The Cognition of Basic Musical Structures

The Cognition of Basic Musical Structures

David Temperley

The MIT Press
Cambridge, Massachusetts
London, England

First MIT Press paperback edition, 2004

This book was set in Sabon on 3B2 by Asco Typesetters, Hong Kong.

Library of Congress Cataloging-in-Publication Data

Temperley, David.
The cognition of basic musical structures / David Temperley.
p. cm.
Includes bibliographical references and index.
ISBN 978-0-262-20134-6 (hc.: alk. paper)—978-0-262-70105-1 (pb.: alk. paper)
1. Music theory—Data processing. 2. Musical perception—Computer simulation. I. Title.
MT6.T35 C6 2001
781—dc21 00-054801

Contents

Preface

This book addresses a fundamental question about music cognition: how do we extract basic kinds of musical information—meter, phrase structure, counterpoint, pitch spelling, harmony, and key—from music as we hear it? My approach to this question is computational: I develop computer models for generating these aspects of structure, with the aim of simply solving the computational problems involved as elegantly and effectively as possible, and with the assumption that this approach may shed light on how the problems are solved in cognition. The models I propose are based on *preference rules*. Preference rules are criteria for evaluating a possible analysis of a piece (in terms of some kind of musical structure). In a preference rule system, many possible interpretations are considered, and the one is chosen that best satisfies the rules.

I begin with an introductory chapter, describing the overall goals and methodology of the project and overviewing the theoretical and implementational strategy. The remainder of the book is then divided into two parts. In part I, I present preference rule systems for generating six basic kinds of musical structure. *Metrical structure* is a framework of levels of beats. *Melodic phrase structure* is a segmentation of the input into phrases; the model I propose is applicable only to melodies, not polyphonic textures. *Contrapuntal structure* is a segmentation of a polyphonic texture into melodic lines. *Pitch spelling*, which I also call the *tonal-pitch-class representation*, involves a labeling of pitch events in a piece with spellings ("tonal-pitch-class" labels) such as A♭ or G♯. *Harmonic structure* is a segmentation of a piece into harmonic segments labeled with roots. The preference rule systems for pitch spelling and harmonic structure are closely integrated, and really represent a single preference rule system. Finally, *key structure* is a segmentation of a piece into larger sections labeled with keys.

A separate chapter is devoted to each preference rule system. In each case, I begin by describing the basic character of the structure in question; I also review any psychological evidence pertaining to it (both the psychological reality of this kind of structure and the way it is inferred in perception). I then discuss earlier computational proposals (if any) for how this structure is inferred. My own preference-rule approach to this problem is then presented in an informal, conceptual way, with discussion of each preference rule and the motivation for it. Next, I discuss the implementation of the model in more technical detail. Finally, I present any formal tests that were done of the model; in each case, at least one such test was performed. I examine flaws in the model revealed by the tests, and consider possible improvements.

A central claim of the current study is that preference rule systems are not merely valuable as proposals for how musical structures are inferred, but also shed light on other aspects of music. The second half of the book attempts to substantiate this claim. I begin in chapter 8 with a discussion of three important aspects of musical experience: ambiguity, retrospective revision, and expectation. The following two chapters explore the possible relevance of preference rule systems to kinds of music outside the Western canon. Chapter 9 applies the metrical, harmonic and key models to rock music; chapter 10 examines the validity of the metrical and phrase structure models for traditional African music. Chapter 11 considers how preference rule systems might be applied to issues of composition and performance, and proposes a framework for the description of musical styles. Finally, in chapter 12, I explore the relevance of preference rule systems to higher-level musical structure and meaning; here I address issues such as motivic structure, musical schemata (gestures or patterns with conventional associations), narrative and dramatic aspects of music, and musical tension.

The content of this book is, in a sense, two-dimensional. With each preference rule system, there are a number of issues to be addressed: basic issues such as psychological evidence, the preference rule system itself, and implementation and testing, as well as more speculative issues such as those addressed in part II. It was difficult to know how to traverse this two-dimensional space in the linear fashion required for a book. I am well aware, however, that not all readers will be interested in all the issues covered here. The sections in part I in which I overview each preference rule system (as well as those relating to psychological evidence and earlier computational approaches) are intended to be interesting and accessible to a broad audience: music theorists and musicians, music psychologists and others in psychology, and workers in music technology and artificial intelligence. In these sections, I try to avoid assuming great knowledge of music theory, and provide at least some explanation of any advanced musical terms that I use. (Even so, these sections will undoubtedly prove more rewarding to those with some knowledge of music; in particular, an ability to imagine simple musical excerpts or play them on a keyboard will be useful, since it will enable readers to compare my claims about musical perception with their own intuitions.) Sections in part II may, obviously, be of special concern to certain audiences with interests in African music, the psychology of performance and composition, and the like, though here again I aim to make the material broadly accessible. The most narrowly aimed sections of the book are those relating to the implementation and testing of each preference rule system (roughly speaking, the final part of each chapter in part I, as well as the section on implementation in the introductory chapter). These are primarily intended for those in the area of computational music analysis, who may wish to learn from or evaluate my implementational approach and compare the performance of my models to their own models or others. Other readers may wish to skip over these sections; they are not essential for understanding the rest of the book.

The computer implementations presented here are publicly available at the website www.link.cs.cmu.edu/music-analysis. (The implementations of the meter, pitch spelling and harmony programs were developed in collaboration with Daniel Sleator.) The programs are written in C, and run on a UNIX platform. The website also provides many of the input files for excerpts discussed in this book. I hope this will encourage others to experiment with the programs, and subject them to further testing; those with alternative models may wish to try their programs on the same input files used in my tests.

Acknowledgments

This project has had the benefit of help and input from a number of people over a period of some six years. The early stages of the project, specifically the harmonic and TPC models, took shape as part of my dissertation at Columbia University. A number of people provided valuable input at this stage, including Ian Bent, Mark Debellis, John Halle, and Carol Krumhansl. Joe Dubiel posed many challenging questions and forced me to think deeply about the goals of the project. Jonathan Kramer was an unfailing source of help and encouragement; my numerous seminars with Jonathan at Columbia gave rise to many of the ideas in this book.

Much of the remainder of the work was done during two years (1997–1999) at Ohio State University, where I was fortunate to receive a postdoctoral fellowship in music cognition from the School of Music, allowing me to focus my full energies on research among a large community of dedicated music cognition enthusiasts. David Huron provided helpful criticism on several sections of the book, and also served as a reviewer in its final stages. Paul von Hippel offered penetrating comments on the counterpoint and key-finding chapters. Numerous discussions with other people at Ohio State, including Bret Aarden, Graeme Boone, Mike Brady, David Butler, Mark DeWitt, Mari Jones, and Caroline Palmer, provided food for thought and helped to solidify my ideas.

Several journal editors (and anonymous reviewers at those journals) provided useful feedback on material that was submitted in articles, including Lucy Green at *Popular Music*, Doug Keislar at *Computer Music Journal*, Bruno Nettl at *Ethnomusicology*, Anthony Pople at

Music Analysis, and Jamshed Bharucha and Robert Gjerdingen at *Music Perception*. In the book's later stages, my editors at MIT Press, Doug Sery and Katherine Almeida, were models of efficiency and professionalism. Thanks are due to Brad Garton at the Computer Music Center at Columbia for providing a hospitable work environment in the 1999–2000 academic year, to Chris Bailey for technical help, and to Robert Rowe for providing feedback on the final draft. Eastman School of Music provided generous institutional support in the final months of the project.

Special recognition is due to Fred Lerdahl. Fred's involvement in the project goes back to its inception, when he was my dissertation advisor at Columbia. Since then he has read every part of the book (sometimes in several different drafts), and has provided guidance and feedback on countless matters large and small, from major theoretical issues to details of musical analysis and writing style. His support and encouragement have been unwavering throughout.

On a personal note, I must thank my wonderful family, my devoted friends, and my endlessly supportive girlfriend, Maya Chaly. My father, Nicholas Temperley, offered valuable comments on several sections of the book.

The final and most important acknowledgment goes to my cousin and collaborator, Daniel Sleator. Danny wrote the code for the meter-finding and harmonic-TPC programs (discussed in chapters 2, 5, and 6). However, his contribution went far beyond mere programming. It was Danny's idea to apply the technique of dynamic programming to the implementation of preference rule systems. Not only did this technique provide a highly effective solution to the search problem, it also offered an elegant model of left-to-right processing and garden-path phenomena. Danny also contributed a number of other important ideas regarding formalization and implementation, for the meter model in particular and also for the TPC-harmonic model. More generally, Danny helped out with the project in a variety of other ways. He provided a stable UNIX environment (at Carnegie-Mellon) for my use, at a time when I was moving around a lot from one institution to another. My own programming efforts benefited greatly from studying and sometimes pillaging his code (any well-written lines in my programs are due to him). Danny also frequently provided technical and debugging help, and patiently answered my naive questions about C and UNIX. In short, Danny's contribution to this project was indispensable; without him it could not have been done. I should also thank Danny's wife Lilya Sleator, who on a number of occasions provided a wonderfully hospitable environment for our work in Pittsburgh.

Source Notes

Thanks are due to the publishers listed below for permission to reprint portions of the musical works indicated.

A Hard Day's Night. From *A Hard Day's Night*. Words and music by John Lennon and Paul McCartney. Copyright © 1964 Sony/ATV Songs LLC. Copyright renewed. All rights administered by Sony/ATV Music Publishing, 8 Music Square West, Nashville, TN 37203. International copyright secured. All rights reserved.

Breathe. Words by Roger Waters. Music by Roger Waters, David Gilmour and Rick Wright. TRO—© Copyright 1973 Hampshire House Publishing Corp., New York, NY. Used by permission.

Day Tripper. Words and music by John Lennon and Paul McCartney. Copyright © 1965 Sony/ATV Songs LLC. Copyright renewed. All rights administered by Sony/ATV Music Publishing, 8 Music Square West, Nashville, TN 37203. International copyright secured. All rights reserved.

Density 21.5 by Edgard Varèse. © Copyright by Casa Ricordi/BMG Ricordi. Copyright renewed. Reprinted by permission of Hendon Music, Inc., a Boosey & Hawkes company, sole agent.

Go Your Own Way. By Lindsey Buckingham. Copyright © 1976. Now Sound Music. All rights reserved. Used by permission.

Here Comes the Sun. Words and music by George Harrison. © 1969 Harrisongs Ltd. Copyright renewed 1998. International copyright secured. All rights reserved.

Hey Bulldog. Words and music by John Lennon and Paul McCartney. Copyright © 1968, 1969 Sony/ATV Songs LLC. Copyright renewed. All rights administered by Sony/ATV Music Publishing, 8 Music Square West, Nashville, TN 37203. International copyright secured. All rights reserved.

Hey Jude. Words and music by John Lennon and Paul McCartney. Copyright © 1968 Sony/ATV Songs LLC. Copyright renewed. All rights administered by Sony/ATV Music Publishing, 8 Music Square West, Nashville, TN 37203. International copyright secured. All rights reserved.

I Can't Explain. Words and music by Peter Townshend. © Copyright 1965 by Fabulous Music Ltd. Copyright renewed. Rights throughout the Western hemisphere administered by Universal-Champion Music Corporation. All rights reserved. Reprinted by permission of Warner Bros. Publications U.S. Inc. and Universal Music Group.

I Heard It Through the Grapevine. Words and music by Norman J. Whitfield and Barrett Strong. © 1966 (renewed 1994) Jobete Music Co., Inc. All rights controlled and administered by EMI Blackwood Music Inc. on behalf of Stone Agate Music (a division of Jobete Music Co., Inc.). All rights reserved. International copyright secured. Used by permission.

Imagine. Words and music by John Lennon. © 1971 (renewed 1999) Lenono Music. All rights controlled and administered by EMI Blackwood Music Inc. All rights reserved. International copyright secured. Used by permission.

Keep on Loving You. Words and music by Kevin Cronin. Copyright © 1981 Fate Music (ASCAP). International copyright secured. All rights reserved.

Let It Be. Words and music by John Lennon and Paul McCartney. Copyright © 1970 Sony/ATV Songs LLC. Copyright renewed. All rights administered by Sony/ATV Music Publishing, 8 Music Square West, Nashville, TN 37203. International copyright secured. All rights reserved.

Mean Mr. Mustard. Words and music by John Lennon and Paul McCartney. Copyright © 1969 Sony/ATV Songs LLC. Copyright renewed. All rights administered by Sony/ATV Music Publishing, 8 Music Square West, Nashville, TN 37203. International copyright secured. All rights reserved.

Mikrokosmos by Bela Bartok. © Copyright 1940 by Hawkes & Son (London) Ltd. Copyright renewed. Definitive corrected edition © Copyright 1987 by Hawkes & Son (London) Ltd.

Proud Mary by John C. Fogerty. © 1968 Jondora Music (BMI). Copyright renewed. Courtesy of Fantasy, Inc. All rights reserved. Used by permission.

Somebody to Love. Words and music by Darby Slick. Copyright © 1967 Irving Music, Inc. Copyright renewed. All rights reserved. Used by permission.

Structures 1A by Pierre Boulez. © 1955 (renewed) Universal Edition (London) Ltd., London. All rights reserved. Used by permission of European American Music Distributors LLC, sole U.S. and Canadian agent for Schott & Co. Ltd., London.

Symphonies of Wind Instruments by Igor Stravinsky. © Copyright 1925 by Hawkes & Son (London) Ltd. Copyright renewed. Revised version © Copyright 1947 by Hawkes & Son (London) Ltd. Copyright renewed.

The Kids Are Alright. Words and music by Peter Townshend. © Copyright 1965 by Fabulous Music Ltd. Copyright renewed. Rights in the U.S. and Canada administered by Songs of Windswept Pacific o/b/o Towser Tunes, Inc. (BMI) All rights reserved. Reprinted by permission of Warner Bros. Publications U.S. Inc. and Songs of Windswept Pacific.

Walking on the Moon. Written and composed by Sting. © 1979 G. M. Sumner. Published by Magnetic Publishing Ltd. and administered by EMI Blackwood Music Inc. in the USA and Canada. All rights reserved. International copyright secured. Used by permission.

The Cognition of Basic Musical Structures

1
Introduction

1.1 An Unanswered Question

The aspects of music explored in this book—meter, phrase structure, contrapuntal structure, pitch spelling, harmony, and key—are well known and, in some ways, well understood. Every music student is taught to label chords, to spell notes correctly, to identify modulations, to identify a piece as being in 3/4 or 4/4, and to recognize the phrases of a sonata and the voices of a fugue. At more advanced levels of musical discourse, these structures are most often simply taken for granted as musical facts. It is rarely considered a contribution to music theory to identify the phrases or the sequence of harmonies in a piece, nor is there often disagreement about such matters. In psychology, too, each of these facets of music has been explored to some extent (some to a very considerable extent), and there are grounds for believing that all of them are important aspects of music cognition, not merely among trained musicians but among listeners in general.

In short, there appears to be broad agreement as to the general character of these structures, the particular form they take in individual pieces, and their reality and importance in music cognition. In another respect, however, our knowledge of these aspects of music is much less advanced. If we assume that harmony, metrical structure, and the like are real and important factors in musical listening, then listening must involve extracting this information from the incoming notes. How, then, is this done; by what process are these structures inferred? At present, this is very much an open question. It is fair to say that no fully satisfactory answer has been offered for any of the kinds of structure listed above; in some areas, answers have hardly even been proposed. I will present a general approach to this problem, based on the concept of *preference rules*, which leads to highly effective procedures for inferring these kinds of information from musical inputs. Because my approach is computational rather than experimental, I must be cautious in my claims about the psychological validity of the models I propose. At the very least, however, the current approach provides a promising hypothesis about the cognition of basic musical structures which warrants further consideration and study.

While exploring processes of information extraction is my main goal, the framework I propose also sheds light on a number of other issues. First of all, music unfolds in time; we do not wait until the end of a piece to begin analyzing it, but rather, we interpret it as we go along, sometimes revising our interpretation of one part in light of what happens afterwards. Preference rule systems provide a useful framework for characterizing this real-time process. The preference rule approach also provides insight into other important aspects of musical experience, such as ambiguity, tension, and expectation. Finally, as well as providing a powerful theory of music perception, the preference rule approach also sheds valuable light on what are sometimes called the "generative" processes of music: composition and performance. I will argue that preference rule systems play an important role in composition, acting as fundamental—though flexible—constraints on the compositional process. In this way, preference rules can contribute not only to the description of music perception, but of music itself, whether at the level of musical styles, individual pieces, or structural details within pieces. The preference rule approach also relates in interesting ways to issues of musical performance, such as performance errors and expressive timing.

An important question to ask of any music theory is what corpus of music it purports to describe. My main concern in this book is with Western art music of the eighteenth and nineteenth centuries: what is sometimes called "common-practice" music or simply "tonal" music.[1] I have several reasons for focusing on this corpus. First, this is the music with which I have the greatest familiarity, and thus the music about which I am most qualified to theorize. Second, common-practice music brings with it a body of theoretical and experimental research which is unparalleled in scope and sophistication; the current study builds on this earlier work in many ways which I will do my best to acknowledge. Third, a large amount of music from the common-practice corpus is available in music notation. Music notation provides a representation which is convenient for study and can also easily be converted into a format suitable for computer analysis. This contrasts with much popular music and non-Western music, where music notation is generally not available. (There are problems with relying on music notation as well, as I will discuss below.) Despite this limited focus, I believe that many aspects of the model I present are applicable to kinds of music outside the Western canon, and at some points in the book I will explore this possibility.

Another question arises concerning the subject matter of this study. No one could deny that the kinds of musical structure listed above are important, but music has many other important aspects too. For example, one could also cite motivic structure (the network of melodic segments in a piece that are heard as similar or related); melodic schemata such as the gap-fill archetype (Meyer 1973) and the $\hat{1}$-$\hat{7}$-$\hat{4}$-$\hat{3}$ schema (Gjerdingen 1988); and the conventional "topics"—musical gestures with extramusical meanings—discussed by Ratner (1980) and others. In view of this, one might ask why I consider only the aspects of music listed earlier. An analogy may be useful in explaining what these kinds of musical structure have in common, and the role they play in music cognition.

Any regular observer of the news media will be familiar with the term "infrastructure." As the term is commonly used, "infrastructure" refers to a network of basic structures and services in a society—largely related to transportation and communication—which are required for the society to function. (The term is most often heard in the phrase "repairing our crumbling infrastructure"—a frequent promise of politicians.) To my mind, "infrastructure" implies two important things. Infrastructure is supposed to be ubiquitous: wherever you go (ideally), you will find the roads, power lines, water mains, and so on that are needed for life and business. Secondly, infrastructure is a means to an end: water mains and power lines do not normally bring us joy in themselves, but they facilitate other things—homes, schools, showers, VCRs—whose contribution to life is more direct. In both of these respects, the aspects of music listed earlier could well be regarded as an "infrastructure" for tonal music. Metrical structure and harmony are ubiquitous: roughly speaking, every piece, in fact every moment of every piece, has a metrical structure and a harmonic structure. Melodic archetypes and topics, by contrast, are *occasional* (though certainly common). Few would argue, I think, that every bit of tonal music is a melodic archetype or a topic. Secondly, while the structures I discuss here may sometimes possess a kind of direct musical value in their own right, they function largely as means to other musical ends. In many cases, these musical ends are exactly the kinds of occasional structures just mentioned. A topic or melodic archetype requires a certain configuration of contrapuntal, metrical, and harmonic structures, and perhaps others as well; indeed, such higher-level patterns are often characterized largely in infrastructural terms (I will return to this point in chapter 12). My aim here is not, of course, to argue for either "ubiquitous" or "occasional" structures as more important than the other—each is important in its own way; my point, rather, is that ubiquitous structures form a "natural kind" and, hence, an appropriate object of exclusive study.

1.2 Goals and Methodology

Discourse about music adopts a variety of methods and pursues a variety of goals. In this section I will explain the aims of the current study and my method of achieving them. It is appropriate to begin with a discussion of the larger field in which this study can most comfortably be placed, a relatively new field known as *music cognition.*

Music cognition might best be regarded as the musical branch of cognitive science—an interdisciplinary field which has developed over the last thirty years or so, bringing together disciplines relating to cognition, such as cognitive psychology, artificial intelligence, neuroscience, and linguistics. Each of the disciplines contributing to cognitive science brings its own methodological approach; and each of these methodologies has been fruitfully applied to music. The methodology of cognitive psychology itself is primarily experimental: human subjects are given stimuli and asked to perform tasks or give verbal reports, and the psychological processes involved are inferred from these. A large body of experimental work has been done on music cognition; this work will frequently be cited below. In theoretical linguistics, by contrast, the methodology has been largely introspectionist. The reasoning in linguistics is that, while we do not have direct intuitions about the syntactic structures of sentences, we do have intuitions about whether sentences are syntactically well-formed (and perhaps about other things, such as whether two sentences are identical in meaning). These well-formedness judgments constitute a kind of data about linguistic understanding. By simply seeking to construct grammars that make the right judgments about well-formedness—linguists reason—we will uncover much else about the syntactic structure of the language we are studying (and languages in general). The introspectionist approach to music cognition is reflected in work by music theorists such as Lerdahl and Jackendoff (1983) and Narmour (1990).

(This is not to say, however, that music theory in general should be regarded as introspectionist cognitive science; I will return to this point.)

The methods of artificial intelligence are also important in music cognition. Here, attempts are made to gain insight into a cognitive process by trying to model it computationally. Often, the aim is simply to devise a computational system which can perform a particular process (for example, yielding a certain desired output for a given input); while there is no guarantee that such a program performs the process the same way humans do it, such an approach may at least shed some light on the psychological mechanisms involved.[2] In some cases, this approach has received empirical support as well, in that neurological mechanisms have been found which actually perform the kind of functions suggested by computational models (see Bruce & Green 1990, 87–104, for discussion of examples in the area of vision). As we will see, this, too, is a widely used approach in music cognition. Finally, cognition can be approached from a neurological or anatomical perspective, through studies of electric potentials, brain disorders, and the like. This approach has not been pursued as much as others in music cognition, though some progress has been made; for example, much has been learned regarding the localization of musical functions in the brain.[3]

Despite their differing methodologies, the disciplines of cognitive science share certain assumptions. All are concerned with the study of intelligent systems, in particular, the human brain. It is widely assumed, also, that cognitive processes involve representations, and that explanations of cognitive functions should be presented in these terms. This assumption is very widely held, though not universally.[4] To appreciate its centrality, one need only consider the kinds of concepts and entities that have been proposed in cognitive science: for example, edge detectors and primal sketches in vision, tree structures and constituents in linguistics, prototypes and features in categorization, networks and schemata in knowledge representation, loops and buffers in memory, problem spaces and productions in problem-solving, and so on. All of these are kinds of mental representations, proposed to explain observed facts of behavior or introspection. A second important assumption is the idea of "levels of explanation." A cognitive process might be described at a neurological level; but one might also describe it at a higher, computational level, without worrying about how it might be instantiated neurologically. A computational description is no less real than a neurological one; it is simply more abstract. It is assumed, further, that a cognitive system, described at a computational level, might be physically instantiated in quite different ways: for example, in a human brain or on a computer. This assumption is crucial for artificial intelligence, for it implies that a computer running a particular program might be put forth as a description or model of a cognitive system, albeit a description at a very abstract level.[5]

This background may be helpful in understanding the goals and methodology of the current study. My aim in this study is to gain insight into the processes whereby listeners infer basic kinds of structure from musical input. My concern is with what Lerdahl and Jackendoff (1983, 3) call "experienced listeners" of tonal music: people who are familiar with the style, though not necessarily having extensive formal training in it. My methodology in pursuing this goal was both introspectionist and computational. For a given kind of structure, it was first necessary to determine the correct analysis (metrical, harmonic, etc.) of many musical excerpts. Here my approach was mainly introspective; I relied largely on my own intuitions as to the correct analyses of pieces. However, I sometimes relied on other sources as well. With some of the kinds of structure explored here, the correct analysis is at least partly explicit in music notation. For example, metrical structure is indicated by rhythmic notation, time signatures, and barlines. For the most part, the structures implied by the notation of pieces concur with my own intuitions (and I think those of most other listeners), so notation simply provided added confirmation.[6] I then sought models to explain how certain musical inputs might give rise to certain analyses; and I devised computational implementations of these models, in order to test and refine them. With each kind of structure, I performed a systematic test of the model (using some source other than my own intuitions for the correct analysis—either the score or analyses done by other theorists) to determine its level of success.

The goals and methodology I have outlined could be questioned in several ways. The first concerns the computational nature of the study. As mentioned earlier, the mere fact that a model performs a process successfully certainly does not prove that the process is being performed cognitively in the same way. However, if a model does *not* perform a process successfully, then one knows that the process is *not* performed cognitively in that way. If the model succeeds in its purpose, then one has at least a hypothesis for how the process might be performed cognitively, which can then be tested by other means. Computer implementations are also valuable, simply because they allow one to test objectively whether a model can actually produce the desired outputs. In the current case, the programs I devised often did not produce the results I expected, and led me to modify my original models significantly.

Another possible line of criticism concerns the idea of "correct" analyses, and the way I arrived at them. It might seem questionable for me, as a music theorist, to take my intuitions (or those of another music theorist) about musical structure to represent those of a larger population of "experienced listeners." Surely the hearing of music theorists has been influenced (enhanced, contaminated, or just changed) by very specialized and unusual training. This is, indeed, a problematic issue. However, two points should be borne in mind. First, it is certainly not out of the question that untrained and highly trained listeners have much in common in at least some aspects of their music cognition. This is of course the assumption in linguistics, where linguists take their own intuitions about syntactic well-formedness (despite their highly specialized training in this area) to be representative of those of the general population. Secondly, and more decisively, there is an impressive body of experimental work suggesting that, broadly speaking, the kinds of musical representations explored here *are* psychologically real for a broad population of listeners; I will refer to this work often in the chapters that follow. Still, I do not wish to claim that music theorists hear things like harmony, key, and so on exactly the same way as untrained listeners; surely they do not. Much further experimental work will be needed to determine how much, and in what ways, music cognition is affected by training.

Quite apart from effects of training, one might argue that judgments about the kinds of structures described here vary greatly among individuals—even among experts (or non-experts). Indeed, one might claim that there is so much subjectivity in these matters that the idea of pursuing a "formal theory of listeners' intuitions" is misguided.[7] I do not deny that there are sometimes subjective differences about all of the kinds of structure at issue here; however, I believe there is much more agreement than disagreement. The success of the computational tests I present here, where I rely on sources other than myself for the "correct" analysis, offers some testimony to the general agreement that is found in these areas. (One might also object that, even for a single listener, it is oversimplified to assume that a single analysis is always preferred to the exclusion of all others. This is certainly true; ambiguity is a very real and important part of music cognition, and one which is considerably illuminated by a preference rule approach, as I discuss in chapter 8.)

An important caveat is needed about the preceding discussion. My concern here is with aspects of music perception which I assume to be shared across a broad population of listeners familiar with tonal music. I must emphasize, however, that I am not at all assuming that these principles are innate or universal. Rather, it is quite possible that they are learned largely from exposure to music—just as language is, for example (at least, some aspects of language). I will argue in later chapters that some aspects of the models I propose have relevance to kinds of music outside the Western canon. However, I will take no position on the questions of universality and innateness; in my view, there is not yet sufficient basis for making claims about these matters.

1.3 Music Cognition and Music Theory

I suggested above that some work in music theory might be regarded as introspectionist cognitive science—work seeking to reveal cognitive processes through introspection, much as linguists do with syntax. Indeed, music theory has played an indispensable role in music cognition as a source of models and hypotheses; much music-related work in cognitive psychology has been concerned with testing these ideas. However, it would be a mistake to regard music theory in general as pursuing the same goals as music cognition. Cognitive science is concerned, ultimately, with describing and explaining cognitive processes. In the case of music cognition, this normally implies processes involved in listening, and sometimes performance; it might also involve processes involved in composition, although this area has hardly been explored. I have argued elsewhere that, while some music theory is concerned with this goal, much music theory is not; rather, it is concerned with enhancing our listening, with finding new structures in pieces which might enrich our experience of them (Temperley in press-b). Many music theorists state this goal quite explicitly. I have called the latter enterprise "suggestive theory"; this is in contrast to the enterprise of "descriptive theory," which aims to describe cognitive processes. Consider Z-related sets, a widely used concept in pitch-class set theory: two pitch-class sets are Z-related if they have the same intervallic content, but are not of the same set-type (related by transposition or inversion). I believe few theorists would claim that people hear Z-related sets (except as a result of studying set theory); rather, Z-related sets serve to enhance or enrich our hearing of certain kinds of music once we are aware of them.

The goal of studying pieces of music in order to understand them more fully, and to enrich our experience of them as much as possible, is an enormously worthwhile one. However, suggesting ways of enhancing our hearing is a goal quite different from describing our hearing. There is a good deal of confusion about this point in music theory, and it is often unclear how specific theories or analyses are to be construed. This is particularly apparent with Schenkerian analysis, a highly influential approach to the study of tonal music. While some theorists have construed Schenkerian theory in a psychological way, others have viewed it as a suggestive theory: a means of enhancing and expanding our hearing of tonal music. Of course, it is possible that a theory could be suggestive in some respects and descriptive in others. My own view is that some aspects of Schenkerian theory are highly relevant to cognition; in particular, Schenkerian analysis draws our attention to subtleties of contrapuntal structure which are often not explicit in notation. (I discuss this further in chapter 8.) With other aspects of Schenkerian theory the relationship to listening is less clear, especially the "reductive" or hierarchical aspect. But to exclude aspects of Schenkerian theory (or any other music theory) from a cognitive theory of tonal music is not at all to reject or dismiss them. Rather, it is simply to maintain that their value is not, primarily, as contributions to a theory of music cognition—a position that many Schenkerian analysts have endorsed.

The psychological, rather than suggestive, perspective of the current study cannot be emphasized too strongly, and should always be kept in mind. For example, when I speak of the "correct" analysis of a piece—as I often will—I mean the analysis that I assume listeners hear, and thus the one that my model will have to produce in order to be correct. I do not mean that the analysis is necessarily the best (most musically satisfying, informed, or coherent) one that can be found. (A similar point should be made about the term "preference rule." Preference rules are not claims about what is aesthetically preferable; they are simply statements of fact about musical perception.) I have already acknowledged that, in assuming a single analysis shared by all listeners, I am assuming a degree of uniformity that is not really present. In making this assumption, I do not in any way mean to deny the importance and interest of subjective differences; such differences are simply not my concern for the moment. I *do* maintain, however, that the differences between us, as listeners, are not so great that any attempt to describe our commonalities is misguided or hopeless.

1.4 The Input Representation

An important issue to consider with any computer model is the input representation that is used. The preference rule systems discussed here all use essentially the same input representation. This is a list of notes,

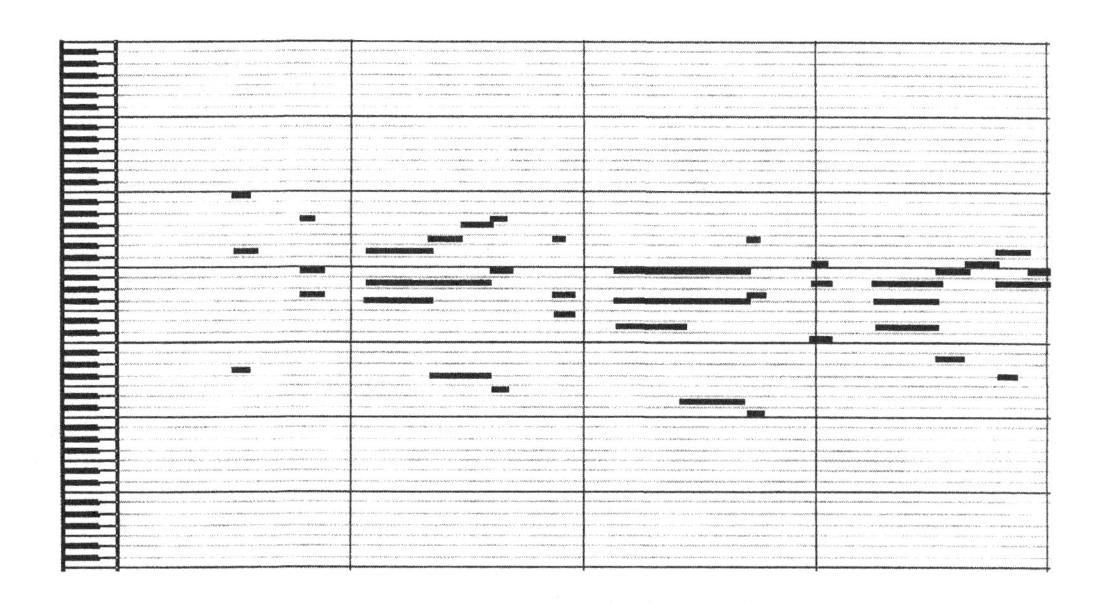

Figure 1.1
A "piano-roll" representation of the opening of the Gavotte from Bach's French Suite No. 5 in G major (generated from a performance by the author on a MIDI keyboard). The score for the excerpt is shown below.

giving the on-time, off-time (both in milliseconds) and pitch of each note—what I will refer to as a "note-list." We can also think of this as a two-dimensional representation, with pitch on one axis and time on the other; each pitch-event is represented as a line segment on the plane, with the length of the line corresponding to the duration of the event. Such a representation is sometimes known as a "piano-roll," since it resembles the representations of pieces used with player pianos in the early twentieth century. Figure 1.1 shows part of the piano-roll representation for a performance of a Bach Gavotte (the score for the excerpt is shown below). Pitches in the input representation are categorized into steps of the chromatic scale; following convention, integers are used to represent pitches, with middle C = 60. In an important sense, then, the pitch axis of the "piano-roll" representation is discrete, not continous. The time axis, however, is essentially continuous; pitch-events are not quantized rhythmically in any significant way (except at the very small level of milliseconds). Other acoustic information such as timbre and amplitude is excluded from the input. (Some of the models also require additional information as input; for example, several of the models require metrical structure. I will discuss this further below.)

In assuming a "piano-roll" representation as input, I am avoiding the problem of deriving pitch information from actual sound. This problem—sometimes known as "music recognition" or "automatic transcription"—has been studied extensively, and proves to be highly complex (Moorer 1977; Foster, Schloss, & Rockmore 1982; Tanguiane 1993). The sounds of the music must be separated out from the other background sounds that are always present in any natural environment; the individual frequencies that make up the sound must be grouped together to form notes; and the notes must be correctly quantized to the right pitch categories, factoring out vibrato, bad intonation, and so on. However, this process is not our concern here; in the following chapters, the existence of an accurate piano-roll representation will simply be taken for granted.

One might wonder what evidence there is that listeners actually form piano-roll representations. Of course very few people could accurately *report* such representations; but this may be because such information is largely unconscious or not easily articulated. Most evidence for the reality of piano-roll representations is indirect, and somewhat inconclusive. For example, the fact that listeners are generally able to learn a melody from hearing it (at least if they hear it enough times), and recognize it later or reproduce it by singing, suggests that they must be extracting the necessary pitch and duration information. Another possible argument for the reality of piano-roll representations is that the kinds of higher-level structures explored here—whose general psychological reality has been quite strongly established, as I will discuss—*require* a piano-roll input in order to be derived themselves. For example, it is not obvious how one could figure out what harmonies were present in a passage without knowing what notes were present. I should point out, however, that several proposals for deriving aspects of the infrastructure—specifically harmony, contrapuntal structure, and key—assume exactly this: they assume that these kinds of structure can be extracted without first extracting pitch information. These proposals will be discussed below, and I will suggest that all of them encounter serious problems. I think a case could be made, then, that the reality of "infrastructure" levels provides strong evidence for the reality of piano-roll representations, since there is no other plausible way that infrastructure levels could be derived.

It was noted above that the input representation does not contain any quantization of events in the time dimension. This is, of course, true to the situation in actual listening. In performed music, notes are not played with perfect regularity; there is usually an *implied* regularity of durations (this will be represented by the metrical structure), but within that there are many small imperfections as well as deliberate fluctuations in timing. In the tests presented below, I often use piano-roll representations that were generated from performances on a MIDI keyboard, so that such fluctuations are preserved. (The piano-roll in figure 1.1 is an example. The imperfections in timing here can easily be seen—for example, the notes of each chord generally do not begin and end at exactly the same time.) However, one can also generate piano-roll representations from a score; if one knows the tempo of the piece, the onset and duration of each note can be precisely determined. Since pieces are never played with perfectly strict timing, using "quantized" piano-roll representations of this kind is somewhat artificial, but I will sometimes do so in the interest of simplicity and convenience.

Another aspect of the piano-roll representation which requires discussion is the exclusion of timbre and dynamics.[8] As well as being important in their own right, these musical parameters may also affect the levels of the infrastructure in certain ways. For example, dynamics affects metrical structure, in that loud notes are more likely to be heard as metrically strong; timbre affects contrapuntal structure, in that timbrally similar notes tend to stream together. Dynamics could quite easily be encoded computationally (the dynamic level of a note can be encoded as a single numerical value or series of values), and incorporating dynamics into the current models would be a logical further step. With timbre, the problem is much harder. As Bregman (1990, 92) has observed, we do not yet have a satisfactory way of representing timbre. Several multidimensional representations have been proposed, but none seem adequate to capturing the great variety and richness of timbre. Studying the effect of timbre on infrastructural levels will require a better understanding of timbre itself.

1.5 The Preference Rule Approach

The approach of the current study is based on *preference rules*. Preference rules are criteria for forming some kind of analysis of input. Many possible interpretations are considered; each rule expresses an opinion as to how well it is satisfied by a given interpretation, and these opinions are combined together to yield the preferred analysis. Perhaps the clearest antecedent for preference rules is found in the Gestalt rules of perception, proposed in the 1920s; this connection will be discussed further in chapter 3.

Preference rules *per se* were first proposed by Lerdahl and Jackendoff in their *Generative Theory of Tonal Music* (1983) (hereafter *GTTM*). Lerdahl and Jackendoff present a framework consisting of four kinds of hierarchical structure: grouping, meter, time-span reduction, and prolongational reduction. For each kind of structure, they propose a set of "well-formedness rules" which define the structures that are considered legal; they then propose preference rules for choosing the optimal analysis out of the possible ones. The model of meter I present in chapter 2 is closely related to Lerdahl and Jackendoff's model; my model of phrase structure, presented in chapter 3, has some connection to Lerdahl and Jackendoff's model of grouping. Lerdahl and Jackendoff did not propose any way of quantifying their preference rule systems, nor did they develop any implementation. The current study can be seen as an attempt to quantify and implement Lerdahl and Jackendoff's initial conception, and to expand it to other musical domains. (I will have little to say here about the third and fourth components of *GTTM*, time-span reduction and prolongational reduction. These kinds of structure are less psychologically well-established and more controversial than meter and grouping; they also relate largely to large-scale structure and relationships, which sets them apart from the aspects of music considered here.)

The preference rule approach has been subject to some criticism, largely in the context of critiques of *GTTM*. The problem most often cited is that preference rules are too vague: depending on how the rules are quantified, and the relative weights of one rule to another, a preference rule system can produce a wide range of analyses (Peel & Slawson 1984, 282, 288; Clarke 1989, 11). It is true that the preference rules of *GTTM* are somewhat vague. This does not mean that they are empty; even an informal preference rule system makes empirical claims that are subject to falsification. If a preference rule system is proposed for an aspect of structure, and one finds a situation in which the preferred analysis cannot be explained in terms of the proposed rules, then the theory is falsified, or at least incomplete. It must be said that very few music theories offer even this degree of testability. The more important point, however, is that preference rule systems also lend themselves well to rigorous formalization. If the parameters of the rules can be specified, the output of the rule system for a given input can be determined in an objective way, making the theory truly testable. This is what I attempt to do here.[9]

Another criticism that has been made of preference rule systems concerns the processing of music over time. Lerdahl and Jackendoff's stated aim in *GTTM* (1983, 3–4) is to model what they call "the final state of [a listener's] understanding" of a piece. Under their conception, preference rules serve to select the optimal analysis for a complete piece, once it has been heard in its entirety. In my initial presentation of the current model (in chapters 2 through 7), I will adopt this approach as well. This "final understanding" approach may seem problematic from a cognitive viewpoint; in reality, of course, the listening process does not work this way. However, preference rule systems also provide a natural and powerful way of modeling the moment-to-moment course of processing as it unfolds during listening. I will return to this in the next section (and at greater length in chapter 8).

One notable virtue of preference rule systems is their conceptual simplicity. With a preference rule system, the rules themselves offer a high-level description of what the system is doing: it is finding the analysis that best satisfies the rules. This is an important advantage of preference rule systems over some other models that are highly complex and do not submit easily to a concise, high-level description. (Some examples of this will be discussed in the chapters that follow.) Of course, preference rule systems require some kind of implementation, and this implementation may be highly complex. But the implementation need not be of great concern, nor does it have to be psychologically plausible; it is simply a means to the end of testing whether or not the preference rule system can work. If a preference rule system can be made to produce good computational results, it provides an elegant, substantive, high-level hypothesis about the workings of a cognitive system.

1.6 The Implementation Strategy

While I have said that details of implementation are not essential to an understanding of preference rule systems, a considerable portion of this book is in fact devoted to issues of implementation. (This includes the present section, as well as the sections of following chapters entitled "Implementation.") While these sections will, I hope, be of interest to some readers, they may be skipped without detriment to one's understanding of the rest of the book. In this section I describe a general implementation strategy which is used, in various ways, in all the preference rule models in this study.

At the broadest level, the implementation strategy used here is simple. In a given preference rule system, all possible analyses of a piece are considered. Following Lerdahl and Jackendoff, the set of "possible" analyses is defined by basic "well-formedness rules." Each preference rule then assigns a numerical score to each analysis. Normally, the analytical process involves some kind of arbitrary segmentation of the piece. Many analytical choices are possible for each segment; an analysis of the piece consists of some combination of these segment analyses. For each possible analysis of a segment, each rule assigns a score; the total score for a segment analysis sums these rule scores; the total score for the complete analysis sums the segment scores. The preferred analysis is the one that receives the highest total score.

As noted above, many of the preference rules used in these models involve numerical parameters (and there are always numerical values that must be set for determining the weight of each rule relative to the others). These parameters were mostly set by trial and error, using values that seemed to produce good results in a variety of cases. It might be possible to derive optimal values for the rules in a more systematic way, but this will not be attempted here.

One might ask why it is necessary to evaluate complete analyses of a piece; would it not be simpler to evaluate short segments in isolation? As we will see, this is not possible, because some of the preference rules require consideration of how one part of an analysis relates to another. Whether an analysis is the best one for a segment depends not just on the notes in that segment, but also on the analysis of nearby segments, which depends on the notes of those segments as well as the analysis of other segments, and so on. However, the number of possible analyses of a piece is generally huge, and grows exponentially with the length of the piece. Thus it is not actually possible to generate all well-formed analyses; a more intelligent search procedure has to be used for finding the highest-scoring one without generating them all. Various procedures are used for this purpose; these will be described in individual cases. However, one technique is of central importance in all six preference rule systems, and warrants some discussion here. This is a procedure from computer science

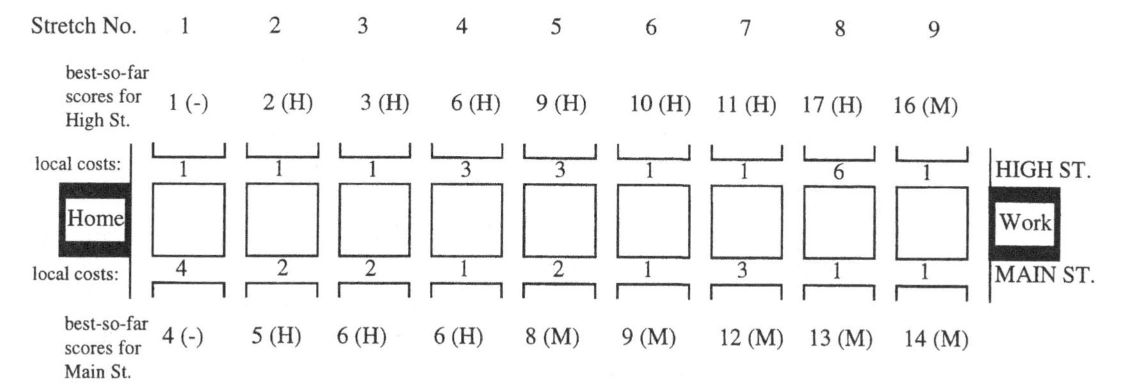

Figure 1.2

known as "dynamic programming." (The idea of using dynamic programming to implement preference rule systems is due to Daniel Sleator.)

Imagine that you are driving through a large city (see figure 1.2). You want to go from home (at the left end of the figure) to work (at the right end). There are two routes for going where you want to go; you can either drive on High Street or Main Street. The two routes are the same in total distance. However, certain stretches of each street are bad (because they have terrible potholes, or construction, or a lot of traffic). You could also switch back and forth between one street and the other at different points, but this carries a cost in terms of time. Suppose that it is worthwhile for you to really sit down and figure out the best route (perhaps because you make this trip every day). You assign each stretch of street a "cost," which is simply the number of minutes it would take you to traverse that stretch. These "local" costs are shown on each stretch of street in figure 1.2. You also assign a cost to any switch between one street and the other; say each such switch costs you 2 minutes. Now, how do you determine the best overall route? It can be seen that there are a large number of different possible routes you could take—2^n, where n is the number of blocks in the east-west direction. You could calculate the cost for every possible route; however, there is a better way. Supposing you compute the cost of all possible routes for the first two stretches that end up on High Street in stretch 2. There are only two, H-H and M-H; the best (i.e. lowest-cost) one is H-H, with a total time of 2 minutes. Then you find the best route ending up on Main Street in stretch 2; it is H-M, with a total time of 5 minutes (local costs of 1 and 2, plus a cost of 2 for switching between streets.) At this point, you do not know whether it is better to end up on High St. or Main St. in stretch 2; that depends on what happens further on. But you do know that no matter what happens later, there will never be any reason to use any route for the first two stretches other than one of the two "best-so-far" routes already identified. Now suppose we want to compute the best way of getting to Main Street in stretch 3. We can use our "best-so-far" routes to stretch 2, continuing each one in stretch 3 and calculating the new total cost; the best choice is H-H-M, with a cost of 6 minutes. Repeating the process with High Street at stretch 3, we now have two new "best-so-far" routes for stretch 3. We can continue this process all the way through to the end of the trip. At each stretch, we only need to record the best-so-far route to each ending point at that stretch, along with its score. In fact, it is not even necessary to record the entire best-so-far route; we only need to record the street that we should be on in the previous stretch. At Main Street in stretch 3, we record that it is best to be on High Street in stretch 2. In this way, each street at each stretch points back to some street at the previous stretch, allowing us to recreate the entire best-so-far route if we want to. (In figure 1.2, the score for the best-so-far route at each segment of street is shown along the top and bottom, along with the street that it points back to at the previous stretch—"H" or "M"—in parentheses.) When we get to the final stretch, either High Street or Main Street has the best (lowest) "best-so-far" score, and we can trace that back to get the best possible route for the entire trip. In this case, Main Street has the best score at the final stretch; tracing this back produces an optimal route of H-H-H-M-M-M-M-M-M.

What I have just described is a simple example of the search procedure used for the preference rule models described below. Instead of searching for the optimal path through a city, the goal is to find the optimal analysis of a piece. We can imagine a two-dimensional table, analogous to the street map in figure 1.2. Columns represent temporal segments; cells of each column represent possible analytical choices for a given segment. An analysis is a path through this table, with one step in each segment. Perhaps the simplest example is the key-finding system (described in chapter 7). Rows of the table correspond to keys, while columns correspond to measures (or some other temporal segments). At each segment, each key receives a local score indicating how compatible that key is with the pitches of the segment; there is also a "change" penalty for switching from one key to another. At each segment, for each key, we compute the best-so-far analysis ending at that key; the best-scoring analysis at the final segment can be traced back to yield the preferred analysis for the entire piece. A similar procedure is used for the harmonic analysis system (where the rows represent roots of chords, instead of keys), the pitch spelling system (where cells of a column represent possible spellings of the pitches in the segment), and the contrapuntal analysis system (where cells represent possible analyses of a segment—contrapuntal voices at different pitch levels), though there are complications in each of these cases which will be explained in due course.

The meter and phrase programs use a technique which is fundamentally similar, but also different. In the case of the phrase program, the table is simply a one-dimensional table of segments representing notes; an analysis is a subset of these notes which are chosen as phrase boundaries. (Choosing a note as a phrase boundary means that a boundary occurs immediately before that note.) Again, each note has a local score, indicating how good it is as a phrase boundary; this depends largely on the size of the temporal gap between it and the previous note. At the same time, however, it is advantageous to keep all the phrases close to a certain optimal size; a penalty is imposed for deviations from this size. At each note, we calculate the best-so-far analysis ending with a phrase boundary at that note. We can do this by continuing all the previous best-so-far analyses—the best-so-far analyses with phrase boundaries at each previous note—adding on a phrase ending at the current note, calculating the new score, and choosing the highest-scoring one to find the new best-so-far analysis. Again, we record the previous note that the best-so-far analysis points back to as well as the total score. After the final note, we compute a final "best-so-far" analysis (since there has to be a phrase boundary at the end of the piece) which yields the best analysis overall. The meter program uses a somewhat more complex version of this approach. The essential difference between this procedure and the one described earlier is that, in this case, an analysis only steps in certain segments, whereas in the previous case each analysis stepped in every segment.

Return to the city example again. Supposing the map in figure 1.2, with the costs for each stretch, was being revealed to us one stretch at a time; at each stretch we had to calculate the costs and best-so-far routes. Consider stretch 7; at this stretch, it seems advantageous to be on High Street, since High Street has the lowest best-so-far score. However, once the next stretch is revealed to us, and we calculate the new best-so-far routes, we see that Main Street has the best score in stretch 8; moreover, Main Street in stretch 8 points back to Main Street in stretch 7. Thus what seems like the best choice for stretch 7 at the time turns out not to be the best choice for stretch 7, given what happens subsequently. In this way the dynamic programming model gives a nice account of an important phenomenon in music perception: the fact that we sometimes revise our initial analysis of a segment based on what happens later. We will return to this phenomenon—which I call "revision"—in chapter 8.

In a recent article, Desain, Honing, vanThienen, and Windsor (1998) argue that, whenever a computational system is proposed in cognitive science, it is important to be clear about which aspects of the system purport to describe cognition, and which aspects are simply details of implementation. As explained earlier, the "model" in the current case is really the preference rule systems themselves. There are probably many ways that a preference rule system could be implemented; the dynamic programming approach proposed here is just one possibility. However, the dynamic programming scheme is not without psychological interest. It provides a computationally efficient way of implementing preference rule systems—to my knowledge, the only one that has been proposed. If humans really do use preference rule systems, any efficient computational strategy for realizing them deserves serious consideration as a possible hypothesis about cognition. The dynamic programming approach also provides an efficient way of realizing a preference rule system in a "left-to-right" fashion, so that at each point, the system has a preferred analysis of everything heard so far—analogous to the process of real-time listening to music. And, finally, dynamic programming provides an elegant way of describing the "revision" phenomenon, where an initial analysis is revised based on what happens afterwards. I know of no experimental evidence pertaining to the psychological reality of the dynamic programming technique; but for all these reasons, the possibility that it plays a role in cognition seems well worth exploring.[10]

I Six Preference Rule Systems

2
Metrical Structure

2.1 Meter

The psychological reality of meter, and several important things about its fundamental character, can be demonstrated in a simple exercise. (This can be done in a classroom, or anywhere else that a group of listeners and a source of music are available.) Play a recording of a piece—preferably a piece that is fairly "rhythmic" in colloquial terms, such as an *allegro* movement from a Classical symphony—and ask people to tap to the music. From trying this in class, my experience is that students are generally able to do this (both music majors and nonmajors) and that there is usually a general consensus about how it should be done. The question is, why do they choose to tap exactly as they do, at a certain speed and at certain moments? One hypothesis might be that listeners simply tap at musical moments that are accented in some obvious or superficial way: at moments of high volume, for example. But this seems unlikely, because in order to keep up with the music, they must anticipate the moment for the next tap before it happens. Listeners may also tap at some moments that are not obviously accented in any way, or even at moments of complete silence in the music. This illustrates an essential point about meter: a metrical structure consists of a series of points in time, or "beats," which may or may not always coincide with events in the music, although they are certainly related to events in the music. Now ask them to tap in another way, more slowly (but still tapping "along with the music"). Again, people can generally perform this task and there is a fair amount of agreement in their responses (although there certainly may be differences; moreover, some in the group may be less certain, and will follow the more certain ones—this is hardly a controlled experiment!). This illustrates a second point about metrical structures: they normally consist of several levels of beats. By convention, beat levels consisting of fewer and sparser beats are known as "higher" levels; beats present only at relatively high levels are known as "strong" beats.

Such informal demonstrations indicate that, at the very least, listeners infer some kind of structure of regular beats from music, which allows them to synchronize their movements with it. This ability has been demonstrated more rigorously in studies in which subjects are asked to tap to repeated patterns (Handel & Lawson 1983; Parncutt 1994). However, the importance of meter goes far beyond this; meter plays an essential role in our perceptual organization of music. Important evidence for this comes from the research of Povel and Essens (1985). Povel and Essens developed a simple algorithm for calculating the implied meter or "clock" of a sequence of events; by this model, a meter is preferred in which events and especially long events coincide with strong beats (this model will be discussed further below). In one experiment, patterns which strongly implied a particular meter (according to Povel and Essens's algorithm) were found to be more accurately reproduced by subjects than those that were metrically ambiguous. In another experiment, subjects were played rhythmic patterns, accompanied by other patterns which simply included the strong beats of various meters, and asked to judge the complexity of pattern pairs. Pattern pairs in which the first pattern implied the meter of the second, according to Povel and Essens's algorithm, were judged to be simpler than others. Thus metrical structure appears to influence the perceived complexity of patterns. Meter has also proven to be a factor in pattern similarity, in that patterns sharing the same meter (duple or triple) are judged as similar (Gabrielsson 1973). In addition, some experimenters have noted in passing that changing the metrical context of a melody, by presenting it with a different accompaniment, can make it sound totally different (Povel & Essens 1985, 432). Even notating a piece in two different ways (i.e. with barlines in different places) can cause performers to regard the two versions as different pieces, not even recognizing the similarity between them (Sloboda 1985, 84). This suggests that the metrical context of a musical passage greatly influences our mental representation of it. Metrical structure also influences other levels of representation such as phrase structure and harmony, as we will explore in later chapters.

Another important function of meter relates to what is sometimes known as *quantization*. Consider the first phrase of the melody "Oh

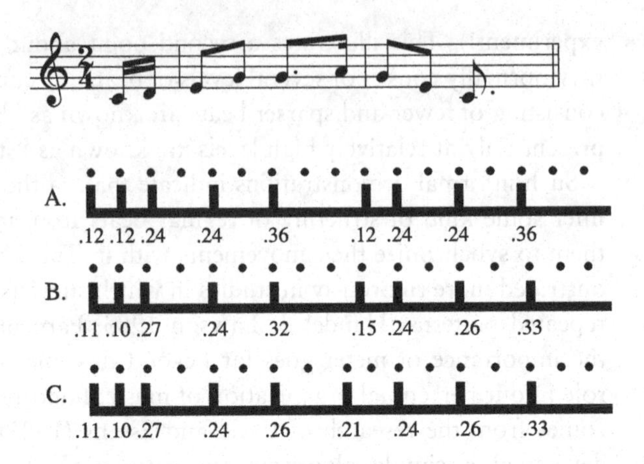

Figure 2.1
The first phrase of the melody "Oh Susannah." Three hypothetical rhythmic performances are shown. The length of each note is represented on a timeline; this is also shown in seconds below. Dots above the timelines show the implied quantized rhythm of each performance. Performance A is perfectly precise; performance B is imprecise, but implies the same quantized rhythm; performance C implies a different quantized rhythm.

Susannah," shown in figure 2.1. An idealized performance of this melody, with perfectly precise durations, is represented as Performance A.[1] However, durational patterns are not usually performed with perfect precision, nor do they need to be in order to be recognized and understood. If we were to hear Performance B, for example—with the dotted eighth-note G a bit too short and the following A a bit too long—we would still understand it as the correct rhythm. However, if the durations were greatly altered, as in Performance C—so that the rhythm of the G-A was more like two eighth notes—then the melody might well sound wrong, as if it was being sung with a different rhythm rather than the correct one.[2] This shows that the perception of rhythm is, in an important sense, categorical: we understand notes as being in one rhythmic category or another, rather than merely perceiving them as continually varying. This process of sorting or "quantizing" notes has been demonstrated experimentally as well; when played patterns of alternating notes whose durations are related by complex ratios (such as 1.5:1 or 2.5:1) and asked to reproduce them, subjects tend to adjust the durations toward simple ratios (such as 2:1) (Povel 1981). Quantization is also reflected in music notation, where notes are categorized as eighth notes, quarter notes, and so on.[3]

Metrical structure—conceived as a framework of rows of beats, as described above—provides a useful way of describing quantization. Imposing a framework of beats on a series of durations can be seen as representing each duration as an integer number of beats. For each rhythmic performance in figure 2.1, the beats that (I suspect) would be perceived are indicated above. In Performances A and B, the dotted-eighth G is given a "length" of three beats and the following A a length of one. In Performance C, however, the G and A are both two beats in length. This allows us to capture the fact that Performance C is qualitatively different from the other two. This does not mean that we do not hear the difference between Performances A and B; quite possibly we do, and such nuances may well be musically important in some ways. But it does mean that we understand the first two performances as in some sense the same, and different from the third. This is another way that metrical structure plays an essential role in our perception of music.

There is a general consensus that meter is an important part of music cognition. However, in modeling the perception of meter, how do we determine what metrical structure is actually heard? In this chapter, as elsewhere in the book, I will rely on my own intuitions about the metrical structures for pieces, and assume that other listeners would generally agree. There is an important source of corroborative evidence, however, namely music notation. The notation for a piece usually indicates a great deal about its metrical structure. For example, a piece in 3/4 meter has one level of quarter-note beats, with every third beat at that level present at the next level up; barlines indicate where these strong beats (or "downbeats") occur. In 3/4, the quarter-note level is generally taken to be the "tactus," or the most salient metrical level (corresponding with the main "beat" of the music in colloquial terms). Lower levels, too, are indicated by rhythmic notation: in 2/4 or 3/4 the tactus is divided duply, with two beats at the next level down for each tactus beat, while in 6/8 or 9/8 the tactus is divided triply. There may also be levels above the measure—so-called hypermetrical levels—so that (for example) odd-numbered downbeats are strong relative to even-numbered ones; and these levels are not usually indicated in notation. There are rarely more than two such levels, however.[4] In most cases the metrical structure indicated by the notation for a piece agrees with my own intuitions, and I think with those of most other listeners as well. (Cases do sometimes arise where the meter indicated by the notation is not the one that seems to be most naturally perceived, but I will generally avoid such cases.)

2.2 Previous Research on Metrical Analysis

It was asserted in the opening sentences of this book that the process of deriving "infrastructural" levels of music has not been widely studied. Meter is a notable exception to this claim. The process of inferring metrical structure from music has, in fact, been explored quite extensively, both from a theoretical perspective and from a computational one. In music theory, a large body of work has been devoted to the study of meter. A number of theorists have explored the way metrical structure interacts with tonal structure, grouping, and other aspects of music (see Schachter 1976; Benjamin 1984; Lester 1986; Berry 1987; Kramer 1988; Rothstein 1989). While this work is mostly informal and humanistic in character, it contains many valuable insights about meter, some of which will be cited below and in following chapters. The most highly developed and formalized theory in this area is Lerdahl and Jackendoff's (1983); their approach forms the starting point for the current study, and will be discussed in greater length in the next section.

Meter has also been the subject of much computational research. Until quite recently, the problem of metrical analysis was generally divided into two smaller problems. One was quantization: taking the events of a live performance and adjusting them so that the time-points (onsets and offsets of notes) are all multiples or divisors of a single common pulse—for example, taking something like Performance B in figure 2.1 and producing something like Performance A. The other problem was higher-level meter-finding: starting with an input which had already been quantized in the way just described, and deriving a metrical structure from that. Recently, people have realized that these problems can really be regarded as two aspects of the same larger problem; however, most research on meter-finding has addressed only one or the other.

The most important work in the area of quantization has been that of Desain and Honing (1992). Desain and Honing propose a connectionist model, consisting of basic units and interaction units. The basic units start out with activation levels representing "inter-onset intervals" or "IOI's" (the time interval between the onset of one note and the onset of the following note) in a live performance. The interaction units connect adjacent basic units, and adjust their relative activation levels to be related by simple ratios such as 1:1, 2:1 or 3:1. In this way, the activation levels of basic units converge to multiples of a common value. The system also has "sum cells," which sum the activations of several basic units, allowing a single unit to represent a time interval containing several notes. Desain and Honing's approach seems elegant and sensible, though it has not been extensively tested. One criticism that might be made of it is that it actually adjusts the durations of input events, apparently abandoning the original duration values. As noted earlier, the small rhythmic irregularities of a performance may convey important expressive nuance, thus it seems that this information should be retained. The model I propose below leaves the original durational values unchanged (except for a very fine level of rounding) and accomplishes quantization by imposing a metrical structure on those values.

A larger body of research has focused on the problem of deriving metrical structures from quantized input. One interesting proposal is that of Lee (1991) (see also Longuet-Higgins & Steedman 1971 and Longuet Higgins & Lee 1982). Lee's model derives a metrical structure for a melody, considering only its rhythmic pattern (pitch is disregarded). The model begins by looking at the interval between the onsets of the first two events (t1 and t2); it then generates a second interval of this length from t2 to a third time-point (t3). This provides a conjectural rhythmic level, which the model then adjusts if necessary (for example, if the interval t2–t3 contains an event longer than the one starting at t3). Higher and lower levels of beats can then be established in a similar fashion. Steedman's (1977) model adopts a similar approach to low-level meter-finding; it then attempts to find higher levels by looking for repeated pitch patterns. This is an interesting attempt to capture the effect of parallelism on meter, something I will discuss further in section 2.7.

In Lee's model (and other similar ones), only a single analysis is maintained at any given time, and is adjusted (or replaced) as necessary. Other systems for meter-finding operate in a rather different manner, by considering a large number of analyses of an entire excerpt and evaluating them by certain criteria. A simple model of this kind is that of Povel and Essens (1985). Their model applies only to short quantized sequences, and derives only a single level of beats. Within such sequences, some events are labeled as accented: long notes (i.e. notes with long IOIs), or notes at the beginning of sequences of short notes. The best metrical structure is the one whose beats best match the accented notes. Parncutt (1994) proposes a similar model; unlike Povel and Essens's system, however, Parncutt's considers the time intervals between beats, preferring levels whose time intervals are close to a value that has been experimentally established as optimal (roughly 600–700 ms).

Finally, we should consider several recent studies which attempt to perform both quantization and meter-finding. One is the system of

Chafe, Mont-Reynaud and Rush (1982), proposed as part of a larger system for digital editing. This system takes, as input, an unquantized "piano-roll" representation of a piece. It begins by comparing all the sequential durations in the piece, and categorizing them as same or different; it then looks for duration pairs in which the second duration is significantly longer, marking these as "rhythmic accents." Adjacent pairs of accents with similar time intervals are located. These pairs, called "local bridges," are taken to indicate probable beat levels in the piece; these are extrapolated over passages where beats are missing. Another system deserving mention is that of Allen and Dannenberg (1990). This system takes as input a series of interonset intervals, and produces just a single level of beats. For each IOI, several possible rhythmic values are considered (quarter, eighth, dotted-eighth, etc.). The system uses a technique known as "beam search"; at any point it is extending and maintaining a number of possible analyses. This is somewhat similar to a preference rule system; however, a number of heuristic pruning methods are used to limit the set of analyses maintained.

An important proposal building on these earlier efforts is that of Rosenthal (1992). Rosenthal's model is based on "noticers." A noticer finds two (not necessarily adjacent) attack-points in a piece; it takes the time-interval between them and uses this interval to generate a rhythmic level in either direction. In choosing the time-point for the next beat, it has a window of choice, and will prefer a point at which there is an event-onset. Even if no event is found, a beat can be placed, creating a "ghost event"; however, only two adjacent ghost events are permitted. Many rhythmic levels are generated in this way. These rhythmic levels are grouped into families of compatible (simple-ratio-related) levels, which are then ranked by salience criteria. One criterion here is IOI (longer events are preferred as metrically strong); another is motivic structure (levels are preferred which support motivic patterns in the music—it is not explained how this is enforced).

A very different approach is reflected in Large and Kolen's connectionist model (1994). In this model, musical input is represented as a continuous function; onsets of events are represented simply as "spikes" in the function. The model involves an oscillator which entrains to both the phase and period of the input. If an event occurs slightly later than the model "expects," then both the phase and the period are adjusted accordingly. Large and Kolen also present a system with several oscillators which entrain to different levels of metrical structure. One interesting feature is that the greater metrical weight of longer notes emerges naturally from the system; the oscillator entrains more strongly to input events which are not closely followed by another event.

These proposals contain many interesting ideas about meter-finding. However, they are also open to several serious criticisms. As mentioned earlier, many of these models assume quantized input, and thus are really only solving part of the meter-finding problem (this is true of the systems of Parncutt, Povel & Essens, Longuet-Higgins & Steedman, Longuet-Higgins & Lee, and Lee). Most of these systems only consider duration information as input, paying no attention to pitch (Longuet-Higgins & Steedman; Chafe, Mont-Reynaud, & Rush; Povel & Essens; Allen & Dannenberg; Lee; Parncutt; Large & Kolen). (Rosenthal's model appears to consider pitch in that it considers motivic patterns; however, it is not explained how this is done.) It will be shown below that pitch is important for meter-finding in several ways; and it is crucial for any system which attempts to handle polyphonic music, something that none of these systems claim to do. It is also difficult to evaluate these systems, due to the lack of systematic testing (at least few such tests are reported).[5] The model proposed below attempts to address these problems.

2.3 A Preference Rule System for Meter

The model of meter-finding described here was developed by myself and Daniel Sleator and was first presented in Temperley & Sleator 1999. It builds on Lerdahl and Jackendoff's theory of meter as presented in *A Generative Theory of Tonal Music* (*GTTM*) (1983), although it also departs from this theory in important ways, as I will explain.

Before we begin, it is necessary to consider the input to the model. As discussed in chapter 1, the input required is a "note list," giving the pitch (in integer notation, middle C = 60) and on-time and off-time (in milliseconds) of a series of notes; we can also think of this as a "piano-roll," a two-dimensional representation with pitch on one axis and time on the other. (Figure 1.1 shows an example of a "piano-roll.")

Lerdahl and Jackendoff's theory of meter begins with the idea that a metrical structure consists of several levels of beats. They propose four well-formedness rules which define the set of permissible metrical structures. Metrical Well-Formedness Rule (MWFR) 1 states that every event onset must be marked by a beat; this goes along with the idea that metrical structure indicates the quantization of events reflected in rhythmic notation. MWFR 2 requires that each beat at one level must also be a beat at all lower levels. This implies that the beats of each level (except the lowest) are always a subset of the beats at some lower level. (One might argue that cases such as hemiolas can feature both a dotted-quarter-note level and a quarter-note level, such that neither level contains all the beats of the other; we will return to such ambiguities in chapter 8.) MWFR 3 states that every second or third beat at one level must be a beat at the next level up. For example, in 4/4 time, it is assumed that there is never just a quarter-note level and a whole-note level; there is always a half-note level in between. Note that the theory is also limited to duple and triple meters, excluding things like 5/4 and 7/8; this is adequate given *GTTM*'s focus on common-practice music, and is adequate for our purposes as well. Finally, MWFR 4 stipulates that beats must be evenly spaced at the tactus level and "immediately higher" levels: that is, levels up to the measure. (We will refer to the levels from the tactus up to and including the measure level as "intermediate levels.") Lerdahl and Jackendoff argue that lower levels may be somewhat irregular—for example, one tactus beat might be divided in two, the next one in three—and higher levels may also sometimes be irregular; but at intermediate levels, perfect regularity is the norm. For the moment, let us accept all of *GTTM*'s well-formedness rules, although we will revise them somewhat later. Lerdahl and Jackendoff's rules say nothing about how many levels a metrical structure should have; they normally assume roughly five levels, with two above the tactus and two below.

Even assuming that a metrical structure must involve several levels of equally spaced beats, one must still determine the duple or triple relationships between levels, the tempo (the time intervals between beats), and the placing of the beats relative to the music. For this purpose, Lerdahl and Jackendoff posit a set of metrical preference rules, stating the criteria whereby listeners infer the correct structure. Consider figure 2.2, the melody "Oh Susannah" in its entirety. The correct metrical structure is shown above the staff. The most important rule is that beats (especially higher-level beats) should whenever possible coincide with the onsets of events. Since MWFR 1 already mandates that every event-onset should coincide with a beat, the thrust of this rule is that the beats coinciding with events should be as strong as possible. For example, in "Oh Susannah," the structure shown in figure 2.2 is preferable to that shown in figure 2.3, since in the latter case most of the beats coinciding with onsets are only beats at the lowest level. Secondly, there is a preference for strong beats to coincide with longer events. In "Oh Susannah," for example, this favors placing quarter-note beats on even-numbered eighth-note beats (the second, fourth, and so on), since this aligns them

Figure 2.2
The traditional melody "Oh Susannah," showing the correct metrical structure.

Figure 2.3

with the dotted eighth-notes. Similarly, this rule favors placing half-note beats on odd-numbered quarter-notes, since this aligns the long note in m. 4 with a half-note beat. We state these rules as follows:

MPR 1 (Event Rule). Prefer a structure that aligns strong beats with event-onsets.

MPR 2 (Length Rule). Prefer a structure that aligns strong beats with onsets of longer events.

(Our wording of the MPR's [Metrical Preference Rules] rules differs slightly from that in *GTTM*. Also, our numbering of the rules will differ from *GTTM*'s, as our model does not include all the same rules.) Note that the preferred metrical structure is the one that is preferred on balance; it may not be preferred at every moment of the piece. For example, the second A in m. 10 is a long note on a fairly weak beat, thus violating the length rule, but on balance, this structure is still the preferred one.

Figure 2.4
Mozart, Sonata K. 332, I, mm. 1–5.

When we turn to polyphonic music, several complications arise. Consider figure 2.4, the opening of Mozart's Sonata K. 332. The correct metrical structure is indicated by the time signature; why is this the structure that is perceived? The eighth-note level is obviously determined by the event rule. The quarter-note level could also be explained by the event rule; on certain eighth-note beats we have two event-onsets, not just one. This brings up an important point about the event rule: the more event-onsets at a time-point, the better a beat location it is. Regarding the dotted-half-note level—which places a strong beat on every third quarter-note beat—one might suppose that it was due to the long notes in the right hand. This raises the question of what is meant by the "length" of an event. The actual length of events is available in the input, and this could be used. However, this is problematic. In the Mozart, the long notes in the right hand would probably still seem long (and hence good candidates for strong beats) even if they were played staccato. Alternatively, one might assume that the length of a note corresponded to its interonset interval (IOI): the time interval between the note's onset and the onset of the following note. (This is the approach taken by most previous meter-finding programs.) While this approach works fairly well in monophonic music, it is totally unworkable in polyphonic music. In the Mozart, the IOI of the first right-hand event is only one eighth-note: it is followed immediately by a note in the left hand. Intuitively, what we want is the IOI of a note within that line of the texture: we call this the "registral IOI." But separating the events of a texture into lines is a highly complex task which we will not address here (I return to this problem in chapter 4). Our solution to this problem, which is crude but fairly effective, is to define the registral IOI of an event as the interonset interval to the next event within a certain range of pitch: we adopt the value of nine semitones. However, actual duration sometimes proves to be important too. In figure 2.5a and b, the pattern of onsets is the same in both cases, but the fact that the G's are longer in figure 2.5a makes

Figure 2.5

Figure 2.6
Schumann, "Von fremden Laendern und Menschen," from *Kinderszenen*, mm. 13–16. The numbers above the staff indicate time intervals (in milliseconds) between tactus beats, in an actual performance of the piece by the author.

them seem metrically stronger, while in figure 2.5b the C's seem like better strong beat locations. Taking all this into account, we propose a measure of an event's length which is used for the purpose of the length rule: the length of a note is the maximum of its duration and its registral IOI.

So far, we have been assuming a "quantized" input, generated precisely from a score. Lerdahl and Jackendoff's model (like many other models of meter) assumes that quantization has already taken place before meter-finding begins; this is reflected in their MWFR 4, stating that beats at intermediate levels must be exactly evenly spaced. In actual performance, however, beats are of course not exactly evenly spaced, at any level. There may be deliberate fluctuations in timing, as well as errors and imperfections. Even so, the performer's intended beats must have a certain amount of regularity for the meter to be correctly perceived. This is illustrated by figure 2.1. Assigning Performance C the same metrical structure as Performance A (that is, making the G three beats long and the A one beat long) would involve too much irregularity in the spacing of beats; thus an alternative analysis is preferred. A more realistic example is shown in figure 2.6. The numbers above the score represent time intervals between tactus (quarter-note) beats, in a performance of this piece by myself. The lengthening of the second beat of m. 14 is not so extreme as to disrupt the intended meter. However, if it was made much longer than this, the listener might be tempted to infer an extra tactus beat. Clearly, we prefer beat levels with relatively evenly spaced beats, though some variation in spacing is tolerated. This suggests that a simple modification of Lerdahl and Jackendoff's model can accommodate unquantized input: make the demand for regularity a preference rule, rather than a well-formedness rule. We express this as follows:

MPR 3 (Regularity Rule). Prefer beats at each level to be maximally evenly spaced.

It is useful in this context to introduce another idea from *GTTM*, the "phenomenal accent." A phenomenal accent is defined as "any event at the musical surface that gives emphasis or stress to a moment in the musical flow" (Lerdahl & Jackendoff 1983, 17). This could include an event-onset, the onset of a long event, or other things that will be discussed below. Rules that pertain to phenomenal accents of one kind or another could be called "accent rules." By the view presented above, meter-finding becomes a search for a metrical structure which aligns beats with the phenomenal accents of the input, thus satisfying the accent rules, while maintaining as much regularity as possible within each level.

While it is clearly necessary to allow flexibility in the spacing of beats, this makes meter-finding much more difficult. If we could assume perfect regularity at intermediate levels, then once these levels were established at the beginning of the piece, they could simply be extrapolated metronomically throughout the piece; meter-finding would then be a problem that arose mainly at the beginning of pieces. Since beats are not exactly regular, meter-finding cannot work this way; rather, the metrical structure must continuously be adjusted to fit the music that is heard. One might suppose that we could at least assume regularity in the relationship between levels; once a piece begins in triple meter (with every third tactus beat strong), it will stay in triple meter. Further consideration shows that we cannot assume this. Pieces do sometimes change time signatures; in fact, they quite often change to a completely different metrical structure, with a different tempo and different relationships between levels. Consider a slow introduction going into an *allegro*, a symphony or sonata with continuous movements, or a segue from one section to another in an opera. Listeners are generally able to adjust to these changes, and do not bullheadedly continue to maintain the previous meter when the evidence

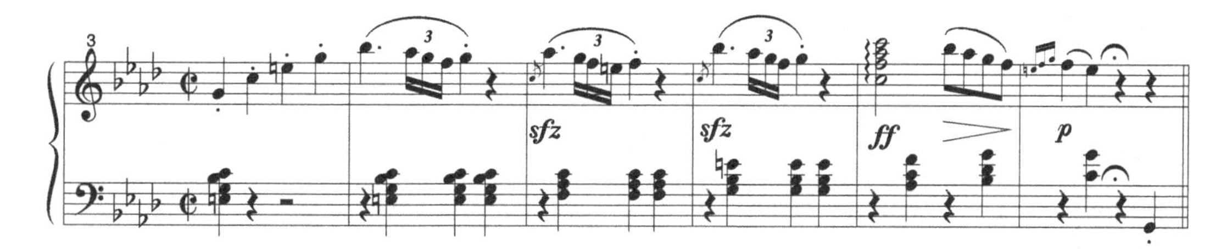

Figure 2.7
Beethoven, Sonata Op. 2 No. 1, I, mm. 3–8.

clearly suggests otherwise. This means that flexibility must be allowed at all levels of metrical structure.

As well as the assumption of perfect regularity, Lerdahl and Jackendoff's model makes another assumption which tends to break down with actual performed music. *GTTM*'s MWFR 1 states that every event-onset must be assigned a beat. For the most part, this rule is valid; as noted earlier, music notation generally assigns each event a position in the metrical structure. However, consider figure 2.7. Measures 5 and 6 each feature "grace notes" before the downbeat of the measure; this is followed by a "rolled chord" in m. 7 and a "turn" figure in m. 8. Each of these ornaments can be represented perfectly well in piano-roll format; figure 2.8 shows my own performance of the passage. However, the notes in these ornaments do not seem to belong to any beat in the metrical structure. Such notes—which Lerdahl and Jackendoff describe as "extrametrical"—are commonplace in all kinds of common-practice music: Baroque, Classical, and Romantic. Lerdahl and Jackendoff's original theory does not really handle extrametrical notes, as they acknowledge (p. 72). The correct way to incorporate them seems to be simply to get rid of MWFR 1. The event rule already provides a preference for aligning beats with events where possible; thus there is strong pressure to avoid making notes extrametrical. The hope is that the model will only label notes as extrametrical when it is appropriate to do so. A trill or turn figure is likely to feature several notes very close together in time, so that it is impossible for the model to assign beats to all of them (without generating implausible extra levels).[6]

Once *GTTM*'s MWFR 1 (requiring every event to coincide with a beat) and MWFR 4 (requiring perfect regularity at intermediate levels) are removed, it can be seen that only two MWFR's remain. We now state these as the two well-formedness rules for the current model:

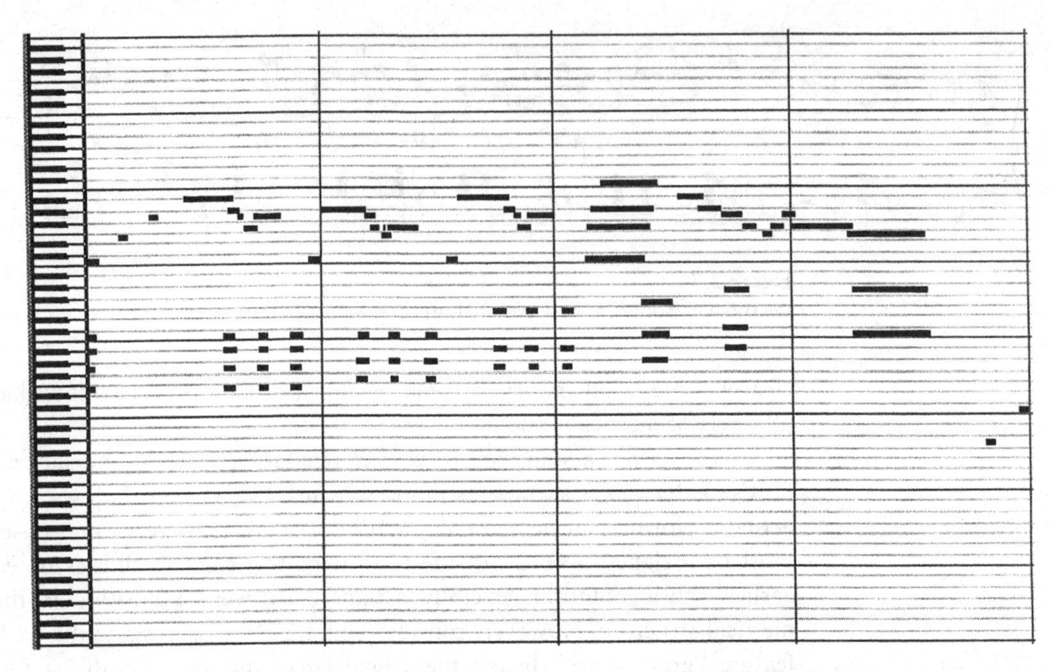

Figure 2.8
A piano-roll representation of the Beethoven excerpt shown in figure 2.7, taken from a performance by the author.

MWFR 1. Every beat at a given level must be a beat at all lower levels.
MWFR 2. Exactly one or two beats at a given level must elapse between each pair of beats at the next level up.

So far, then, our model has these two well-formedness rules, as well as the three preference rules described above: the event rule, the length rule, and the regularity rule.

In initial computational tests, the model as just described proved quite effective in generating good metrical analyses. However, one problem frequently arose; this concerns the highest level, that is, the level two levels above the tactus. (We adopt the convention of calling the tactus level 2; higher levels are then called 3 and 4, and lower levels are 1 and 0. As we discuss below, our program generally only generates these five levels.) Given only the three preference rules listed above, the program will generally identify level 4 as duple, which it usually is, but it often chooses the incorrect phase (this occasionally happens with level 3 as well). In "Oh Susannah," for example (figure 2.2), the program assigns level 4 beats to even-numbered measures, rather than odd-numbered ones (as seems correct). The reason for this is clear: there are often long notes at the ends of phrases, which makes the program prefer them as strong, although they are often weak at higher metrical levels. The solution to this problem lies in phrase structure, or what Lerdahl and Jackendoff call "grouping structure." Grouping structure involves a segmentation of events, with grouping boundaries occurring after long notes and rests. (There are other factors in grouping as well; this is discussed further in the next chapter.) As Lerdahl and Jackendoff note, grouping affects meter: there is a preference for locating strong beats near the beginning of groups. In the first phrase of "Oh Susannah," the beginning of the melody is clearly a group boundary; this exerts pressure for the first measure downbeat to be strong, since it is near the beginning of the group. The long note on the downbeat of m. 4 suggests the end of a group, exerting pressure for the downbeat of m. 5 to be strong as well.[7] Ideally, it would be desirable to incorporate grouping as a factor in meter, as Lerdahl and Jackendoff suggest, at least for higher levels of meter, exerting pressure for strong beats near the beginning of groups. However, getting a computer to recognize grouping boundaries is a difficult problem, especially in polyphonic music (see chapter 3); and meter is also a factor in grouping, creating a chicken-and-egg problem. Instead, we have adopted a cruder solution. At level 4, the program ignores the length rule, and simply prefers beat locations which hit the maximum number of event-onsets. In addition, we give a slight bonus at level 4 for placing a level 4 beat on the first level 3 beat of the piece (rather than the second or third), thus encouraging a level 4 beat near the beginning of the piece. This *ad hoc* solution improves performance somewhat, but incorporating grouping structure would clearly be more satisfactory. For completeness, we state the grouping rule here:

MPR 4 (Grouping Rule). Prefer to locate strong beats near the beginning of groups.

A second problem in our initial tests concerned the relationship between levels. While both duple and triple relationships are common, it appears that there is a slight preference for duple relationships. Consider an undifferentiated sequence of chords, at a typical tactus rate of quarter = 90. Obviously a beat will be heard on each chord. The first one will probably be heard as strong; this is predicted by the grouping rule above. In addition, however, the level above the tactus will most likely be heard as duple rather than triple, with a strong beat on every second chord. (If pressed to supply a lower level, also, we will most likely assume a duple division of the beat rather than a triple one.) We state this preference as follows:

MPR 5 (Duple Bias Rule). Prefer duple over triple relationships between levels.

This is obviously a weak rule, frequently overruled by other factors—as proven by the wealth of music with some kind of triple relationship between levels.

The five rules discussed above—the event rule, the length rule, the regularity rule, the grouping rule (or rather our very limited implementation of it), and the duple bias rule—constitute the preference rules used in our meter program. (Lerdahl and Jackendoff propose a number of other preference rules, some of which will be discussed in a later section.) It remains to be explained how the model is formalized and implemented.

2.4 Implementation

As explained in chapter 1, the basic idea for implementing preference rule systems is straightforward. The system must consider all "well-formed" analyses of a piece. We regard a well-formed metrical structure as one consisting of several levels of beats, such that each beat at one level is a beat at all lower levels, and exactly one or two beats at one level elapse between each pair of beats at the next level up. We assume metrical structures of five levels, with two below the tactus and two above. In many cases, this seems quite sufficient; in some pieces there may be more than two levels above the tactus (perhaps three or four), but we will not consider these higher levels here. (While Lerdahl and Jackendoff only include levels below the tactus where they are needed—that is, where there are event-onsets—our program generates beats even when they are not needed.) According to the procedure outlined in chapter 1, the system must then evaluate all possible well-formed analyses according to the preference rules. The current system essentially works this way, with one important qualification. Rather than considering complete metrical analyses, the system generates one level at a time, starting with the tactus level and proceeding to the upper and lower levels.

The first step is that the input representation (the "note list" described earlier) is quantized into very short segments of 35 ms, which we call pips (the value of 35 ms was simply found to be optimal through trial and error). Every note onset or offset is adjusted to the nearest pip. Beats may also occur only at the start of a pip. There are several reasons for performing this step. One is purely a matter of implementation: in order for the dynamic programming scheme outlined in chapter 1 to work, the input must be quantized at some level; and since the speed of the program depends on the number of segments, it is better to avoid a very small level of quantization. A more substantive reason for quantization is that, in live performance, notes that are intended as simultaneous are not usually played simultaneously; it is very common for the notes of a chord to be 10 or 20 ms apart. The event rule and length rule favor beats at locations where there are event-onsets, but in fact, the notes of a chord may not be at exactly the same location. By quantizing to pips, we attempt to make the notes of each chord exactly simultaneous. (Note that this quantization is quite different from the "quantization-to-beats" seen earlier in the work of Desain and Honing; 35 ms is much smaller than the smallest level of beats.)[8]

We now turn to the problem of deriving the tactus level. A well-formed tactus level is simply a single row of beats. We arbitrarily limit the time intervals between beats to a range of 400 to 1600 ms (this is the typical range for the tactus level). However, there is no "well-formedness" constraint on the variation between beat intervals within this range; a tactus level could in principle have one beat interval of 400 ms, followed by one of 1600, followed by another of 400. An analysis can then be evaluated in the following way. Each pip containing a beat is given a numerical score called a "note score." This is a complicated score which reflects both the number of events that have onsets at that pip, as well as their lengths; pips with more note onsets, and onsets of longer notes, receive higher note scores.[9] Pips with no event-onsets have a note score of zero. In summing these note scores for all the pips that are beat locations, we have a numerical representation of how well the analysis satisfies the event rule (MPR 1) and the length rule (MPR 2). Each pip containing a beat is also given a score indicating how evenly spaced it is in the context of the previous beats in that analysis (MPR 3). There are various ways that this could be quantified, but we have found a very simple method that works well: the regularity of a beat B_n is simply given by the absolute difference between the interval from B_n to B_{n-1} and that from B_{n-1} to B_{n-2}. This beat-interval score therefore acts as a penalty: a representation is preferred in which the beat-interval scores are minimized. The preferred tactus level is the one whose total note score and (negative) beat-interval score is highest. One complication here is that simply defining the note score for a beat level as the sum of the note scores for all its beats gives an advantage to analyses with more beats. Thus we weight the note score of each beat by the square root of its beat interval (the interval to the previous beat). The reason for using the square root here will be explained further in section 2.8.

The beat-interval scores capture the role of context in metrical analysis. The best location for a tactus beat within (for example) a one-second segment depends in part on the beat-interval penalty for different pips, but this in turn depends on the location of previous beats, which in turn depends on the beat-interval scores for previous pips (and their note scores), and so on. In theory, then, all possible analyses must be considered to be sure of finding the highest-scoring one overall. However, the number of possible analyses increases exponentially with the number of pips in the piece. There is therefore a search problem to be solved of finding the correct analysis without actually generating them all.

The search procedure uses a variant of the dynamic programming technique discussed in chapter 1. Imagine moving through the piece from left to right. At each pip, we consider each possible beat interval (the interval from the current pip to a previous one), within the allowable interval range for the tactus; we find the best analysis of the piece so far ending with that beat interval. To do this, we must consider each possible beat interval at the previous pip and the associated best-so-far score (which has already been calculated). This allows us to calculate beat-interval scores for the current pip, since this depends only on the difference between the current beat interval and the previous one. (The note scores for pips can of course be easily calculated; they do not depend on what happens elsewhere in the analysis.) For each beat interval at the current pip, we choose the previous beat interval leading to the optimal score; this gives us a new set of "best-so-far" analyses for the current pip. When we reach the end of the piece, the pip and beat-interval with the highest total score yields the preferred analysis for the piece. (Since there is no guarantee that the final pip of the piece is a beat location, we must consider all pips within a small range at the end of the piece.)

The preferred tactus level for a piece, then, is the one that emerges from the procedure just described. It remains to be explained how the other levels are derived. We derive the other levels "outward" from the tactus (level 2); first we derive the upper levels, 3 and then 4, then the lower levels 1 and then 0. There are, first of all, "well-formedness" constraints here that did not exist with the tactus level. Because of MWFR 1, the possible locations for level 3 beats are limited to level 2 beat locations; and level 2 beat locations are obligatory locations for level 1 beats. The same rule applies to the levels 0 and 4. MWFR 2 also comes into play: exactly one or two level 2 beats must occur between each pair of level 3 beats, and the same rule applies between every other pair of adjacent levels.

The application of the preference rules to the outer levels is similar to that described for the tactus level, though there are some differences. The event rule and the length rule apply, and are quantified in exactly the same way.[10] Here the duple bias rule (MPR 5) is also incorporated: we simply provide a small bonus for duple rather than triple relationships between the level being added (higher or lower) and the existing one. The regularity rule (MPR 3) applies here as well, but in a rather different way than at the tactus level. At the upper levels, rather than imposing a penalty for differences between adjacent beat intervals (as it does with the tactus level), the regularity rule imposes a penalty for changes in the number of beats that elapse between upper level beats. If one level 2 beat elapses between a pair of level 3 beats, and two level 2 beats elapse between the next pair, a penalty is imposed. At level 4, the rule works the same way with respect to level 3. At level 1, similarly, a penalty is imposed depending on regularity in relation to level 2: if one level 2 beat is divided duply, and the next triply (or vice versa), a penalty is imposed. However, the beat-interval measure of regularity also imposes a penalty here, to ensure that the division of each tactus beat is relatively even. Similarly, the regularity of level 0 is evaluated in relation to level 1. In this way, other things being equal, there is a preference for maintaining either a consistent duple or triple relationship between each pair of levels, rather than frequently switching back and forth. The logic of this is as follows. It is the job of the regularity rule at the tactus level to penalize excessive changes in tempo (that is, to favor some other interpretation rather than a radical change in tempo). Using a "beat-interval" measure of regularity at upper and lower levels would simply be adding another, redundant, penalty for changes in tempo. Rather, it seemed to us that, at the higher and lower levels, the regularity rule should function to penalize changes in relationships between levels—roughly speaking, changes in *time signature*—not changes of tempo.

These quantitative expressions of the preference rules can be used to evaluate possible analyses at each outer level. A dynamic programming procedure similar to that described for the tactus is then used to find the best analysis for each level.

2.5 Tests

Two tests were done of the program described above. First, it was tested on quantized input files generated from a corpus of excerpts from the

Kostka and Payne theory workbook (1995b). Second, it was tested on MIDI files generated from performances by an expert pianist of certain excerpts from the same corpus (the ones for solo piano).

The Kostka-Payne workbook was judged to be an appropriate source of testing material, since it contains excerpts from a variety of different composers and genres within the common-practice period. In addition, the workbook provides harmonic and key analyses, making it a suitable source for testing the harmonic and key programs as well (as discussed in later chapters).[11] The corpus of excerpts used consisted of all excerpts from the workbook of eight measures in length or more for which harmonic and key symbols were provided. (Shorter excerpts were not used, since it seemed unfair to expect the program to infer the meter or key for very short excerpts.) This produced a corpus of forty-six excerpts. Nineteen of these were for solo piano; the remainder were mostly chamber pieces for small ensembles. The same corpus was also used for testing the pitch-spelling, harmonic, and key programs.

For the quantized test, quantized input files were generated for the forty-six excerpts in the corpus. The tempi of course had to be determined; I simply chose tempi that I thought were reasonable. Once this was done, the duration of each note could be determined objectively. One problem concerned extrametrical notes, such as trills and grace notes. Due to the difficulty of deciding objectively on the timing for these, all such notes were excluded.

The unquantized files were generated from performances on a MIDI keyboard by a doctoral student in piano at Ohio State University. The student was given the excerpts beforehand, and asked to practice them. She was then asked to play them on the keyboard as naturally as possible, as if she was performing them. She was allowed to play each excerpt repeatedly until she was satisfied with the performance. The student was discouraged from using the sustain pedal (since the program does not consider sustain pedal), although she was allowed to use the pedal and sometimes did so. No other instructions or input were given; in particular, the student was not discouraged from using expressive timing or extrametrical notes in her performance.

The input files for both the quantized and unquantized performances were then analyzed by the program. As mentioned above, the metrical structure for a piece is usually explicit in notation up to the level of the measure. Given the program's output for a piece, then, it was possible to determine objectively whether this analysis was correct.[12] In the evaluation of the program's output, each metrical level was examined individually. For each measure of each excerpt, it was determined whether the program's output was correct for each metrical level: that is, whether beats were placed on exactly the right event-onsets, with the right number of beats elapsing in between onsets. Recall that the program generates five levels, numbered 0 through 4, with level 2 being the tactus level. (The program did not always choose the correct level as the tactus; we will return to this problem below. In this test, however, we simply compared each of the program's levels with whichever level of the correct metrical structure it matched most closely.) Since the scores for the excerpts contained no information about metrical levels above the measure, hypermetrical levels—levels above the measure—could not be scored. In twenty-one of the excerpts, level 4 was hypermetrical; these excerpts were excluded in the evaluation of level 4. (Two excerpts were excluded for level 3 for the same reason.) Also, for sixteen of the excerpts, the lowest metrical level was not used at all (that is, there were no notes on any level 0 beats, either in the correct analysis or in the program's output); these excerpts were excluded from the evaluation of level 0. (One excerpt was excluded for level 1 for the same reason.)

The results of the test are shown in table 2.1. It can be seen that the program performed somewhat better on the quantized files than on the unquantized files. On the quantized files, it achieved the best results on the tactus level; on the unquantized files, its results were best on the tactus level and level 4. However, its success rate did not vary greatly between different levels.

2.6 Problems and Possible Improvements

While the performance of the program on the tests described above seems promising, it is far from perfect. We have examined a number of the errors that the program makes (in these tests and others), and considered possible solutions. We will begin by discussing the quantized files first, and then consider the unquantized files. It should be noted, first of all, that some of the errors made by the program should probably not be regarded as errors. For example, in the quantized file for the Chopin passage in figure 2.9 (a segment of a larger excerpt), the sixteenth-note in the right hand in m. 41 was treated as being simultaneous with the triplet eighth-note just before (likewise the sixteenth-note in m. 42). Though strictly incorrect, it is very likely that it would be heard this way, if played with perfect precision (though of course it never would be). Thus the program's performance is somewhat better than it appears.

Table 2.1
Test results for metrical program on Kostka-Payne corpus

Metrical level	0	1	2	3	4
Quantized input files	88.7	89.6	94.4	86.2	93.5
Unquantized input files	71.5	77.0	85.5	83.0	85.6

Note: The numbers in each cell show the percentage of measures in the corpus for which the program's analysis was correct at a given metrical level. For the quantized corpus, there were 553 measures total; for the unquantized one there were 235 measures.

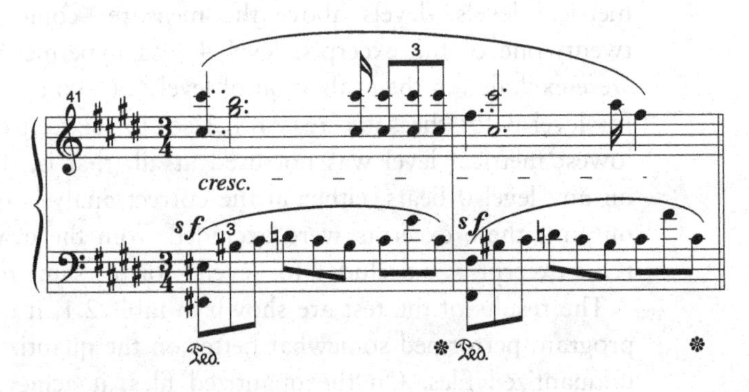

Figure 2.9
Chopin, Nocturne Op. 27 No. 1, mm. 41–2.

Having said that, the program did make a considerable number of real errors on the quantized files. Consider figure 2.10a, from a Chopin Mazurka; the tactus here is the quarter note. Initially, the metrical program produced a 3/4 metrical structure, but "out of phase" with the correct analysis; that is, level 3 beats were placed on the second and fifth quarter-note beats rather than the first and fourth. A similar error was made in the Schubert song shown in figure 2.10b; here, level 3 (the quarter-note level in this case) was identified as duple, but again out of phase, with beats on the second and fourth eighth-notes of each measure. Perceptually, the important cue in these cases seems to be harmony. In the Chopin there are changes of harmony on the first quarter-note of each notated measure (with the possible exception of the second); in the Schubert, too, the downbeats of the first and third measures feature clear chord changes. (The factor of harmony is noted by Lerdahl and Jackendoff in their MPR 5f.) However, this presents a serious chicken-and-egg

Figure 2.10
(A) Chopin, Mazurka Op. 63 No. 2, mm. 1–4. (B) Schubert, "Auf dem Flusse," mm. 13–17.

problem; as we will see in chapter 6, meter is an important input to harmony as well. One solution would be to compute everything at once, optimizing over both the metrical and harmonic rules, but we have not attempted this; we have, however, experimented with an alternative solution to the problem. First we run the basic metrical algorithm, but generating only the tactus level and lower levels. Then we run the output through the harmonic program (described in chapter 6), in what we call "prechord" mode. In this mode, the harmonic program processes the piece in the usual way, but outputs only a list of "prechord" events, which are simply chord changes. (All that matters here is the timepoints of the chord changes, not the actual roots.) In other words, the harmonic program determines what it thinks is the best location for chord changes, based on very limited metrical information. These prechord statements are added to the original note list. Now this is fed back into the basic metrical program, which has been modified so that it recog-

Figure 2.11
Beethoven, Sonata Op. 2 No. 1, I, mm. 41–5.

nizes prechord events and has a preference for strong beats at prechord locations. In effect, the metrical algorithm is now taking harmony into account in choosing the metrical structure. This allows us to get better results in several cases, including the Chopin and Schubert examples.[13]

In both the examples discussed above, one might point to another factor as influencing our perception of the meter. In both cases, the bass notes are in positions that should be considered metrically strong. One might get the right results in such cases, then, by exerting special pressure for strong beats on bass notes. This principle is reflected in Lerdahl and Jackendoff's MPR 6, which states "prefer a metrically stable bass." We experimented with an implementation of this rule. The program first identified bass notes, where a bass note was defined as a note below G3 in pitch, and not followed within 500 ms by any other note below it or less than six half-steps above it. The program then gave a bonus to the note score for bass notes, so that there was a greater incentive to make them metrically strong than for other notes. However, two problems arose. In the first place, defining bass notes proves to be a hard problem; it is not difficult to find notes that are incorrectly included or excluded by the definition posed above. (Here again, it might be necessary to analyze polyphonic lines in order to identify bass notes correctly.) Secondly, we found that this modification created new errors; quite often, bass notes actually *do* occur on weak beats, and the program needs to be able to allow this. Examples of this can easily be found; figure 2.11 shows one, from a Beethoven sonata.[14] Our tests suggest, then, that harmonic information is more useful than bass-note information as a cue to metrical structure.

We should also consider the program's performance on the unquantized files. It can be seen that the program's success rate on the unquantized files is somewhat lower that on the quantized ones. Most of the errors the program made on the quantized input files were made on the corresponding unquantized files as well; the program also made a number of additional errors on the unquantized files. The most common kind of error was what might be called "smudged chord" errors, where the notes of a chord, intended to be simultaneous, were not analyzed that way; either some of the notes of the chord were placed on a different beat from the correct beat, or they were analyzed as "extrametrical" (not on any beat at all). (Notes that were intended as extrametrical often caused problems here; in some cases a beat was placed on an extrametrical note instead of the intended metrical note.) Other errors were caused by expressive fluctuations in timing. Not suprisingly, fermatas and other major fluctuations in tempo usually confused the program, causing it to put in extra beats. However, there were relatively few errors of this kind; the slight ritards that typically occurred at phrase-endings did not usually cause errors.

2.7 Other Factors in Metrical Structure

Several other possible factors in metrical analysis should be mentioned here. One is loudness, or what Lerdahl and Jackendoff call "stress." We have not incorporated loudness as a factor, although it would be quite easy to do so, allowing a loudness value for each note and factoring this into the "note scores." Loudness does not appear to be a centrally important cue for meter; consider harpsichord and organ music, where loudness can not be varied (see Rosenthal 1992 for discussion). Still, it can be a significant factor when other cues are absent.[15] Anther kind of stress that is important in vocal music is linguistic stress; there is a tendency to align strong beats with stressed syllables of text (see Halle & Lerdahl 1993). This rule will assume considerable importance in chapters 9 and 10.

Little has been said about the role of melodic accent in metrical analysis. It has sometimes been argued that a melody can convey points of accent simply by virtue of its shape; one might wonder whether these melodic accents could serve as metrical cues. A number of hypotheses have been put forward regarding what exactly constitutes a melodic accent. Various authors have suggested that melodic accent can be conveyed by pitches of high register, pitches of low register, pitches following large intervals (ascending, descending, or both), or pitches creating a change in melodic direction (see Huron & Royal 1996 for a review). Inspection of the program's errors on the Kostka-Payne corpus gives little indication that incorporating a factor of melodic accent would improve results. There are other reasons, as well, to be skeptical about the idea of melodic accent. Huron and Royal (1996) analyzed melodic databases (folk songs and instrumental melodies) and examined the correlation between the various kinds of melodic accent cited above and actual metrical strength (as indicated by the notation). In all cases the correlation was very small. (The best correlation found was with a model proposed by Thomassen [1982], defining melodic accent in a rather complex contextual way, but even here only a small correlation was found.) This is not to deny that there might be some other useful and psychologically valid notion of "melodic accent" which is unrelated to meter. The evidence for melodic accent as a cue to meter, however, is slim.

A final factor that should be mentioned is parallelism. Parallelism refers to repeated patterns of pitch and rhythm; as Lerdahl and Jackendoff observed, there is a tendency to align metrical levels with such patterns, so that repetitions of a pattern are assigned parallel metrical structures. While parallelism is certainly a factor in meter, it is not so easy to tease out the role of parallelism from other metrical cues. In a passage with a simple repeating rhythmic pattern, such as the Mozart excerpt in figure 2.4, one might suppose it is parallelism that causes us to assign the same metrical analysis to each instance of the pattern. However, this may simply be due to the fact that each instance of the pattern, considered independently, naturally yields the same metrical analysis. A clearer demonstration of parallelism is found in a repeating pitch pattern such as figure 2.12. In this case, there is no apparent factor favoring any particular interpretation of the three-note pattern itself. (There is perhaps a slight tendency to put a strong beat at the beginning of the entire excerpt, due to the grouping rule, but this is clearly a weak factor; it is perfectly possible to hear the third note of the excerpt as metrically strong.) Even so, there is a definite tendency to hear some kind of triple metrical structure here, with every third eighth-note strong. I say "some kind of" triple structure, because importantly, the factor of parallelism itself does not express a preference between different triple structures. As stated earlier, parallelism relates to the alignment of the metrical struc-

Figure 2.12

ture with instances of a repeated pattern; repetitions of the pattern should be assigned parallel structures. This means, essentially, that the *period* of the metrical structure should be the same as the period of the pattern. (More specifically, *one level* of the metrical structure should have the same period as thc pattern, so that the interval between beats at that level is the same as the length of the pattern.) Parallelism in itself says nothing about *phase*, that is, where in the pattern the strong beats should occur. This, of course, is governed by the event rule and other rules.

As well as the tendency to assign parallel metrical structures to parallel segments, there is another aspect to parallelism, which affects higher levels of meter in particular. Consider figure 2.13, from the first movement of Beethoven's Piano Sonata Op. 2 No. 1. The meter is clear enough up to the level of the measure, but which are stronger, odd-

Figure 2.13
Beethoven, Sonata Op. 2 No. 1, I, mm. 20–32.

numbered or even-numbered measures? The first part of the passage suggests odd-numbered measures as strong (the fact that m. 21 is the beginning of a melodic group is a major factor here). However, m. 26 seems stronger than m. 27. The reason, I submit, is the parallelism between m. 26 and m. 27: the repetition in rhythmic pattern, and partly in pitch pattern as well. As already noted, we tend to choose a metrical structure with the same period as the parallelism, but this rule expresses no preference between the "odd-measure-strong" and the "even-measure-strong" hearing. Another factor is involved here: in cases where a pattern is immediately repeated, we prefer to place the stronger beat in the *first* occurrence of a pattern rather than the second; this favors m. 26 as strong over m. 27. This "even-measure-strong" hearing persists for several measures, but in m. 31 we are again reoriented, as the parallelism between mm. 31 and 32 forces us to switch back to an odd-measure-strong hearing. This factor of "first-occurrence-strong" is an important determinant of higher-level meter, and can frequently lead to metrical shifts.[16]

Despite the clear role of parallelism in meter, it would be a very difficult to incorporate parallelism into a computational model. The program would have to search the music for patterns of melodic and rhythmic repetition. Since this seems to me to be a huge and complex problem, I am not addressing it formally in this book. However, we will give some attention to the issue of musical pattern recognition in chapter 12. (For an interesting attempt to handle parallelism, see Steedman 1977.)

The previous sections have listed several factors which seem important to metrical analysis, though we have not included them in our implemented model. It will be useful to list them here, since I will refer to them in later chapters.

MPR 6 (Harmony Rule). Prefer to align strong beats with changes in harmony.

MPR 7 (Stress Rule). Prefer to align strong beats with onsets of louder events.

MPR 8 (Linguistic Stress Rule). Prefer to align strong beats with stressed syllables of text.

MPR 9 (Parallelism Rule). Prefer to assign parallel metrical structures to parallel segments. In cases where a pattern is immediately repeated, prefer to place the stronger beat on the first instance of the pattern rather than the second.

2.8 Choosing the Right Tactus

A metrical structure does not just consist of several levels of equal strength and importance. As noted earlier, there is generally one level, the tactus, which corresponds to the main "beat" of the music. There is psychological evidence that the tactus level serves a special cognitive function. In performance, there is less variance of beat intervals in the tactus level than in other levels, suggesting that the tactus serves as the internal "timekeeper" from which other levels are generated (Shaffer, Clarke, & Todd 1985). For a metrical model, then, it is important not only to identify all the levels of the metrical structure, but to identify the correct level as the tactus.

In the current case, the program indicates its choice of tactus as level 2—the first level it generates. It was noted earlier that the program's choice of tactus intervals is limited to a range of 400 to 1600 ms. Thus the program is really being forced to choose a tactus in the appropriate range. However, in a piece with a constant tempo and a regular structure of duple and triple levels, there will always be at least two levels within this range that the program can choose from. The program's choice here is guided by the same preference rules noted earlier: most importantly, it prefers the level that hits the most note-onsets (especially onsets of longer notes). (Recall that "note scores" are added to the score of a tactus level for every note that it hits.) Consider a simple sequence of quarter notes. If this sequence is given to the program with a tempo of quarter = 120, it will choose the quarter note as the tactus (with tactus intervals of 500 ms). If the same sequence is given with a tempo of quarter = 60, it will *still* choose the quarter note level as the tactus, although this level is now half as fast, with tactus intervals of 1000 ms. In both cases, the alternative beat level (500 or 1000 ms) is also found, but it is treated as a higher or lower level than the tactus level. The program's behavior here is surely correct; however, attaining this result is not trivial, and depends on a subtle feature of the scoring. Suppose the note scores for a given tactus analysis were simply summed. Then, given a sequence of quarter notes at a tempo of quarter = 60, the program would have no incentive for choosing the 1000 ms level; the score for the 500 ms level would be just as high, since it hits just as many notes as the 1000 ms level. On the other hand, suppose that note scores were weighted according to the beat intervals. Then, given a quarter-note sequence at quarter = 120, the program would have no incentive to choose the 500 ms level; the 1000 level would hit half as many notes, but would score twice as much for each one. This is why note scores are weighted by the *square root* of the

Figure 2.14
Beethoven, Sonata Op. 13, II, mm. 1–4.

beat interval. This allows the program to choose intelligently between possible tactus levels over a wide range of time intervals. It is an attractive feature of the current approach that the same criteria the model uses in choosing a set of beat levels also lead it to choose the correct level as the tactus.

Having said this, it should be noted that the program is only moderately successful in choosing the right level as the tactus. In the Kostka-Payne corpus, it almost always chooses the fastest level within the possible range, though this is not always the most plausible tactus. In figure 2.14, for instance, it chooses the sixteenth level (with intervals of 600 ms), though the eighth or even the quarter note level would be a better choice. It is interesting to ask, what is it that makes us favor these higher levels as the tactus here? Simply from the point of view of the preference rules, the sixteenth note level seems like a reasonable choice, since there are notes on every sixteenth note beat. There are many unaccompanied melodies—with (at most) one note on each tactus beat—in which the perceived tactus is in the range of 600 ms or faster; for example, "Oh Susannah" (figure 2.2) might easily be performed at this tempo (quarter = 100). One might point to the relatively slow harmonic rhythm in figure 2.14 as the reason for the slower tactus, or the fact that the sixteenth-note line is clearly accompanimental. I believe there is another factor, however, which is that there is no support in figure 2.14 for any metrical level *lower* than the sixteenth-note level. It may be that we prefer to choose a tactus level which allows lower levels to score well also (rather than just hitting all the same events as the tactus level). In "Oh Susannah," a tactus of 600 ms is more reasonable than in figure 2.14, since there are many notes in between the tactus beats to support the lower levels.

This brings up an important general point about the current program. We have assumed that metrical structures can be generated by finding the tactus level first, and then the upper and lower levels. This seems to work well in a large majority of cases. Occasionally, however, it seems that the program needs to consider the lower levels that will be allowed by a given tactus level. In such cases, having the program search for entire metrical structures—generating all the levels at once—might result in better performance. While this approach could cause computational problems, as it would greatly increase the number of analyses to be considered, it might in some cases produce improved results.

3
Melodic Phrase Structure

3.1 Musical Grouping and Phrase Structure

The study of musical grouping in psychology can be traced back to the Gestalt school—a group of psychologists active in Germany in the 1920s. These psychologists proposed a set of principles governing the grouping of elements in perception. One principle is similarity: we tend to group together things that are similar. Another is proximity: we tend to group things that are close together (in space or in some other dimension). A third rule, "good continuation," states that elements which follow each other in a linear pattern are grouped in this fashion; for example, two lines crossing in an X will tend to be seen as two crossing lines, rather than (for example) as a V over an upside-down V.[1] These principles were seen to interact and compete with one another. In figure 3.1a, similarity seems to be the decisive factor, so that the squares form one group and the circles form another; in figure 3.1b similarity is overruled by proximity. Gestalt rules are similar to preference rules, in that each rule expresses an opinion as to how well it is satisfied by a given analysis of the input, and these opinions are combined together in some way to yield the preferred analysis.

Gestalt principles were held to apply to perception generally, both visual and auditory; and the Gestalt psychologists were well aware of their possible application to music.[2] The rules may seem both somewhat obvious and somewhat vague, and the Gestalt psychologists themselves had little success in developing them into a more rigorous theory. But they have provided a useful starting point for much of the recent work on grouping, as we will see below.

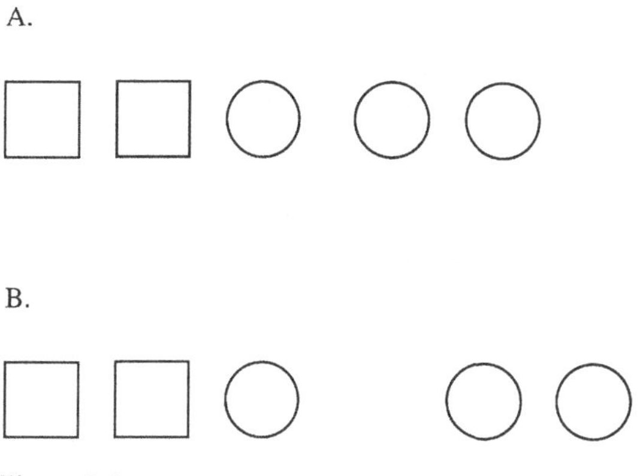

Figure 3.1

Grouping in music is an extremely complex matter. Generally, a piece of music can be divided into sections and segments at a number of different levels. Lerdahl and Jackendoff proposed the term *grouping* to describe the general process of segmentation at all levels (and the multi-leveled structure that results). The term *phrase* generally refers to basic, low-level groups, containing a few measures and a handful of notes. In this chapter we will mostly restrict our attention to this lower level of grouping, for reasons that I will explain. Most previous psychological and computational work, too, has focused on lower-level grouping structure.[3]

3.2 Studies of Musical Grouping in Psychology

It is worth considering the *purpose* of grouping—in musical perception and in perception generally. In some cases, grouping serves a clear function of parsimony. If elements can be grouped into larger familiar elements, then the grouping reduces the number of elements that must be stored: for example, it is much easier to remember a list of 10 words than a list of 70 letters. However, there seems to be a strong tendency to group smaller elements into larger ones even in cases when the larger elements formed are not familiar—as in the Gestalt-type patterns of figure 3.1. Undoubtedly this is because certain conditions—such as the proximity of elements in space (or time)—suggest that the elements are part of the same larger entity, which should be encoded now to aid further processing (for example, because it is likely to recur in the future). Indeed, grouping structure plays an important role in the recognition of repeated patterns or "motives" in music, as I will discuss in chapter 12.

Figure 3.2
From Deutsch 1980.

The grouping of events as a phrase may also affect our processing of the events themselves. It may be, for example, that we are are more sensitive to small units such as intervals or motives if they occur within phrases rather than across them (and there is indeed evidence for this, as I discuss below). To put it another way, it may be that a phrase serves as a kind of "psychological present," such that all the events within the phrase are particularly salient and available for processing while the phrase is in progress.[4]

If we assume that grouping events together involves storing them in memory as a larger unit, this suggests a possible approach to the psychological study of grouping. If a sequence of notes is being perceived as a group, it should be more easily identified and recognized than other sequences. In one experiment, Dowling (1973b) presented subjects with a series of notes interrupted by pauses; listeners were much more able to learn and identify sequences of notes separated by pauses than sequences that spanned pauses. Several other experiments have reached similar conclusions, using melodies with repeated intervallic patterns. In such cases, it is generally found that subjects can learn the melody more easily when pauses occur between repetitions of the pattern, as in figure 3.2a, rather than when the pauses break up instances of the pattern, as in figure 3.2b (Deutsch 1980; Boltz & Jones 1986). Apparently, the presence of pauses between repetitions of the pattern helps people to identify instances of the pattern (or perhaps the presence of pauses within pattern instances hinders this), and this allows people to encode the melody more efficiently.[5] Such experiments demonstrate both the reality of grouping and the importance of pauses in the way groups are perceived.

While pauses are perhaps the most well-established cue in grouping, Deliege (1987) has shown that a number of other factors play a role as well. Deliege examined the role in grouping of a number of factors cited by Lerdahl and Jackendoff (whose theory of grouping will be discussed at greater length below). Lerdahl and Jackendoff's rules state that there is a preference for grouping boundaries after rests (or at the end of a slur), at long intervals between attack-points, and at changes of articulation, dynamics, note length, and register (the possibility of a rule pertaining to changes in timbre is mentioned but not pursued [1983, 46]). Deliege categorized these factors into two classes: the rules pertaining to rests and attack-points relate to the Gestalt principle of proximity (in time), whereas the rules pertaining to changes in dynamics and other things relate to the Gestalt principle of similarity. Deliege identified a number of passages from the common-practice literature which seem to feature grouping boundaries according to one or more of these rules. Subjects were then played recordings of these excerpts, and asked to indicate where they thought the grouping boundaries were; there was a strong tendency to place them at the locations predicted by the rules. This suggests that all of the factors cited by Lerdahl and Jackendoff play a role in grouping in some circumstances. A second experiment involved artificially constructed nine-note melodies, such as the one in figure 3.3a, in which two grouping cues were placed at different locations. For example, a change in dynamics might occur after note 3 and a rest after note 4. Again, subjects had to indicate the grouping boundaries they perceived. Timbre proved to be the most "preferred" cue (the one that determined grouping in the highest proportion of cases); the rest-slur, attack-point, register, and dynamics rules were also important. As Deliege points out, we could not expect all the rules to be consistently important, since some rules are in conflict with others. In figure 3.3b, the change in length after the third note could be

Figure 3.3
(A) From Deliege 1987.

regarded as a cue for a grouping boundary there, but one might also say that the greater attack-point interval after the fourth note (relative to the previous notes) demands a grouping boundary there.

The main conclusion that can be drawn from these experiments is that all of the cues described by Deliege can sometimes be cues to grouping. More specific conclusions are difficult to draw. In the first experiment, Deliege chose excerpts that contained cues for each rule, but it is not clear what qualified as a cue; how large does a rest or a change in register have to be to qualify as a grouping cue? (One also wonders if the cue in question was always the only cue at that point in the excerpt.) The second experiment is of greater interest, since it tells us something about the relative importance of cues of particular magnitudes. It seems problematic, however, to claim—as Deliege does—that one *kind* of cue is more important or—in Deliege's terms—more "preferred" than another. If we assume that all the factors cited above might sometimes be cues, and that increasing the magnitude of the factor increases its strength as a cue, then probably a very large presence of one factor could always override a very small presence of another. What is needed is to establish the magnitude of one cue that is equal to the magnitude of another.

An experiment by Tan, Aiello and Bever (1981) suggests that tonal factors may also play a role in grouping. Subjects (both musicians and nonmusicians) were played short sequences of tones; they were then played a two-tone probe and asked if the probe had occurred in the sequence. The reasoning was that, if the two-tone probe spanned a perceived phrase boundary, it should be more difficult to identify. (This experiment employs the reasoning—noted earlier—that the identification of notes as a phrase should promote the identification of smaller units *within* the phrase, not just the phrase itself.) If a point of harmonic closure occurred between the two tones of the probe, the probe was less easily recognized, suggesting that subjects were inferring a segment boundary there. In figure 3.4, for example, subjects had difficulty identifying the probe when it spanned the tenth and eleventh notes; the authors' explanation for this was that subjects interpreted notes 7

Figure 3.4
From Tan, Aiello & Bever 1981.

through 10 (D-G-E-C) as implying a V-I cadence, and therefore inferred a phrase boundary after the C. Another study, by Palmer and Krumhansl (1987), explored the influence of both tonal and metrical factors on segmentation. An excerpt from a Mozart sonata was performed, stopping at various points, and listeners were asked to indicate which stopping points sounded more good or complete. Excerpts were generally judged as more complete when the final event was more tonally stable in terms of Krumhansl's key-profiles (discussed further in chapter 7) and more metrically stable (i.e. aligned with a relatively strong beat).

3.3 Models of Grouping Structure

As with meter, attempts to model the grouping process include both speculative theoretical proposals and more rigorous computational studies. However, there has been far less work of either kind on grouping than on meter. One difficulty with studying grouping lies in determining the "correct" analysis—the one listeners would perceive—for a given input. With meter, the correct analysis for a piece is usually clear, at least up to the level of the measure. (Not only is it indicated by music notation, but there usually seems to be agreement that the structure indicated by the notation is perceptually correct.) With grouping, the correct analysis is often much less certain. In much notated music, there is no explicit indication of grouping structure (at least none provided by the composer); in other music, low-level structure is indicated by slurs, but only in certain passages, where it also indicates "legato" articulation. (The connection between legato articulation and phrase structure will be discussed further in section 11.6.) The problem is compounded by the fact that our intuitions about grouping are often ambiguous and vague, much more so than meter—a point I will discuss below. Despite these problems of ambiguity and indeterminacy, there seems to be enough agreement about grouping to make the modeling of grouping analysis a worthwhile endeavor.

In music theory, grouping has long been a central concern. As early as 1793, the German theorist Heinrich Koch proposed a complex hierarchical system of musical segmentation, assigning terms for each hierarchical level.[6] However, progress in this area has often been hindered by confusion between grouping structure and metrical structure. In their groundbreaking book *The Rhythmic Structure of Tonal Music* (1960), Cooper and Meyer proposed a single hierarchical structure which captured both meter and grouping. Groups of notes were classified in terms of rhythmic units borrowed from poetry—iamb, trochee, and the like—

implying both a segmentation and an accentual structure. (For example, an iamb consists of two elements in the relation weak-strong.) These units then entered similar relations with other units in a recursive manner. More recently, however, a consensus has emerged that meter and grouping are best regarded as independent structures (Schachter 1976; Lerdahl & Jackendoff 1983; Rothstein 1989). Meter involves a framework of levels of beats, and in itself implies no segmentation; grouping is merely a segmentation, with no accentual implications. However, this is not to say that there is no interaction between meter and grouping; we return to this issue below.[7]

The first computational study of musical grouping—and a seminal study in the computational analysis of music generally—was Tenney and Polansky's model of grouping in monophonic music (1980). Like other work on grouping, Tenney and Polansky cite two main kinds of grouping factors, arising out of the Gestalt rules: proximity—the preference for grouping boundaries at long intervals between onsets, and similarity—the preference for group boundaries at changes in pitch and dynamics. Tenney and Polansky also point out, however, that the distinction between proximity and similarity factors is somewhat arbitrary. In some cases one could regard something either as an "interval" in some kind of space or as a "difference" in some parameter: this is the case with both pitch and onset-time, for example. The important point is that intervals and differences can each be quantified with a single number, and these values can then be added to produce an overall interval value between each pair of adjacent events. Tenney and Polansky observe that whether a certain value in some parameter (an interonset-interval, for example) is sufficient to be a grouping boundary depends on the context. For example, an IOI is unlikely to be sufficient for a grouping boundary if either the preceding or the following IOI is longer. This leads to the following proposal: an interval value in some parameter tends to be a grouping boundary if it is a local maximum, that is, if it is larger than the values of intervals on either side. The same applies when values for different parameters are added together. One nice effect of this is that it prevents having groups with only one note, something which clearly should generally be avoided. (Since two successive intervals cannot both be local maxima—one must have a higher value than the other—it is impossible to have a single note with grouping boundaries on either side.) The authors describe an implementation of this algorithm. The program searches for smallest-level groups by identifying local maxima in interval values in the way just described; it then finds larger-level groups by looking for local maxima among the intervals between smaller-level groups. The program was tested on several monophonic twentieth-century pieces, and its output was compared to segmentation analyses of these pieces.

Tenney and Polansky's model is an interesting one in the current context. It bears some resemblance to a preference rule system, in that it takes several different factors into account and also considers context in a limited way, but it does so in a very computationally efficient manner that does not require a lot of searching. Clearly, the authors were more concerned with twentieth-century music than with common-practice music. As I will discuss, their algorithm sometimes does not produce such good results on more traditional melodies, but this may be largely due to differences in the corpora being studied.

An alternative computational approach is proposed by Baker (1989a, 1989b). Unlike Tenney and Polansky's model, Baker's model is intended for tonal music; moreover, its analysis is based largely on tonal criteria. The system begins by performing a harmonic analysis; from this, it constructs a hierarchical structure using phrase structure rules, such as I → V-I. This produces a tree structure, similar to what Lerdahl and Jackendoff call a "time-span reduction"; the final segmentation is then generated by searching for high-level motivic parallelisms. Baker's system is highly complex; he also presents no examples of the output of his system, making it difficult to judge. Also worthy of mention is Todd's "primal sketch" model of grouping (1994). In this system, auditory input is analyzed by units which integrate energy over different time scales; identifying peaks of energy in different units produces a hierarchical grouping structure. The use of sound input means that the system can take account of fluctuations in expressive timing and dynamics. Todd's model is intriguing and promising, though I will suggest that such a "statistical" approach to grouping may encounter problems, especially with polyphonic music.

In *GTTM*, Lerdahl and Jackendoff (1983) propose a model of grouping in tonal music. The model is designed to handle polyphonic music; however, it imposes only a single hierarchical grouping structure on a piece, and does not allow for different groupings in different contrapuntal lines. Lerdahl and Jackendoff's well-formedness rules state that each group must consist of events contiguous in time; groups must be non-overlapping (with exceptions which I will discuss); and any group containing a smaller group must be exhaustively partitioned into smaller groups. The authors then propose a series of preference rules. As mentioned earlier, some of Lerdahl and Jackendoff's preference rules relate to proximity in time: the "slur-rest" rule, preferring grouping boundaries at rests or at the ends of slurs, and the "attack-point" rule, preferring boundaries at relatively widely separated attack-points. Others relate to similarity: grouping boundaries are preferred at changes of register, dynamics, articulation, and length. Several other rules are proposed. One states that very short groups should be avoided; another prefers grouping structures in which larger groups are subdivided into two parts of equal length; another states that two segments "construed as parallel" should preferably form parallel parts of groups (no precise formulation of this rule is provided—an important gap in the theory, as the authors admit). A final rule prefers grouping structures that result in more stable reductional structures; for example, a tonally complete phrase should preferably be regarded as a group. In addition, Lerdahl and Jackendoff propose mechanisms for handling "grouping overlaps." In some cases, a single event is heard as being both the last event of one group and the first event of the next; in such cases the two groups are allowed to overlap.

Lerdahl and Jackendoff's attempt to devise a theory of grouping adequate to the complexity of common-practice music is laudable. However, there are several serious problems with it. One is the fact, readily acknowledged by the authors, that the theory can only accommodate "homophonic" music in which a single grouping structure applies to the entire texture. Thus it works fairly convincingly for things like Bach Chorales. In much music, however, one feels that different parts of the texture demand different grouping boundaries. Examples of this abound, even in Lerdahl and Jackendoff's own examples; two of these are shown in figure 3.5. In figure 3.5a, the grouping of the bass line seems to overlap with that of the soprano; in figure 3.5b, the accompaniment and the melody seem to have different groupings. These problems would multiply even further in highly contrapuntal music such as fugues. One solution here would be to allow groups to have boundaries that were not exactly vertical; for example, one could well argue for the grouping boundaries shown with dotted lines in figure 3.5b. At the very least, however, examples such as these suggest that grouping analysis requires identification of grouping boundaries within different parts of the texture; and this requires identification of the individual parts themselves, a complex problem in itself which I will address in the next chapter.

Another problem, perhaps even more fundamental, lies with the nature of grouping itself. Lerdahl and Jackendoff's analyses suggest a concep-

Figure 3.5
Two passages with conflicting grouping in right and left hands. (A) Beethoven, Sonata Op. 22, III, mm. 4–6. (B) Schubert, Waltz in A Major, *Valses Sentimentales,* Op. 50, mm. 1–6.

tion of grouping in which a piece is divided exhaustively into small units (often one or two measures), and these are grouped into larger units in a hierarchical fashion. In many passages, the identity of these small units is clear enough. In many other cases, however, it is extremely ambiguous, even considering just the main melody. Consider a passage like figure 3.6. The beginning of the run of sixteenth-notes at the downbeat of m. 3 encourages us to begin a group there, and then—preserving the parallelism—to begin groups on each downbeat, or perhaps each half-note beat (grouping A). However, one could just as easily regard the units as beginning on the following note (grouping B). Under grouping B (unlike grouping A), the ending of each group is metrically strong and tonally stable; more importantly (as I will argue), grouping B creates a rhythmic parallelism with the opening, as it begins each group in the same metrical position as the opening melodic phrase.[8] I am quite unable to decide which of these two groupings I hear more strongly. (Which of these groupings we actually perceive could be studied in the manner of the experiment by Dowling cited earlier. Is group "a" a more readily identifiable unit than group "b"—would we more easily recognize it as being from this piece?) Passages such as this are commonly found in

Figure 3.6
Bach, Invention No. 1, mm. 1–4, showing two alternative grouping analyses of mm. 3–4.

all kinds of common-practice music, intermixed with passages of very clear and unambiguous phrase structure. Indeed, an important aspect of common-practice composition is the contrast between sections of very clear phrasing (this is often found in thematic passages), and passages of ambiguous phrasing (found in transitions and development sections).[9]

Preference rule systems are certainly capable of handing ambiguity; indeed, this is one of their important strengths, as I will discuss in a later chapter. However, it seems reckless to attempt a computational model of an aspect of perception in which it is so often unclear what the "correct" analysis would be. Rather than attempt this, I will address a smaller problem.

3.4 A Preference Rule System for Melodic Phrase Structure

As discussed above, phrase structure in common-practice music presents many complexities: ambiguities of grouping and conflicting groupings in different lines. In designing a computer model of grouping, it seemed wise to begin with a repertoire where these problems do not arise. One such repertoire is Western folk melodies. Folk melodies are of course monophonic, and they tend to have a fairly transparent phrase structure. Moreover, several anthologies of folk songs are available which have phrases marked in the form of slurs. This serves as a convenient source of "correct" phrase structure against which the program can be tested.[10] I will return later to the question of how this smaller problem relates to the larger goal of modeling grouping in polyphonic music.

I will simplify the problem, also, by only attempting to derive one level of groups—the level of the "phrase." This seems to be the level that is clearest and least ambiguous in perception (though it, too, is sometimes susceptible to ambiguity). The derivation of lower-level and higher-level groups will not be considered.

As with the metrical program, the main input required by the model is a "note-list" (or "piano-roll"), showing the on-time, off-time, and pitch of a series of notes. (Metrical information is also required; this simply consists of a framework of rows of beats aligned with the piano-roll representation, exactly as proposed in the previous chapter.) It is not implausible to suppose that listening to a performed folk melody involves representing the pitch, on-time and off-time of each note. As described earlier, we can imagine a piano-roll representation as being derived either from a live performance or from a score. In this case I will mostly use quantized piano-roll representations, generated from scores. One reason for this is that it is difficult to generate the piano-roll representation that would correspond to a vocal performance; it certainly cannot be easily done from a MIDI keyboard. We should note, however, that a quantized piano-roll representation may differ significantly from one that would be implied by a live performance. Consider the melody shown in figure 3.7a (the numbers above the staff will be discussed further below). A quantized representation—in which each note is given its full notated duration—would have no temporal break between the end of the first G in m. 7 and the beginning of the following note. However, a live performance might well contain a short break here. As we will see, such "offset-to-onset" gaps can be important cues to phrase structure. Nuances of tempo might also affect grouping, particularly the tendency to slow down at the end of phrases. (The role of performance nuance in the perception of infrastructural levels will be discussed further in section 11.6.) A live performance of a melody might also contain other cues to phrase structure which are not present in a piano-roll representation at all, even one corresponding to a live performance. In particular, folk melodies (and vocal melodies generally) usually have text, and the syntactic structure of the words might be a factor in the phrases that were perceived. In short, there are a variety of possible cues to phrase structure which are absent from a quantized piano-roll representation. One might wonder, indeed,

Figure 3.7
Two melodies from Ottman 1986. Numbers above the staff indicate gap scores (assuming a measure length of 1.0). Local maxima in gap scores are circled. (A) Melody No. 103 (Hungary). (B) Melody No. 265 (France), mm. 1–4.

whether it is even *possible* to infer the correct phrase structure from a quantized piano-roll input alone. However, I will suggest that it is indeed possible. Intuitively, it usually seems fairly clear where the phrase boundaries in a folk melody are, simply from the notes alone. The tests I will report here, in which a computer model has considerable success in identifying the correct phrase boundaries, provide further evidence that phrase structure in folk melodies is largely inferable from quantized piano-roll information, though not completely.

Figure 3.7a, a Hungarian melody from Ottman's *Music for Sight Singing* (1986), illustrates several basic principles of phrase structure. (*Music for Sight Singing* contains a large number of folk melodies, which I used for informal testing and experimentation while developing the phrase structure model. Generally the melodies are not identified, except by country of origin.) Ottman indicates the phrase structure with slurs, as is customary. The reader is encouraged to sing the melody and examine his or her intuitions about the phrase structure. To my mind, the phrase structure shown by Ottman is the most intuitive and natural one. (One might also end a phrase after the first D in m. 5, but this seems more dubious.) What are the reasons for these intuitions? The most obvious factor is rhythm. Both the last note of the first phrase and the last note of the second are long notes (relative to their neighbors), not closely followed by another note. All students of grouping agree that an important factor in grouping is temporal proximity. However, it is not clear exactly how the proximity between two notes should be measured. It could be done by measuring the time interval between their onsets: the "inter-onset interval," or IOI. This accounts for both the first phrase ending and the second in figure 3.7a; in both cases, the IOI at the final note of the phrase is longer than that of any nearby notes. However, consider figure 3.7b. Here, the logical phrase boundary is after the second A♭ in m. 2, though the IOI here is only two eighth-note beats, shorter than the previous IOI of three eighth-note beats. In this case, the important cue seems to be the rest between two notes. As mentioned earlier, a very short break between two notes (not reflected in notation) can also be a cue to phrase structure in performance. What is important here, then, is the time interval from the offset of one note to the onset of the next; we could call this the "offset-to-onset interval," or OOI. Significant (non-zero) OOI's tend to be strong cues for phrase boundaries.

In short, both IOI and OOI are important cues to phrase structure. (Both IOI and OOI are reflected in Lerdahl and Jackendoff's rules, and in Deliege's experiments: the IOI in the "attack-point" rule, and the OOI in the "slur-rest" rule.) This leads to the first of our Phrase Structure Preference Rules (PSPR's):

PSPR 1 (Gap Rule). Prefer to locate phrase boundaries at (a) large inter-onset intervals and (b) large offset-to-onset intervals.

Of course, IOI and OOI are related: the IOI between two notes is always at least as long as the OOI. One way to think about this is that each IOI contains a "note" portion and a "rest" portion (the OOI), but the "rest" portion carries more weight as a grouping cue than the "note" portion. Thus we could express the goodness of a grouping boundary between two notes as a weighted sum of the note part and the rest part of an IOI. A value that works well is simply to sum the note part of the IOI and two times the rest part. This is equivalent to adding the entire IOI to the OOI. We will call this value a "gap score"; the gap score for a note refers to the interval between it and the following note.

The next question is, how big does a "gap" have to be (both in terms of IOI and OOI) in order to qualify as a phrase boundary? It will not work to simply define a certain magnitude of gap as sufficient for a phrase boundary. In figure 3.7a, a quarter-note IOI (with no OOI) is sufficient for a phrase boundary after the second phrase, but other gaps of similar magnitude are not phrase boundaries (e.g., the two quarter notes in m. 2). Tenney and Polansky's model offers one solution: a gap is a phrase boundary if it is larger than the gaps on either side. Consider this solution as applied to figures 3.7a and b, using the formula for gap scores proposed earlier. In figure 3.7a, this produces exactly the right solution, except for the extra phrase boundary after the D in m. 5. In figure 3.7b, the results are much less good; there are far too many phrase boundaries. It can be seen that any melody with a repeated long-short rhythm will have this problem. In general, Tenney and Polansky's solution leads to an excess of phrase boundaries.[11]

An alternative solution takes advantage of a striking statistical fact: phrases are rather consistent in terms of their number of notes. In the Ottman collection, the mean number of notes per phrase is 7.5; over 75% of phrases have from 6 to 10 notes, and less than 1% have fewer than 4 or more than 14. (The reasons for this are unclear. It may be due in part to constraints on performance: phrases normally correspond to vocal breaths, and it is difficult to sing more than a certain number of notes in one breath. On the other hand, it may also be due to information-processing constraints in perception: 8 is a convenient number of items to group together into a "chunk.") We can incorporate this regularity by assigning a penalty to each phrase depending on its deviation from the "ideal" length of 8 notes, and factoring this in with the gap scores. We state this as follows:

PSPR 2 (Phrase Length Rule). Prefer phrases to have roughly 8 notes.

We can assume that phrases whose length is close to 8 (between 6 and 10) will be penalized only very slightly by this rule; more extreme deviations will be more highly penalized.

Figure 3.8a illustrates the need for a further rule. What is the motivation for the phrase structure here? Only the phrase boundary after the F♯ in m. 6 can be accounted for by the gap rule; the phrase-length rule prefers some kind of segmentation of the first 6 measures, but does not express a strong opinion as to exactly where the boundaries occur. The apparent reason for the first and second phrase boundaries is parallelism: the first phrase begins with a descending three-eighth-note figure, and when this figure repeats in a metrically parallel position there is pressure to treat that as a phrase boundary as well. In the third phrase the eighth-note figure is ascending, but it is rhythmically and metrically the same as in the previous two instances. As observed in chapter 2, recognizing

Figure 3.8
(A) Melody No. 82 (Germany), from Ottman 1986. (B) Melody No. 115 (England), from Ottman 1986.

motivic parallelisms—repeated melodic patterns—is a major and difficult problem which is beyond our current scope. Another solution suggests itself, however, which is to incorporate a preference for phrases that begin in the same position relative to the metrical structure. Given that the first phrase clearly begins on the fourth eighth-note of a measure, there is pressure for subsequent phrases to do so as well. It can be seen that this rule, along with the phrase-length rule, can achieve the right result in figure 3.8a. This rule proves to be extremely useful in a variety of situations: consider figure 3.8b, where parallelism is the only apparent factor favoring a boundary before the last note in m. 6. (It also provides added support for the first phrase boundary in figure 3.7b; note that the gap rule is indecisive here.) We state this rule as follows:

PSPR 3 (Metrical Parallelism Rule). Prefer to begin successive groups at parallel points in the metrical structure.

Assuming metrical structure as input to grouping is problematic. As noted in chapter 2, grouping is a significant factor in meter, in that there is a preference for strong beats near the beginning of groups. It appears that meter and grouping both affect one another. There is a preference for strong beats to occur early in groups, and also for groups to begin at parallel points in the metrical structure; and both meter and grouping structures may be adjusted to achieve these goals.[12]

Once the effect of meter on grouping is allowed, it can be seen that the influence of motivic parallelism on grouping can be accounted for as well (at least in principle). Parallelism affects meter, as discussed in chapter 2, favoring a metrical structure with the same period as the parallelism; PSPR 3 then favors a grouping structure with this period also. For example, in figure 3.8a, it is possible that motivic parallelism is one factor leading to the 3/4 metrical structure; once the meter is determined, a grouping strucure aligned with this metrical structure is then preferred.

3.5 Implementation and Tests

An implementation was devised of the grouping system described above. The usual "note-list" representation of the input is required, though it should be borne in mind that only monophonic input is allowed. (There may be breaks between notes—i.e., rests—but notes may not overlap in time.) In addition, the system requires a specification of the metrical structure. This consists of a list of beats; each beat has a time-point and a level number, indicating the highest metrical level at which that time-point is a beat. (As in chapter 2, we label the tactus level as level 2.) The system considers all "well-formed" grouping analyses of the input. A well-formed grouping analysis is simply some segmentation of the melody into phrases; every note must be contained in a phrase, and the phrases must be non-overlapping. It seemed reasonable to exclude the possibility of a phrase beginning in the middle of a note. Also, in cases where phrases are separated by a rest, it seemed unnecessary to worry about which phrase contained the rest. (If phrases are indicated by slurs, a rest at the boundary between two phrases is usually not included under either slur.) Given these assumptions, choosing a phrase analysis simply means choosing certain note-onsets as phrase boundaries, meaning that these notes are the first note of a phrase. The first note of the melody must of course be the beginning of a phrase, and the last note must be the end of a phrase.

The system searches for the well-formed grouping analysis that best satisfies the three rules presented above: the gap rule, the phrase length rule, and the parallelism rule. The first step in devising the implementation was to determine a way of evaluating analyses by these three rules. As noted earlier, the gap rule looks at the size of the gaps between pairs of notes at each phrase boundary in the analysis, considering both IOI and OOI; a simple formula which worked well was simply to take the sum of the IOI and OOI. The total gap score for the analysis sums these values for each pair of notes at a phrase boundary. One problem that arises is how gaps should be measured; should they be calculated in terms of absolute time, or in some other way? If they were calculated in absolute time, this would mean that gap scores would be larger at slower tempos; since the scores for the other two rules do not depend on the tempo, the gap rule would carry more weight at slower tempos. This seems counterintuitive; the perceived phrase structure of a melody does not seem to depend much on the tempo. To avoid this problem, the gap scores are adjusted according to the length of the notes in the melody so far; a "raw" gap score for an interval produced by summing the IOI and OOI is divided by the mean IOI of all the previous notes.

For the phrase length rule, we impose a penalty for each phrase, depending on the number of notes it contains. The following formula is used:

$$|(\log_2 N) - 3|$$

where N is the number of notes in the phrase. This formula assigns a penalty of 0 for a phrase length of 8. The penalty grows as the length deviates in either direction from this ideal value; it grows according to the ratio of the actual value to the ideal value. For example, a phrase length of 16 and a phrase length of 4 both carry a penalty of 1.

Finally, the parallelism rule is quantified by assigning a penalty according to whether each pair of adjacent phrases begin at parallel points in the metrical structure. We calculate this by assigning each note a "level 3 phase number." (Recall that level 3 is one level above the tactus.) This is the number of lowest-level beats elapsing between the beat of that note (i.e. the beat coinciding with the onset of the note) and the previous level 3 beat. If the first notes of two successive phrases do not have the same level 3 phase number, a penalty is assigned. We also calculate a level 4 phase number, and assign a smaller penalty if the two notes do not have the same level 4 phase number. The intuition is that it is best for successive phrases to be in phase at both level 3 and 4, but some credit should be given if they are in phase at level 3 only. (We assume a perfectly regular metrical structure, so that two beats which are the same number of beats after a level 3 beat are guaranteed to be at parallel positions in the metrical structure. If a melody switched back and forth between duple and triple divisions of the beat, this strategy would not work. We also assume that every note-onset coincides with a beat; "extrametrical" notes will not be handled, or at least cannot be considered as phrase boundaries.)

In this way it is possible to evaluate any grouping analysis according to the three preference rules. (Correct weights for the three rules relative to one another were determined by a trial-and-error process. For a given phrase, each kind of score was weighted by the length of the phrase, so as to prevent analyses with many phrases from having higher penalties.) Given this way of evaluating possible analyses, a search procedure was devised for finding the best analysis, using the dynamic programming approach outlined in chapter 1. The program proceeds in a left-to-right manner; at each note, it finds the best possible analysis of the piece so far ending with a phrase boundary immediately before that note. This entails considering each prior note as the location of the previous phrase boundary, adding the score for the current phrase on to the "best-so-far" score for the prior note, and choosing the prior note that leads to the best new score. The program continues to build up these "best-so-far" analyses through the piece; when the end of the piece is reached (where there is assumed to be a phrase boundary), the best analysis can be traced back through the piece.

A quantitative test of this program was performed. The test used the Essen folksong database, compiled by Schaffrath (1995). This is a database of European folksongs, transcribed in conventional notation, with phrase markings added. The phrase structures in the transcriptions are "well-formed" by the current definition, in that they exhaustively partition each melody into phrases, with no overlapping phrases. The folksongs were gathered from a range of ethnic groups across Europe. For this test, I used a random sample of eighty songs from the database. A stratified sample was used, including four songs from each of the twenty ethnic groups featured in the database.[13] Note-list representations of each melody were generated, showing the pitch, on-time, and off-time of each note. (Since the program's output does not depend on tempo, the exact length of notes was unimportant; only the relative length mattered, and this could easily be determined from the notation.)

Since the program requires metrical information, the question arose of how this information should be generated. I could have had the program proposed in chapter 2 figure out the metrical structure. This would be problematic, for two reasons. First, it would sometimes get the metrical structure wrong, which would be an irrelevant distraction in terms of the current goal (testing the grouping program). Secondly, the Essen folk song collection provides no information about tempo, which the metrical program needs in order to operate. Instead, metrical structures for the pieces were determined from the musical notation in the Essen collection.[14] Following the usual convention, the tactus level was taken to be the quarter note in 4/4, 2/4, and 3/4; in 3/8 and 6/8, the tactus level was taken as the dotted quarter. Most often (in songs in 2/4, 3/4 or 6/8), level 3 was the measure level; this too was indicated by the notation. In most cases, level 4 of the meter was a "hypermetrical" level—the level just above the measure level—and thus was not explicit in the notation. It was assumed in such cases that level 4 was duple with beats on odd-numbered level 3 beats. (In a few cases this seemed incorrect, but it seemed problematic to adjust the hypermeter in individual cases, given the possibility of experimenter bias.) Fifteen of the songs in the sample had irregular metrical structures; these were excluded, given the difficulty of determining the phase numbers of notes in this situation. This left a sample of 65 songs.

The test involved feeding the 65 songs to the program, and comparing the program's output with the phrase boundaries as shown in the database. First, a "fair" test was done, before any adjustment of the parameters to improve performance on the database. (The parameters had been set through informal testing on songs from the Ottman collection.) On this test, the program found 316 boundaries, compared to the 257 in the database. (Since every song includes a phrase boundary at the beginning and end of the melody, both in the transcriptions and in the program's output, these were disregarded; only "internal" phrase boundaries were counted.) The program made 115 "false positive" errors (finding phrase boundaries at locations not marked by boundaries in the transcriptions) and 56 "false negative" errors (omitting phrase boundaries at locations marked by boundaries in the transcriptions). Clearly, the program found significantly too many phrase boundaries overall. Recall that, according to the phrase length rule, the program prefers phrases with 8 notes, imposing a penalty for phrases either longer or shorter; this value was deemed to be optimal for the Ottman collection. Analysis of the Essen collection showed that the mean phrase length there is slightly larger: 9.2 notes. It was found that giving the program a preferred length of 10 notes yielded the optimal level of performance. With this value, it produced 66 false positives and 63 false negatives. If we consider only the false negatives (which seems reasonable, given that the number of false positives and negatives were roughly the same), the program on this latter test achieved a rate of 75.5%; that is, it correctly identified 75.5% of the phrase boundaries in the transcriptions.

The results were inspected to determine why errors were made. In a number of cases, the program's errors seemed quite understandable, and

Figure 3.9
Three excerpts from the Essen folksong collection. The phrasing given in the collection is shown in slurs; the program's phrasing is shown in square brackets. (A) "Vater und Tochter Schoenes Maedchen, was machst du hier" (Czech), mm. 1–8. (B) "Die Schlangenkoechin Dove si sta jersira" (Italian), mm. 1–4. (C) "Die Naehterin 's sass a Naehterin und sie naeht" (Czech).

well within the range of reasonable phrasings. Sometimes, phrases marked in the transcription were divided into two by the program; other times, pairs of phrases in the transcription were grouped into larger phrases (figure 3.9a). In such cases, one could argue that the program was simply finding a different level of grouping from the transcriber (though in figure 3.9a I would agree with the transcriber's intuitions as to the "primary" level of phrasing). In a few melodies, the transcriber's phrasing seemed bizarre; why put a phrase boundary after the A in m. 2 of figure 3.9b? We should bear in mind that phrase structure may sometimes depend on factors such as text syntax which are not apparent in notation; this may be the case in figure 3.9b, though there seemed to be relatively few cases of this. In several cases, the program's output definitely seemed wrong. A few of the program's errors were due to tonal factors. In figure 3.9c, it clearly seems better to put a phrase boundary after the third C in m. 5, rather than after the following F as the program did. The main factor seems to be that C (the fourth degree of the scale) is a more stable tonal destination than F (the seventh degree). (One might argue that C is actually the tonal center here; in this case, too, C is clearly more stable than F.)

Beyond the occasional role of tonality, it was difficult to identify any other important factors in phrase structure in this corpus. This might seem surprising, since Lerdahl and Jackendoff's model (and Deliege's experiments) put forth a number of other factors as significant cues to grouping, among them changes in timbre, articulation, dynamics, note length, and register. The current test has little to say about timbre, articulation, and dynamics, since this information is not represented in the Essen database; in any case, it seems unlikely that these factors would be major cues to grouping in folk melodies. Changes in note length or register *might* have proven to be important factors in the current test, but I could find almost no cases in the current corpus where such information seemed necessary or useful. However, all of these factors might well prove important in other kinds of music, particularly in cases where the factors discussed earlier (such as the gap rule) did not yield clear preferences.

3.6 Grouping in Polyphonic Music

Unlike the preference rule systems proposed in other chapters, the system proposed here does not even aspire to suffice for common-practice instrumental music. I argued earlier that, given the complexity and ambiguity of grouping in much common-practice music, attempting a formal model was premature. However, it is instructive to consider what would be needed for such a model. The Mozart quartet excerpt in figure 3.10 illustrates a number of points about grouping in polyphonic music.

It seems clear that grouping in polyphonic music must involve some kind of identification of gaps, both in terms of IOI and OOI. Many of the clear examples of grouping boundaries in polyphonic music are marked with rests in all the voices—the end of m. 2 of the Mozart, for example. Such rests could be easily identified. In other cases, however, a phrase ending is marked only by a gap in one voice (normally the melody), while the accompaniment continues. Almost any "melody-and-accompaniment" texture contains examples of this; for example, see figure 2.10b. One might wonder whether gaps could be identified in some statistical way—for example, by a decrease in the density of attack-points in the entire polyphonic texture. However, extensive consideration of this idea (and

Figure 3.10
Mozart, String Quartet K. 387, I, mm. 1–55. Vertical dotted lines show my own analysis of the phrase structure.

Figure 3.10 (continued)

Figure 3.10 (continued)

Figure 3.10 (continued)

Figure 3.10 (continued)

some computational playing around) convinces me that it will not work.[15] At a minimum, grouping requires the identification of voices, particularly the main melody.

Once the voices of a texture are identified, the grouping problem becomes somewhat more tractable. The grouping of the melody is often a primary determinant of the phrase structure of the entire texture. Indeed, one approach to grouping would simply be to identify the melody, subject it to the melodic grouping algorithm proposed earlier, and impose that grouping on the entire piece. Consider how this approach would fare with the Mozart quartet shown in figure 3.10. (I have identified what I consider to be the correct phrase boundaries; boundaries which seem vague or debatable are indicated with question marks.) In a number of cases, the groups of the melody are clearly identified by the factors described earlier. The gap rule is of frequent and obvious importance. Consider also the factor of metrical parallelism; the clear melodic phrase beginning on the fourth beat of m. 38 favors a boundary at the parallel place in m. 40. In many cases the phrase boundaries in the melody seem to indicate the primary phrase structure of the texture as a whole—although it is sometimes not clear how to continue these groups across the other voices (mm. 13–15, 20–1). In some cases the phrasing of the melody itself is unclear. The phrase boundaries in mm. 42 and 44 appear to be good examples of overlaps; the D in the melody is both the end of one phrase and the beginning of the next. Here, tonality is perhaps a factor: the clear tonic (D) on the third beat of m. 42 makes it a good candidate both as a phrase ending and a phrase beginning. Similarly, tonal stability is the main thing qualifying the D in m. 49 as a phrase ending. The role of tonal factors is also apparent in mm. 15–20. In terms of purely rhythmic factors, the most plausible points for grouping boundaries here are after the first eighth of m. 16 and after the first eighth of m. 19; metrical parallelism favors phrase boundaries before the second eighth of the measure, as in mm. 14 and 15, while the gap rule favors a boundary after the long note of mm. 17–18. However, both of these are points of harmonic instability: V6 in the first case, ii6 in the second case. Because of this, one is disinclined to hear any clear phrase boundary in this passage—a common feature of transition sections, as I suggested earlier.

The second theme (mm. 25–30) presents a typical case of phrasing ambiguity. It could certainly be regarded as one long six-measure phrase. However, it could also be split into three smaller phrases. The question then is whether to include the last melody note of the second measure (e.g. the F♯ at the end of m. 26) in the previous phrase or the following one. The gap rule would favor including it in the following phrase (the IOI before the F♯ is longer than the one after it), as would metrical parallelism, since the first phrase of the second theme clearly begins on the last eighth of a measure. On the other hand, the lower voices do not favor a phrase boundary before the last eighth. The slur across the last two eighths of m. 26 in the viola line exerts pressure against a boundary between them; in the case of the cello line, such a boundary would entail splitting a note between two phrases—something which we clearly prefer to avoid. For both of these lines, then, placing the phrase boundary at the bar line is strongly preferred. (If the cello and viola had eighth-note rests on the last eighth of m. 26, a phrase boundary before the last eighth note would be decidedly more plausible.) One other subtle factor is that there is a slight preference to place a phrase boundary so that strong beats occur as close as possible to the beginning of a phrase: this means, ideally, placing the boundary right at the barline. While I stated earlier that this factor primarily affects meter, rather than grouping, it can also affect grouping in cases where other factors are indecisive.

One rule of the current model seems to be of little value in the case of the Mozart quartet: the phrase length rule. Many phrases here have far more than 8 notes, even considering just the melody (the phrase beginning at the end of m. 38, for example). Is the idea of "optimal phrase length" irrelevant for a piece such as this, or does it simply require modification? One possible solution would be to express the "optimal length" of a phrase in terms of absolute time, rather than number of notes; in those phrases containing many notes, the notes are usually very fast. This distinction was not of great importance in the Essen database, since the density of notes per unit time was fairly consistent. However, this idea requires further study.

This example suggests that identifying the grouping of the melody in a polyphonic piece is an important step towards analyzing the grouping of the entire texture. But melodic phrase structure in common-practice instrumental music is much more complex than it is in folk songs, due largely to the role of tonal factors. In some cases (as in the second theme of the Mozart), accompanying voices can play a role in grouping as well. And these problems are only magnified by the fact that our intuitions about phrase structure are so often indecisive. Thus grouping in polyphonic music presents formidable problems. It is apparent, also, that polyphonic grouping cannot be done without first grouping the notes into contrapuntal lines. This is the topic of the next chapter.

4
Contrapuntal Structure

4.1 Counterpoint

An essential feature of common-practice music is *counterpoint*: the simultaneous combination of multiple melodic lines. When we think of counterpoint, we think first, perhaps, of compositions such as fugues, in which a small number of clearly identifiable voices are maintained throughout the piece. However, a great deal of other Western music is essentially contrapuntal also, in that it must be understood, at least in part, as being constructed from simultaneous melodic lines. In many cases, these lines are made explicit by being assigned to different instruments; and they are often perceptually reinforced by the timbres of those instruments. But even in music for a single instrument—most notably piano music—multiple contrapuntal lines are often present and quite easily heard. Consider figure 4.1, the opening of a Mozart piano sonata. The first four measures present one melodic line in the right hand and another in the left. (The left-hand line is less melodic than the right-hand one, and there are other possible ways of understanding it, as I will discuss below.) The right-hand melody continues in mm. 5–8, and is now imitated by the left-hand line. In m. 8 the main melody is joined by another melody in the right hand. Beginning in m. 12, the left hand features two simultaneous lines (the top one simply doubles the right-hand melody an octave lower); in m. 15, a second right-hand melody enters, making four simultaneous lines in all. All this seems clear enough; in some respects, however, the identity and extent of melodic lines in this passage—what we might call its "contrapuntal structure"—is not so clear. Does the left-hand line in mm. 1–5 continue to the left-hand line in

Figure 4.1
Mozart, Sonata K. 332, I, mm. 1–16.

m. 7, or are these two separate lines? How do the three lines of the downbeat chord of m. 12 connect with the three lines of the chord two beats later? One might well connect the main melody across these two chords; but to say that the middle line of the first connects with the middle line of the second seems rather arbitrary. This brings up an important fact about contrapuntal structure: in much contrapuntal music, there is not necessarily a fixed number of voices that endure throughout the piece; voices may come and go as they are needed. A contrapuntal voice, therefore, is best viewed as something bounded in time as well as pitch.[1]

If the identification of contrapuntal lines is assumed to be a real and important part of musical listening, then music perception must involve grouping notes into lines in the appropriate way. This process, which I will call *contrapuntal analysis*, is the subject of the current chapter.

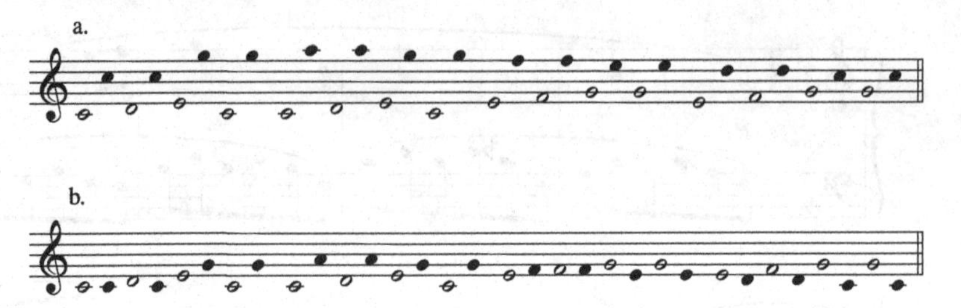

Figure 4.2
From Dowling 1973a.

4.2 Sequential Integration in Auditory Psychology

A great deal of experimental work in auditory psychology has focused on contrapuntal analysis (though not usually under that name). The grouping of notes into lines is an important aspect of "stream segregation," the larger process of analyzing and segmenting sound input (see Bregman 1990 for a review). Bregman (1990, 30–1) suggests that stream segregation can be divided into two processes, simultaneous integration and sequential integration. In musical terms, simultaneous integration involves the grouping of component frequencies or "partials" into notes; sequential integration involves the grouping of notes into lines.[2] It is only the latter process that concerns us here; we will assume, as we do elsewhere in this book, that notes have been fully identified before sequential integration begins. It should be emphasized at the outset, however, that the interaction between sequential integration and simultaneous integration is highly complex; to assume that the identification of notes is completely independent of, and prior to, their sorting into lines is undoubtedly an oversimplification.

Perhaps the most important factor in sequential grouping is pitch proximity. When a sequence of two alternating pitches is played, the two pitches will seem to fuse into a single stream if they are close together in pitch; otherwise they will seem to form two independent streams (Miller & Heise 1950). The importance of pitch proximity has been demonstrated in more musical contexts as well. Dowling (1973a) played subjects interleaved melodies (in which notes of one melody alternate with those of another); he found that the melodies could be recognized much more easily when the pitches were separated in register, as in figure 4.2a, rather than overlapping in register, as in figure 4.2b. As well as showing the role of pitch proximity, this experiment nicely demonstrates the musical importance of sequential integration. Once notes are assigned to the same stream, they form a coherent musical unit, which can then be

Figure 4.3
From Deutsch 1975. Notes marked "R" were heard in the right ear, those marked "L" in the left ear.

given further attention and analysis. (A similar point was made in the previous chapter with regard to phrase structure.) A related finding is that a rhythmic pattern is harder to identify if it involves crossing between two registrally separated streams; this, too, suggests the grouping together of registrally proximate events (Fitzgibbons, Pollatsek, & Thomas 1974).

An experiment by Deutsch (1975) sheds further light on the role of pitch in sequential grouping. Subjects heard rising and falling scales simultaneously, with the notes of each stream presented to alternating ears, as shown in figure 4.3a. Rather than grouping events by spatial location, or hearing crossing scales in opposite directions (analogous to the "good continuation" principle found in visual perception), subjects most often perceived the sequences indicated by beams in figure 4.3b; Deutsch dubbed this the "scale illusion." This suggests that there is a particular perceptual tendency to avoid crossing lines, a finding that has been confirmed in other studies as well (Bregman 1990, 418–9).

Another important factor in sequential integration is tempo. If two tones are played in alternation, they will tend to segregate into separate streams if they are more than a certain interval apart in pitch; however, the crucial interval depends on the speed of alternation. If the two tones are alternating rapidly, they will segregate at quite a small interval; if played more slowly, a larger interval is needed (van Noorden 1975).[3] The large-scale temporal structure of patterns can affect sequential grouping as well: given a pattern of two alternating tones, the tendency to hear two separate streams increases with the number of repetitions of the pattern (Bregman 1978).

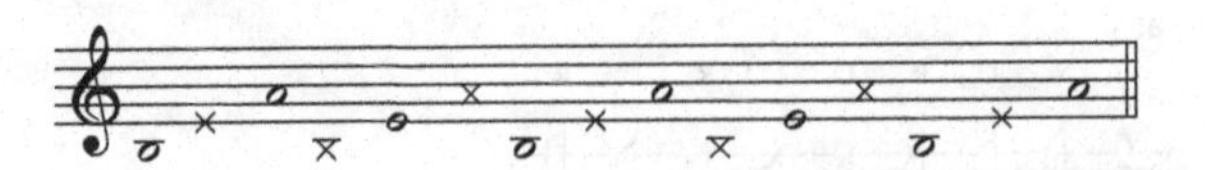

Figure 4.4
From Wessel 1979. Notes indicated with crosses and ovals were assigned contrasting timbres.

Other factors in sequential integration have also been studied. One might suppose that spatial location would be an important factor; however, it proves to play a limited role. In Deutsch's "scale illusion" experiment, the notes of each scale were played to alternating ears (as shown in figure 4.3a), yet subjects did not generally group the events according to their spatial location. However, spatial location may affect sequential integration in certain circumstances (Bregman 1990, 73–83). The role of amplitude has also been studied, but with inconclusive results (Bregman 1990, 126–7). One factor which does seem important is timbre, though here, too, the evidence is inconclusive. In one study, subjects were played the pitch sequence shown in figure 4.4 (Wessel 1979). When all the pitches were given the same timbre, a single line was heard; when the pitches were assigned different timbres as indicated in the figure, they tended to be grouped according to timbre. Other experiments have obtained similar findings (Bregman 1990, 83–103); however, a study by Butler (1979) produced the opposite result. Butler used stimuli such as that in figure 4.5a, with the top and bottom lines coming out of left and right speakers respectively. Subjects generally heard two lines as shown in figure 4.5b, with pitch proximity overriding spatial location; moreover, even when the spatial location grouping was reinforced by differences in timbre, pitch proximity still prevailed. Perhaps a fair conclusion would be that timbre can sometimes affect sequential grouping, but is easily overridden by pitch factors.[4]

We should note three basic assumptions that underlie much of the experimental work discussed here. One is that there is a preference to include each note in only one stream. Consider the simple case of a pattern of alternating tones. When the two tones are far apart in pitch, they tend to be heard as two static streams; bringing the tones close together in pitch causes them to be heard as a single alternating stream, but it also weakens the perception of the two static streams. The obvious explanation is that the one-stream and two-stream interpretations are competing with each other; once a group of notes are assigned to one stream, we tend not to assign them to another stream as well. (Other

Figure 4.5
From Butler 1979.

studies have also demonstrated this kind of competition between streams; see Bregman 1990, 166–9.) Secondly, there is a tendency to prefer an interpretation with fewer streams. Return to the Deutsch scale pattern; if we heard a separate stream for each pitch (one stream for the low C's, another for the D's, and so on), then pitch proximity would be maximal within each stream; but we do not hear the pattern this way. Apparently we prefer to limit the number of streams, if a reasonable degree of pitch proximity can still be maintained. Finally, there is a strong preference to have only one note at a time in a stream. Consider Butler's stimuli, as they are normally perceived (figure 4.5b). By pitch proximity alone, it would be preferable for the final pair of simultaneous notes (F and A) to be included in the lower stream, since both are closer in pitch to the previous A than to the previous C. Presumably this interpretation was not favored because it would result in the lower stream containing two simultaneous notes.

A final issue to be considered is temporal grouping. In most of the experiments discussed above, the stimuli are short enough that it can be assumed that any streams present will persist throughout the stimulus. As noted earlier, however, in extended pieces it often seems arbitrary to insist that streams must span the entire piece. Rather, one frequently has the sense of streams beginning and ending within the piece; a stream is thus a group of notes that is bounded both in pitch and time. This may bring to mind the preference rule system for phrase structure discussed in the previous chapter. As discussed there, an essential factor in temporal grouping is temporal separation; a sequence of notes tends to form a group when bounded by pauses. (See section 3.2 for a discussion of relevant experimental work.) There are other factors involved in temporal grouping besides pauses, but these will not concern us here. The connection between contrapuntal structure and phrase structure will be discussed further below.

Let us summarize the important points about sequential integration that emerge from this survey. (1) We prefer to group together events that are close in pitch. (2) Grouping together events far apart in pitch is harder at faster tempos. (3) The tendency to hear a pattern of alternating pitches as two separate streams increases with the number of repetitions of the pattern. (4) There is a preference to avoid crossing streams. (5) We generally avoid including a single note in more than one stream. (6) We generally avoid streams with multiple simultaneous notes. (7) We generally try to minimize the number of streams. (8) Pitch-events are also grouped temporally; temporal separation (by pauses) is an important factor here. (9) Although timbre, amplitude, and spatial location may sometimes affect sequential integration, they are often outweighed by pitch and temporal factors.

4.3 Computational Models of Contrapuntal Analysis

Several computational models have been proposed which relate to contrapuntal analysis. One interesting proposal is Huron's algorithm (1989) for measuring pseudo-polyphony. Pseudo-polyphony occurs when a single line, through the use of large leaps, creates the impression of multiple lines. Huron's model predicts the number of concurrent streams that are perceived to be present in a passage. The model begins with a piano-roll representation; every event is given a "tail," a short time interval following the event in which the event is still perceptually salient. If another event closely follows the first that is close to it in pitch, the "tail" of the first event is assumed to be cut off; if the second event is further away in pitch, however, the tail of the first event is not cut off (figure 4.6). The number of streams is then indicated by the mean number of simultaneous events (including their tails) present at each point in time. Consider again a case of two alternating pitches. If the pitches are close enough in register to cut off one another's tails, then only one event is present at each moment, implying a single stream; otherwise, the events will overlap in time, implying two streams. Calcu-

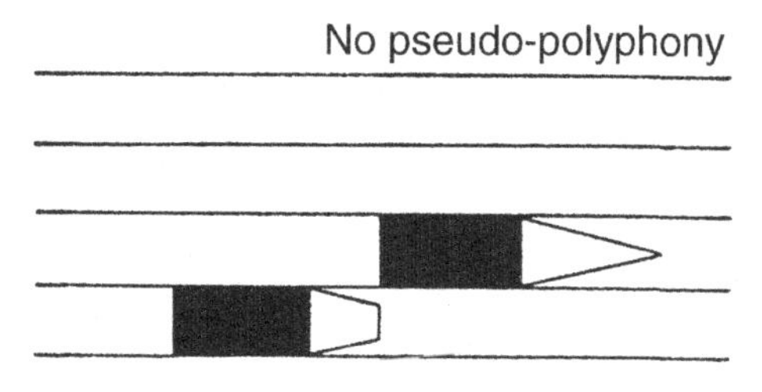

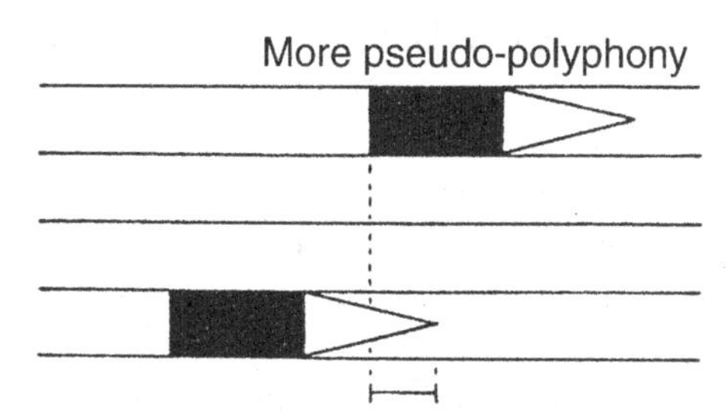

Figure 4.6
From Huron 1989.

lating the mean number of events present at each moment produces an aggregate measure of the number of concurrent streams in the piece. Huron tested his model by creating a set of 29 monophonic musical passages and having music theorists rank them according to their degree of pseudo-polyphony. The model's judgments of pseudo-polyphony correlated well with the theorists'; indeed, it correlated better with the theorists' judgments than the theorists' judgments did with each other.

As Huron notes, his model does not actually produce a contrapuntal analysis; converting it into such a model would require some further work. For example, if the tail of an event is cut off by two subsequent events (perhaps one higher in pitch and one lower), the model makes no decision about which note the first note will connect to. One fundamental problem with the model is that it is entirely "left-to-right"; it has no capacity for revising its analysis of one segment based on what happens afterwards. This point is worth elaborating, since it arises with other models of contrapuntal analysis as well. Consider m. 5 of the Mozart sonata (see figure 4.1); the initial F in the melody clearly connects to the

Figure 4.7
A recomposition of mm. 5–8 of the Mozart sonata shown in figure 4.1.

following A. However, if these two notes had been followed by the context shown in figure 4.7, the F and A would be heard as being in separate streams. Thus the correct contrapuntal interpretation of a segment can only be determined by looking ahead.

A highly original approach to contrapuntal analysis is proposed by Gjerdingen (1994), based on an analogy with vision. Imagine a field of units laid out in two dimensions, corresponding to pitch and time. A single pitch event activates not only units at that pitch but also (to a lesser degree) neighboring pitch units as well, and also neighboring units in time; the result is a kind of "hill" of activation. Another event closely following in time and nearby in pitch may generate an overlapping hill. If one then traces the points of maximum activation for each point in time, this produces a line connecting the two pitches; the melodic line is thus made explicit. Tracing maxima of activation across the plane can produce something very like a contrapuntal analysis. The model can also track multiple simultaneous lines; an example of its output is shown in figure 4.8. A number of phenomena in stream segregation are captured by the model, including the failure of melodic lines to connect across large intervals, the greater ease of hearing large intervals melodically connected at slower tempi, and the greater salience of outer voices as opposed to inner ones. The main problem with the model is that its output is difficult to interpret. At the left of the diagram in figure 4.8, we can discern an upper-voice line and a lower-voice line, but these are not actually fully connected (our tendency to *see* lines there is an interesting perceptual phenomenon in itself!); towards the end of the diagram, the upper line appears to break up into several streams. It is also sometimes not clear which pitches the lines correspond to; the lower line in the diagram lies between the two actual lower lines of the texture. One might argue that the highly ambiguous nature of the model's output is true to human perception, but it does make the model difficult to test.

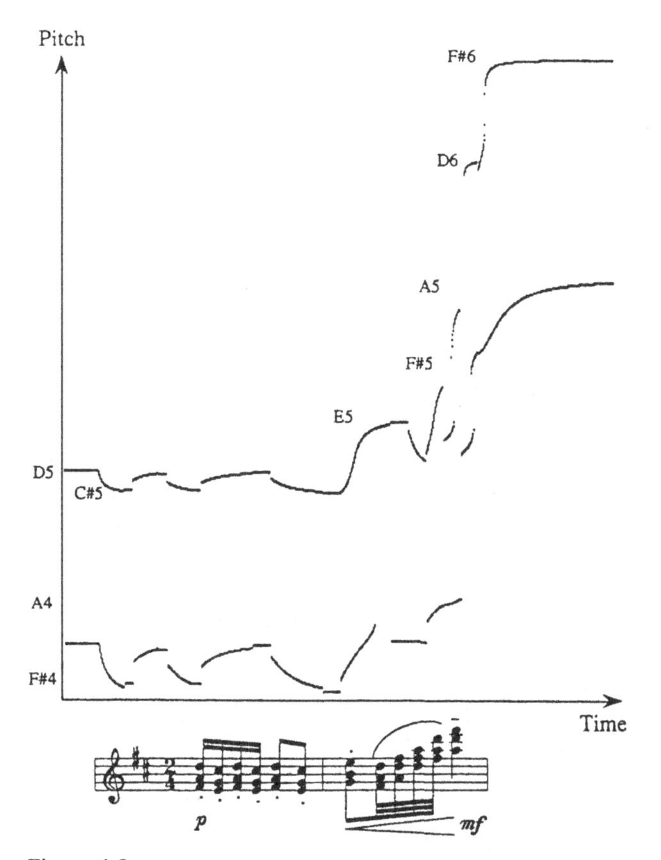

Figure 4.8
From Gjerdingen 1994. Reprinted by permission of the Regents of the University of California.

Another approach somewhat similar to Gjerdingen's is that of McCabe and Denham (1997). Like Gjerdingen's model, the McCabe-Denham model consists of units arranged two-dimensionally in pitch and time; however, two networks are used, corresponding to two streams, "foreground" and "background." Activated units in one stream activate nearby units in the stream and inhibit corresponding units in the other stream; thus acoustic input is segregated into the two streams. The model is able to account for a number of the experimental findings discussed above. In a way, the McCabe-Denham model is more ambitious than the model I will propose (or Gjerdingen's), since it begins with acoustic input rather than notes; in a sense, it bypasses simultaneous integration and

Figure 4.9

goes straight from acoustic input to sequential integration. On the other hand, the model does not appear to have been tested on any complex musical inputs. Like Gjerdingen's model, the output of McCabe and Denham's system is difficult to interpret; a given input may result in certain units in a stream being activated to some degree, but it is not clear whether this means that a certain note is being included in that stream.

The final model to be considered, developed by Marsden (1992), is the closest to the current model in its perspective and goals. This model begins with a "piano-roll" representation, is designed to handle musical inputs, and produces an explicit contrapuntal analysis with pitch events connected into contrapuntal voices. Marsden begins by presenting a simple model in which each note is connected to the note in the following chord which is closest to it in pitch. As he observes, this model cannot handle cases like figure 4.9. The model would connect the first B to the second B here; it would be preferable to connect the first B to the G♯3, and the G♯4 to the second B. (Another possibility would be to connect each note to the closest note in the *previous* chord, but as Marsden points out, that does not help in this case.) To address such problems, Marsden proposes another version of the model in which connections between notes (called "links") compete with one another. A weight is assigned to each link, which reflects the size of the pitch interval it entails; this value may also reflect other factors, such as whether the link crosses other links. Links then activate other compatible links and inhibit incompatible ones, in a manner similar to a neural network. In figure 4.9, the link from the first B to the second would be inhibited by the links from G♯4 to the second B, and the first B to G♯3. This leads to a way of connecting adjacent chords so that small intervals within streams are favored. Marsden discusses the possibility of incorporating motivic parallelism into his model; he also shows how the findings of Deutsch's "scale illusion" experiment could be accommodated.

Marsden's study is an important contribution to the contrapuntal analysis problem, and the current model builds on it in several ways. In particular, the idea of giving each link a score (based on several criteria), with competition between them, is closely related to the preference rule approach presented below. It is somewhat difficult to evaluate Marsden's model, since several versions of it are presented, each with acknowledged strengths and weaknesses. Early version of the model have limited power —for example, they do not allow rests within voices (they are only able to connect notes that are perfectly adjacent). Later versions become extremely complex, so much so that their behavior is difficult to predict or even "non-deterministic." (Another general problem with all versions of the model is that they do not allow the subsequent context to influence the analysis; as noted earlier, this appears to be essential for contrapuntal analysis.) As Marsden discusses, there is often a trade-off in computational models between clarity and predictability on the one hand, and level of performance on the other. The following model attempts to achieve a high level of performance within a reasonably simple and predictable design.

4.4 A Preference Rule System for Contrapuntal Analysis

The input required for the current model is the usual piano-roll representation, with pitch on one axis and time on the other. As usual, pitches are quantized to steps of the chromatic scale; for the purposes of the model, it is also necessary to quantize pitch-events in the temporal dimension. It is convenient to use lowest-level beats of the metrical structure for this. (Here I am assuming a metrical structure along the lines of that proposed in chapter 2.) Since pitch-events are quantized in both dimensions, we can think of the piano-roll representation as a grid of squares; each square is either "black" (if there is a note there) or "white" (if there is not). In addition, the input representation must show where events begin and end; given two black squares of the same pitch in adjacent time segments, the input must show whether they are part of the same event or not. (The reason for this will be made clear below.) This can be done simply by indicating the onset of each note; the offset is then indicated either by the following event-onset at that pitch, or by a white square. (Overlapping notes of the same pitch are disallowed in the input representation.) Figure 4.10 shows a graphic representation of the input for the opening of Mozart's Sonata K. 332 (the first few measures of the excerpt shown in figure 4.1); here temporal units correspond to eighth-notes. The onset of a note is indicated by a white bar.

As discussed in chapter 1, the use of a "piano-roll" representation as input means that several important aspects of musical sound are disregarded, including timbre, amplitude, and spatial location. As noted

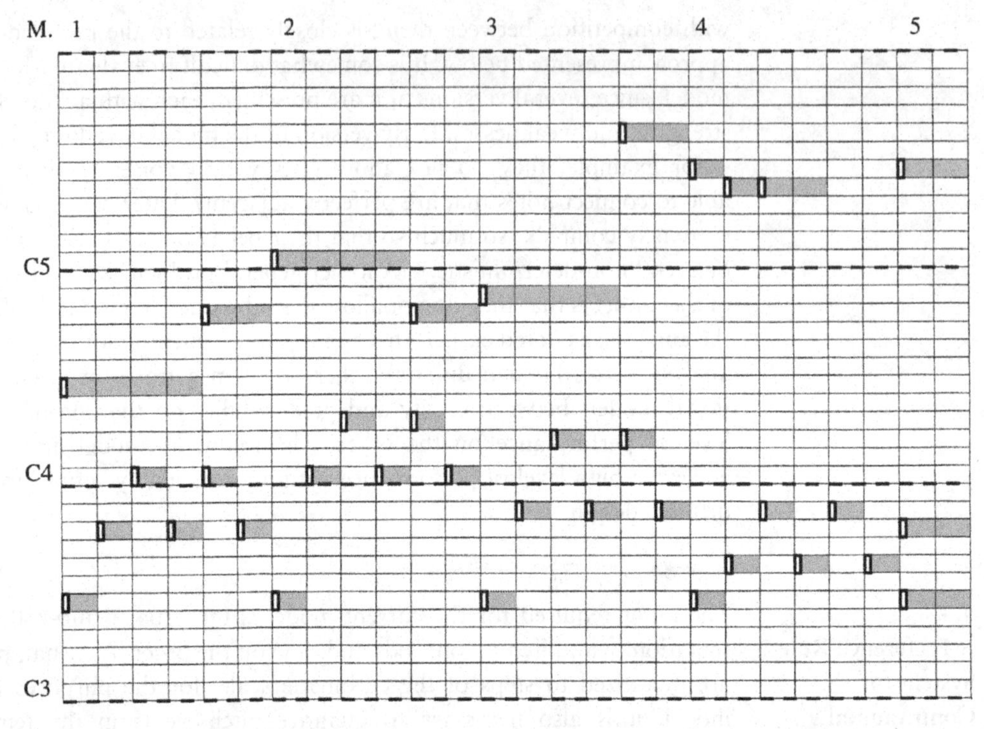

Figure 4.10
The input representation for mm. 1–4 of the Mozart sonata shown in figure 4.1.

earlier in this chapter, the experimental evidence for these factors as determinants of sequential integration is equivocal, but it seems likely that they could play a role under some circumstances. However, some genres of polyphonic music—such as solo piano and string quartet music—allow little differentiation in timbre and spatial location (though not in amplitude). The fact that sequential integration appears to be possible with these genres suggests that a "piano-roll" representation may be largely sufficient. Whether a computer model can successfully perform stream segregation using only pitch and time information may shed further light on this question.

A contrapuntal analysis is simply a set of streams, subject to various constraints. We begin, as usual, by setting forth some basic well-formedness rules, stating criteria for a permissible analysis.

CWFR 1. A stream must consist of a set of temporally contiguous squares on the plane.

That is to say, a stream may not contain squares that are unconnected in the temporal dimension. A stream may contain two black squares that are separated by white squares (i.e., two notes separated by a rest), but then it must contain any white squares that are between them. The fact that streams can contain white squares greatly increases the number of possible well-formed streams. This is clearly necessary, however, since streams often do contain rests. (Examples can readily be seen in the experimental stimuli discussed above, as well as in the Mozart excerpt in figure 4.1.) Note that it is not required for streams to persist throughout a piece; a stream may be of any temporal length.

CWFR 2. A stream may be only one square wide in the pitch dimension.

In other words, a stream may not contain more than one simultaneous note. This seems to be the usual assumption in musical discourse; it is also assumed in most auditory perception work.[5]

CWFR 3. Streams may not cross in pitch.

As discussed earlier, this rule is well-supported by experimental findings. It is true that crossing-stream situations are not unheard of, either in psychology or in music; thus including this as a well-formedness rule rather than a preference rule may seem too strict. I will return to this question.

CWFR 4 (first version). All black squares must be contained in a stream.

Again, treating this as a well-formedness rule may seem overly strict, in view of the fact that there are sometimes notes—in big chords, for example—which do not seem to belong to any larger stream. But this is not really a problem. In such cases, there may be a stream containing only one note, which is no disaster.

As well as ensuring that all black squares are contained in streams, it is also desirable to maintain the integrity of notes. It is difficult to imagine hearing different parts of a note in two separate streams. We could enforce this with a rule which required that each note must be entirely included in a single stream. Notice, however, that this makes the first version of CWFR 4 unnecessary; if every note is entirely included in a single stream, then every black square will be included in a stream as well (since every black square is included in a note). Thus we can replace the old CWFR 4 with this:

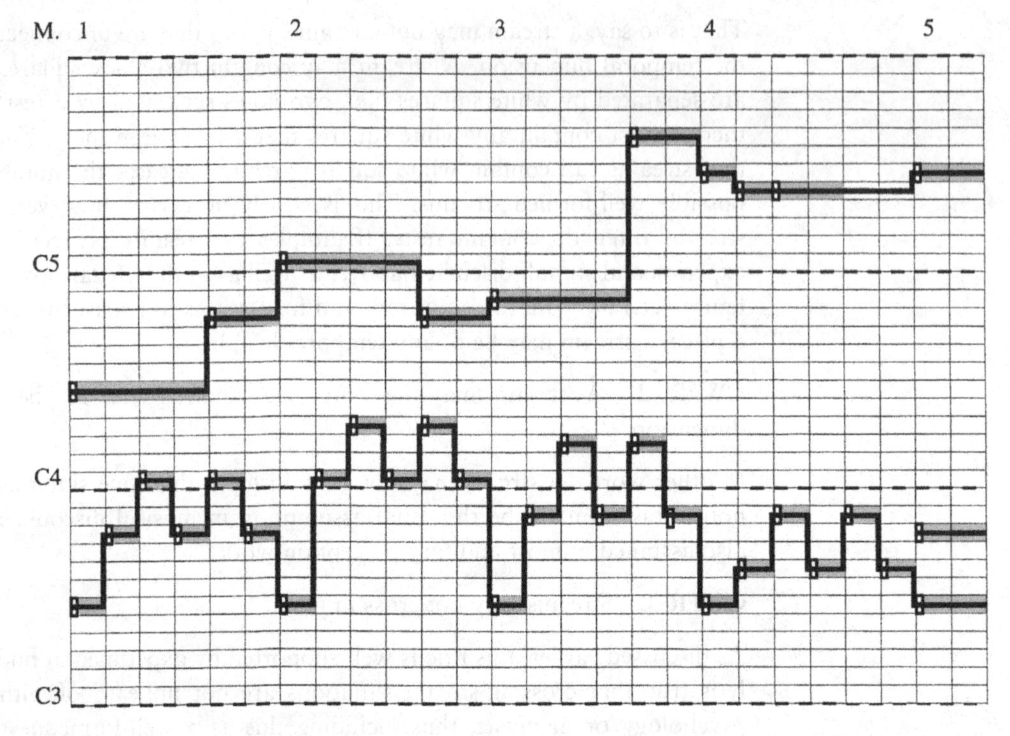

Figure 4.11
The correct contrapuntal analysis of mm. 1–4 of the Mozart sonata shown in figure 4.1. Streams are indicated by thick black lines.

CWFR 4 (final version). Each note must be entirely included in a single stream.

It is because of this rule that the input representation must contain information about where notes begin and end, and not just about which squares are black. Given two black squares of the same pitch in adjacent time segments, if they are both part of the same note, they must be included in the same stream. (Notice that it *is* permissible for a note to be contained in more than one stream.)

Though this may seem like quite a limiting set of well-formedness rules, it in fact permits an unimaginably huge number of possible contrapuntal analyses, most of them quite ridiculous, even for a short segment of music. Figure 4.11 shows the correct analysis for the opening of the Mozart sonata; figure 4.12 shows a much less plausible analysis, which reveals how licentious the above set of well-formedness rules really

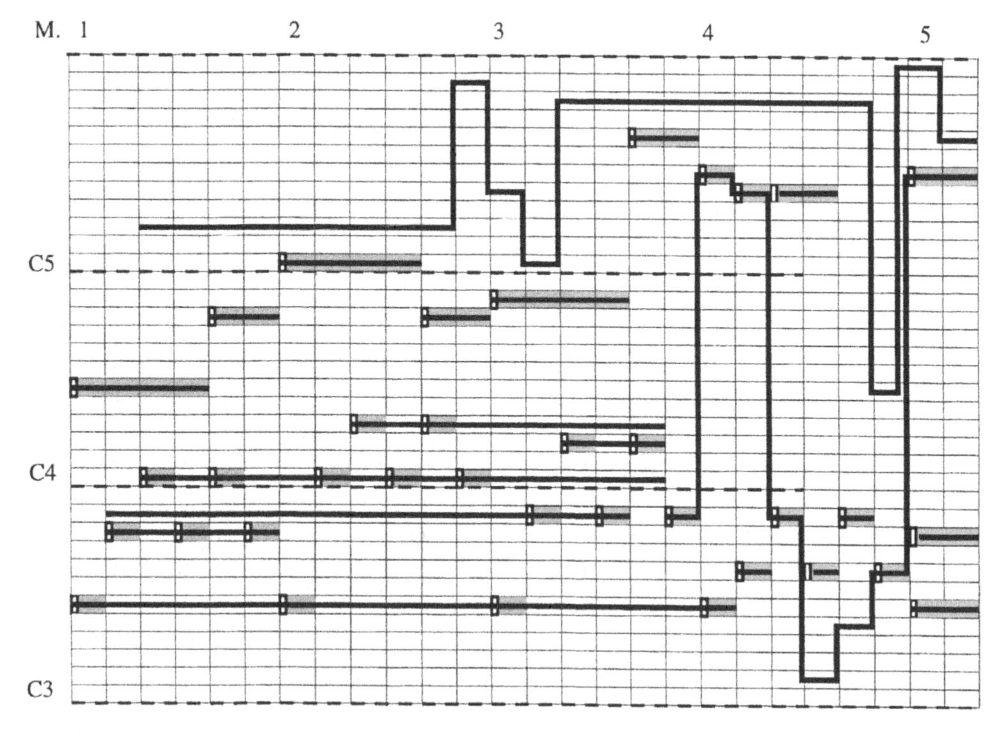

Figure 4.12
An absurd (but well-formed) contrapuntal analysis of mm. 1–4 of the Mozart analysis shown in figure 4.1. Streams are indicated by thick black lines.

is. There is nothing in them to prevent a situation where each note of a piece occupies a separate stream (as in the opening notes of the right-hand melody), or where a stream forms for each pitch, consisting mostly of empty space (see the opening notes of the left hand). Indeed, there is nothing to exclude streams, like the uppermost stream in figure 4.12, which contain no notes at all. Streams may also leap wildly from one note to another, as seen in one stream towards the end of the passage.

Three simple preference rules solve these problems.

CPR 1 (Pitch Proximity Rule). Prefer to avoid large leaps within streams.

This first CPR (contrapuntal preference rule) is simply the principle of pitch proximity, well-known from auditory psychology. This would exert strong pressure against the large leaps seen in the "absurd" analysis of the Mozart.

CPR 2 (New Stream Rule). Prefer to minimize the number of streams.

This principle is reflected in much auditory psychology, though it is usually taken for granted. Without it, it would be preferable to form a separate stream for each note.

CPR 3 (White Square Rule). Prefer to minimize the number of white squares in streams.

As noted earlier, it is clear that streams must be allowed to contain white squares. However, we do not continue to maintain streams perceptually over long periods of silence. If the gap between one note and another is long enough, we will hear one stream ending and another one beginning.

In effect, the white square rule exerts pressure to separate a sequence of notes into two temporal segments, when there is a large pause in between. The reader may note a connection here with the model proposed in the previous chapter. There, I proposed a preference rule system for identification of melodic phrases; pauses (or "offset-to-onset intervals") were a major factor in this model, though there were several others as well. However, the temporal grouping proposed in chapter 3 is of a slightly different kind from what is needed here. A stream is not the same thing as a melodic phrase; one often has the sense of a single stream that is clearly divided into multiple phrases. The kind of temporal group we are interested in here is somewhat larger than a phrase. It seems reasonable to treat pauses as the main factor in such large-scale temporal units, as opposed to the other factors (such as phrase length and metrical parallelism) involved in low-level phrase structure.

A fourth preference rule is also needed:

CPR 4 (Collision Rule). Prefer to avoid cases where a single square is included in more than one stream.

The tendency to avoid including a note in two streams is reflected in the auditory perception research discussed above. However, the model does not *forbid* two streams from including the same note; as we will see, such "collisions" actually arise quite often in actual music. If two streams do occupy the same note, it is still understood that one is higher than the other. Thus it is not permissible for two voices to cross by occupying the same note; when they diverge from the note, they must have the same height relationship that they had before.[6]

4.5 Implementation

A computer implementation of this preference rule system was devised. Like the grouping program described in chapter 3, the current program requires a "note-list" input (indicating the onset, offset, and pitch of each event), as well as a "beat-list" specifying the metrical structure. In this case, the only reason metrical information is needed is that the lowest-level beats of the meter are used to generate the indivisible segments the program requires. Time-points of notes are adjusted by the contrapuntal program so that every note both begins and ends on a beat. This then produces the kind of grid described earlier, where each square is either completely "black" or "white."

As with the metrical and grouping programs discussed earlier, the implementation problem can be broken down into two sub-problems. The first problem is how the program is to evaluate individual analyses. The program does this by assigning a numerical score to each analysis which reflects how well it satisfies the four preference rules. Consider first the pitch proximity rule (CPR 1). For each stream in the analysis, the program assigns a penalty to each pitch interval within the stream, which is proportional to the size of the interval (it assigns a penalty of 1 point for each chromatic step; so a leap of a perfect fifth would receive a penalty of 7). Turning to the new stream rule (CPR 2), the program assigns a penalty of 20 points for each stream. For the white square rule (CPR 3), the program assigns a penalty for each white square in any stream. In this case, it seemed logical that the penalty should depend on the actual duration of the segment (which may vary, since segments are determined by the metrical structure); the penalty used is 20 points per second. For segments of 0.2 second, then, the penalty will be 4 points for each white square. Finally, for the collision rule (CPR 4), a penalty is assigned for each square included in more than one stream. Here, too, the penalty is proportional to the duration of the segment; again, a value of 20 points per second is used. (The numerical parameters for these rules were set simply by trial-and-error testing on a number of examples.)

Given this means of evaluating analyses, the program must then find the analysis with the highest score. (Since all the scores are penalties, they are in effect negative scores; the highest-scoring analysis is then the one with the smallest total penalty.) In theory, the program must consider all possible well-formed analyses of the piece it is given. As usual, this creates a search problem, since the number of possible analyses even for a single temporal segment is extremely large, and grows exponentially with the number of segments. It is worth considering why the program must consider global analyses, rather than simply choosing the best analysis

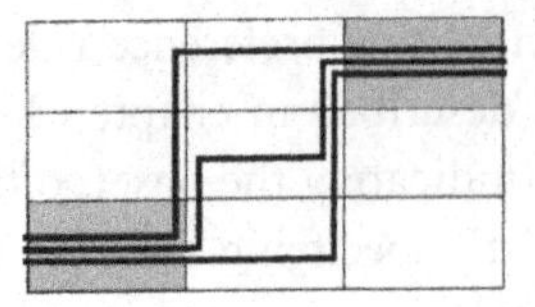

Figure 4.13
Three possible ways are shown of connecting two events across a rest.

for each segment in isolation. The reason is that the best analysis for a segment depends on both the previous and subsequent context. Figure 4.7 gave one example of this; another is seen in the third quarter-note segment of m. 4 of the Mozart (figure 4.1). The optimal analysis of this segment has a stream at E5, even though there is no note there. If we considered only this segment, it would be preferable to have no stream there, since this would avoid penalties from both the new stream rule and the white square rule. The reason that a stream is preferred is that there are notes in both previous and subsequent segments, and it is preferable to connect them with a stream, rather than starting a new stream at the subsequent note. But we can only know this by seeing both the previous and subsequent context. Cases like this show that it is necessary to search for the analysis that is preferred overall, rather than analyzing segments in isolation.

In essence, then, the program must consider all possible analyses of each segment, and find the best way of combining them. A possible analysis of a segment is simply some subset of the squares labeled as stream locations. Because of CWFR 4, we know that any "black squares" are stream locations, so the only question is which white squares (if any) are to be considered as stream locations. However, it is not necessary to consider all white squares as possible stream locations. Consider figure 4.13. Three possible ways are shown of connecting two events across a rest: up-up-right, up-right-up, or right-up-up. Since the pitch proximity penalty is linear, the total pitch proximity penalty for the three analyses will be the same. This means that only one of them needs to be considered. In general, it can be seen that any gap between two notes can be bridged in an optimal way by simply maintaining the stream at the pitch of the first note and then moving to the second note when it occurs. This is important, because it means that there is never any reason for a stream to move vertically to a white square; we only need to consider putting a stream on a white square if it is already at that pitch level.[7] There is also a limit on the number of consecutive white squares that need to be considered as stream locations. If the new stream rule imposes a penalty of 20, and the white square rule imposes a penalty of 5 points per square (at a given tempo), this means that after 4 consecutive white squares, it will be advantageous to begin a new stream; it will never be advantageous to have 5 consecutive white squares within a stream. Thus we proceed as follows. For each black square in the piece, we label a small number of subsequent white squares at that pitch level (depending on the tempo) as "grey squares." Grey squares and black squares are then the only stream locations that need to be considered. More precisely, the possible analyses for a segment are the analyses which have streams on (1) all black squares and (2) all possible subsets of any grey squares. (The fact that an analysis may have multiple streams on one square complicates this somewhat, but I will not discuss this here.)

Given that only certain squares need to be considered as grey squares, the program is able to construct a relatively small set of possible segment analyses. In deciding which of these segment analyses to use, the program adopts the dynamic programming technique described in chapter 1. The input is analyzed in a left-to-right manner; at each segment, for each analysis of that segment, the program records the best global analysis ending with that segment analysis (known as a "best-so-far" analysis). At the next segment, the program only needs to consider adding each possible analysis of the new segment on to each "best-so-far" analysis at the previous segment, creating a new "best-so-far" analysis for each segment analysis at the new segment. In this way the program is guaranteed to find the best global analysis for the whole piece, while avoiding a combinatorial explosion of possible analyses.

When considering going from one segment analysis to another, the program must choose the "transition" between them: i.e., which streams to connect to which other streams. It is here that the pitch proximity rule and new stream rule are enforced (as well as CWFR 3, prohibiting crossing streams). The best transition between two segment analyses can be calculated in a purely local fashion; this is done whenever the program is considering following one segment analysis with another.

An example of the program's output is shown in figure 4.14. This is a portion of the program's analysis of a Bach fugue; the score for the excerpt is shown in figure 4.15. The output represents the grid of squares discussed earlier. Time is on the vertical axis and pitch on the horizontal axis (unlike in the figures shown above). The time segments used are eighth-notes, unless a change of note happens within an eighth-note beat,

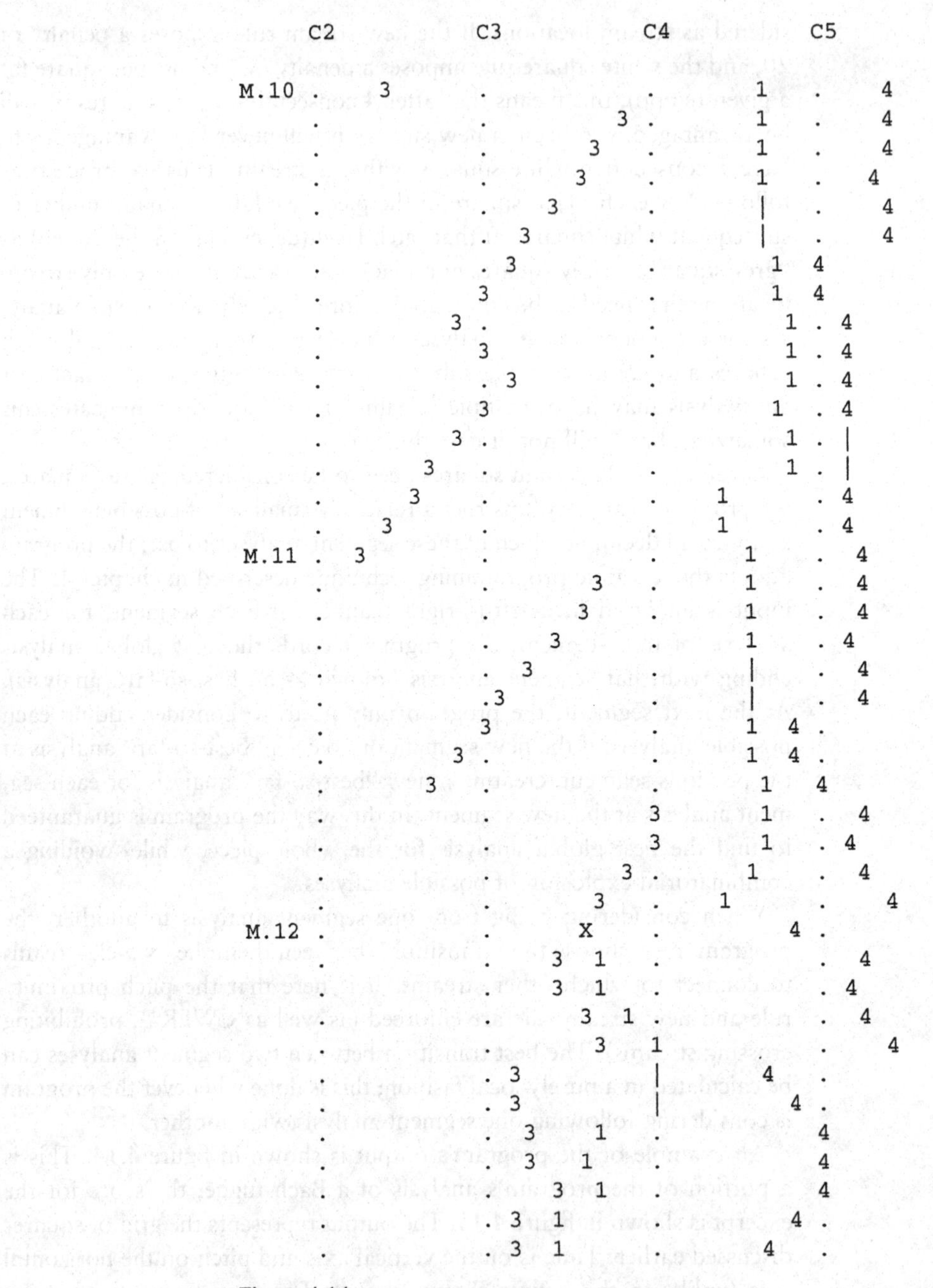

Figure 4.14
An excerpt from the contrapuntal program's analysis of Fugue No. 2 in C Minor from Bach's *Well-Tempered Clavier* Book I.

Figure 4.15
The Bach fugue excerpt analyzed in figure 4.14.

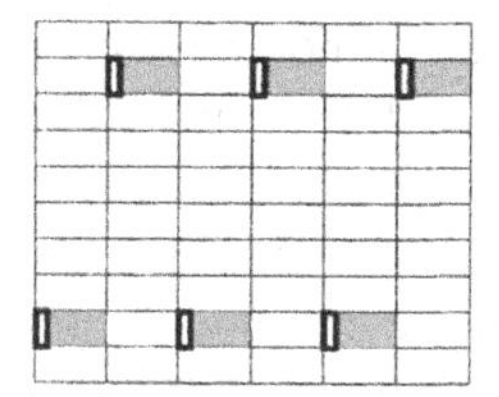

Figure 4.16
A pattern of two alternating pitches. Assume that segments are 0.25 second in length.

in which case smaller units are used. Each square containing a note (or "black square") is given a number indicating which stream is present there. (Recall that every black square must be contained in a stream.) The streams are numbered in the order of their appearance in the piece; streams beginning simultaneously are numbered from bottom to top in pitch. (It may seem odd that the streams are numbered 1, 3, and 4; this will be explained below.) If two streams are present in a square, this is marked by an "X"; figure 4.14 contains one example of this (this is in fact an error on the program's part, as I will discuss). If a stream contains a white square, this is shown with a vertical bar.

4.6 Tests

Two ways of evaluating the model will be discussed here. First, I will examine how the model accommodates the experimental results discussed earlier. Secondly, I will consider the model's handling of actual musical inputs.

The model is able to account for a number of the experimental findings on auditory stream segregation. Consider a pattern of two pitches in alternation, as shown in figure 4.16; assume that the segments here are

0.25 second in length, so that the white square rule (CPR 3) imposes a penalty of 5 points per segment. There are two possible analyses of this pattern (at least, two that are plausible): it could be heard as a single oscillating stream, or as two static streams. (Imagine that the pattern has continued for some time; the new stream rule [CPR 2] imposes a slightly larger penalty for the two-voice interpretation, but this penalty is only imposed once and is therefore negligible.) In figure 4.16, the one-stream analysis receives a penalty of 14 (2×7) from the pitch proximity rule (CPR 1) for each cycle of the pattern (a cycle being two notes); the white square rule imposes no penalty. The two-stream analysis receives a penalty of 10 from the white square rule (2×5) for each cycle of the pattern; the pitch proximity rule imposes no penalty. Thus on balance the two-stream analysis is preferred. If the pitch distance between the two streams were reduced to 3, then the penalty from the pitch proximity rule for the one-stream analysis would only be 6 per cycle, while the score for the two-stream analysis would be unchanged; the one-stream analysis would then be preferred. Now suppose that the speed of the pattern in figure 4.16 was cut in half, so that the length of each note was doubled. The one-stream analysis still receives a penalty of 14 per cycle; but now the two-stream analysis receives a penalty of 20 per cycle, since there are now 4 white squares per cycle. Thus slowing the tempo favors the one-stream analysis; this accords well with the experimental findings discussed above.

It was noted above that the new stream rule imposes a higher penalty for the two-stream interpretation in figure 4.16. Once the pattern has repeated many times, this difference becomes negligible. When it has only been heard a few times, however, the difference is not negligible. After two occurrences of the pattern, the one-stream analysis receives a penalty of 21 from the pitch proximity rule and a penalty of 20 from the new stream rule, for a total of 41; the two-stream analysis is penalized 20 points by the white square rule and 40 points by the new stream rule (20 points for each stream), for a total of 60. Thus, after only two occurrences of the pattern, the one-stream analysis will be preferred; as the pattern continues, though, the two-stream analysis will gain favor. As noted earlier, this is exactly what happens perceptually; with such stimuli, there is a greater tendency for stream segregation as the number of repetitions of the pattern increases.

Other experimental findings are accommodated by the model quite straightforwardly. The perceptual abhorrences for crossing lines, for including a single note in two streams, and for having multiple simultaneous notes in a single stream are explicitly reflected in the model's rules. The preference for minimizing the number of voices is captured in the new stream rule. Finally, the factor of temporal separation in the grouping of notes is reflected in the white square rule; this penalizes white spaces within a stream, thus favoring the beginning of a new stream instead.

An effort was also made to test the model on real pieces of music. A number of pieces were encoded in the appropriate format, and submitted to the model for analysis. It would clearly be desirable to subject the model to some kind of systematic, quantitative test. However, this is problematic. As noted with respect to figure 4.1, the correct "contrapuntal analysis" for a piece is often not entirely clear. (It is usually *mostly* clear, but there tend to be many points of ambiguity and uncertainty.) One case where the correct contrapuntal analysis is explicit is Bach fugues (and similar pieces by other composers). In that case, the separate voices of the piece are usually clearly indicated by being confined to particular staffs and notated with either upward or downward stems. The program was tested on four Bach fugues: fugues 1 through 4 from Book I of the *Well-Tempered Clavier*. The output was examined, and the program's errors were analyzed. The results are shown in table 4.1.

Four kinds of errors were recorded. One kind is a "break," in which the program breaks what should be a single voice into two or more streams. It should be noted that such errors are not necessarily errors. The voices of a Bach fugue are assumed to persist throughout the piece, though they may contain long periods of silence; in such cases the program tends to split them into a number of temporally separated streams. For example, the three-part C Minor Fugue was analyzed as having 11 streams. (This explains why the streams in figure 4.14 are labeled as 1, 3, and 4, rather than 1, 2, and 3, as one might expect.) By the understand-

Table 4.1
Test results of contrapuntal program on the first four fugues of Book I of Bach's *Well-Tempered Clavier*

	Breaks	Missed collisions	Incorrect collisions	Misleads	Total errors
Fugue 1 (C major, 4 voices)	10	2	1	9	22
Fugue 2 (C minor, 3 voices)	7	2	2	5	16
Fugue 3 (C♯ major, 3 voices)	25	12	4	75	116
Fugue 4 (C♯ minor, 5 voices)	36	22	5	21	84

ing of the term proposed here, one might argue that several streams are indeed present in each voice. A second kind of error was a "missed collision," where the program failed to identify a collision between two voices; in a third kind of error, an "incorrect collision," the program analyzed two voices as colliding when they really were not. (Figure 4.14 shows one such error: the G on the downbeat of m. 12 is incorrectly assigned to two voices at once.) The fourth kind of error was a "mislead," where the program simply failed to connect the notes in the correct way between one segment and the next.

A number of the errors were due to voice-crossings. In such cases, however, it was often quite unclear that the voice-crossings would actually be perceived (an example of an imperceptible voice-crossing will be given below). Other errors also seemed like ones that a listener might easily make. One case is shown in figure 4.17; here, the program interpreted the D on the last eighth of m. 10 as a continuation of the tenor line (second from the bottom) rather than the alto line (third from the bottom), as shown by the dotted line. This connection was favored partly because it resulted in a slightly smaller leap (7 steps instead of 9), and also because it results in fewer white squares (i.e. a shorter rest) between the two connected notes. This "error" seems quite understandable, if not actually perceptually correct. However, there were also numerous cases where the program's interpretation was really implausible; two of these will be discussed below.

The program's performance on the four fugues varies considerably. The larger number of errors on Fugue 4, compared to Fugues 1 and 2, may be due to its greater length, and also to the fact that it has five voices, which leads to a greater density of voices and hence a greater difficulty of tracking the voices by pitch proximity (as well as a larger number of collisions, which the program is only moderately successful in

Figure 4.17
Bach, *Well-Tempered Clavier* Book I, Fugue No. 1 in C Major, mm. 10–11.

Figure 4.18
Bach, *Well-Tempered Clavier* Book I, Fugue No. 3 in C♯ Major, mm. 1–3. The program analyzed this passage (incorrectly) as two separate voices, as indicated by the upward and downward stems.

identifying). However, the very large number of errors on Fugue 3, a three-voice fugue, is surprising. Inspection of the errors showed that this was due to two factors. First, the fugue contains a large number of voice-crossings, mostly quite imperceptible. Secondly, the fugue subject itself was consistently misinterpreted by the program as containing two voices, not one; for example, the opening of the fugue is analyzed as shown in figure 4.18. While this is strictly incorrect, such subordinate streams within a melody might well be regarded as a real and important part of perception; I will discuss this possibility in section 8.6.

The model was also tested on a number of other keyboard pieces from the Baroque and Classical periods. In these cases, quantitative testing was not possible, since no "correct" contrapuntal analysis was available. However, the results could still be inspected in an informal way. A significant problem emerged that had not arisen in the Bach fugues; the Mozart sonata discussed earlier provides an example (figure 4.1). It seems clear that the right-hand line in mm. 12–14 continues as the top line in m. 15; it is then joined by a second right-hand line underneath. However, the program connected the right-hand line in mm. 12–14 with the lower right-hand line in m. 15. The reason is clear: connecting with the lower line involves only a one-step leap in pitch (F to E), whereas connecting with the upper line requires a two-step leap (F to G). A number of other cases occurred in this piece and others where the upper voice incorrectly connected with an inner voice, or where the upper voice ended, so that an inner voice became the upper voice. These problems were largely solved by the addition of a fifth preference rule:

CPR 5 (Top Voice Rule). Prefer to maintain a single voice as the top voice: try to avoid cases where the top voice ends, or moves into an inner voice.

Presumably there is a tendency to regard whichever voice is highest as "the melody," and to accord it special status. Given that tendency, it is

Figure 4.19
Haydn, String Quartet Op. 76 No. 1, I, mm. 29–31. The dotted lines indicate the written instrumental lines.

not surprising that we prefer to keep this attention focused on a single voice rather than switching frequently from one voice to another. Interestingly, while this rule generally improved the model's performance, it did not improve performance on the fugues, and actually caused new errors. This is in keeping with the conventional view that the voices in a fugue are all of equal importance; the top voice does not require special attention as it does in much other music.

Finally, the model was tested on several excerpts from Classical-period string quartets. String quartets (or other ensemble pieces) might appear to be good pieces for testing, since the "correct" contrapuntal analysis is explicit in the instrumentation. However, it is often problematic to take the score as the correct contrapuntal analysis. It is common in string quartets for voices to cross in ways which are not perceptually apparent; in figure 4.19, the viola and second violin parts cross as indicated by the dotted lines, but these crossings would surely not be heard. Even in the case of a string quartet, then, determining the correct analysis from a perceptual point of view—the one that would be heard by a competent listener—is somewhat subjective. There are also occasional cases in string quartets where the most plausible analysis does not appear to be in any way inferable from a piano-roll representation. In figure 4.20, the notes alone suggest that all three melodic phrases belong to a single stream (certainly it would be heard this way if played on a piano). In a live performance, however, other cues—in particular, *visual* information about who is playing what—would make it clear that the three phrases belong to different voices. However, cases where such information is needed appear to be fairly rare in string quartet music.

As noted above, many voice-crossings in string quartet music are not actually perceived as such; this is of course in keeping with CWFR 3,

Figure 4.20
Mozart, String Quartet K. 387, I, mm. 4–6.

Figure 4.21
Mozart, String Quartet K. 387, I, m. 34. The dotted lines indicate points where one line crosses another.

prohibiting crossing streams. Other cases can be found, however, where voices are actually perceived to cross. Figure 4.21 shows an example; it is not difficult to hear the second violin line crossing over the viola here. Cases like these suggest that the prohibition against voice crossings cannot really be treated as an inviolable rule. An important point should be made about this example, which applies to most other cases I have found of perceptible voice-crossing. In Deutsch's "scale illusion" stimuli —where there is a strong tendency not to hear crossing voices—an alternative "bouncing" percept is available (see figure 4.3): at the point of crossing, it is possible for the lower voice to take over the descending scale and the upper voice to take over the ascending one, so that both voices are maintained and crossing is avoided. In figure 4.21, however—where the second violin's A3–A4 crosses the viola's E4—this interpretation is not possible. Maintaining both voices and avoiding a cross would mean splitting the long viola note between two voices, which would violate CWFR 4. Another alternative is maintaining one stream on the E4, ending the lower stream on A3, and starting a third stream on A4; but this incurs an extra penalty from the new stream rule. In this case, then,

Figure 4.22
Bach, *Well-Tempered Clavier* Book I, Fugue No. 1 in C Major, m. 12.

Figure 4.23
Bach, *Well-Tempered Clavier* Book I, Fugue No. 2 in C Minor, m. 20.

there is no palatable alternative to the crossing-stream interpretation. In general, crossing-stream interpretations only seem to arise as a "last resort," when no reasonable alternative is available.

Two other factors should be mentioned which occasionally arose in these tests. One is motivic parallelism. Consider figure 4.22 (m. 12 from the Bach C Major Fugue). Here the model began the upper stream of the left hand on the F♯3 rather than the previous E3. Beginning it on the E would be permissible, but is undesirable due to CPR 4. Perceptually, however, one would probably hear the upper stream as beginning on the E. One reason for the latter interpretation is that it would identify the melody as an instance of the main fugue subject. Getting the model to recognize motivic parallelisms of this kind would be quite difficult (Marsden 1992 has an interesting discussion of this problem). Another factor favoring the second interpretation of m. 12 is tonal: it might make more tonal sense to begin a voice on E rather than F♯, given that E major is the underlying harmony at that moment. A clearer case of the influence of tonal factors on contrapuntal analysis is given in figure 4.23, from the C Minor Fugue. Here, the model ends the middle voice on the G3 halfway through the measure, and then begins a new stream on A♭4 on the sixth eighth of the measure. The model is reluctant to continue the middle voice as it should, since this would involve a collision on F3 followed by a fifteen-step leap up to A♭4. But ending a voice on a secondary dominant harmony (V/iv) without continuing it through to its resolution (iv) feels quite wrong. Meter may also be a factor here; the model's interpretation involves ending a stream on a weak beat. Tonal, motivic, and metrical factors appear to arise in contrapuntal analysis relatively rarely; these are the only two cases in Fugues 1 and 2 where such information seemed necessary.

The pieces used for testing in this study were deliberately confined to genres with little timbral differentiation: solo piano and string quartet music. It would be interesting to apply the model to other genres, such as orchestral music, where a variety of timbres are used. In such genres, timbre might prove to play an important role in contrapuntal analysis; or perhaps it would not. Of course, addressing this issue would raise the very difficult question of how to encode timbre in the input representation.

Although it requires refinement and further testing, the model proposed here offers a promising computational solution to the problem of contrapuntal analysis. It performs creditably on actual musical inputs, and accords well with the psychological evidence on sequential integration. Still, the reader may feel somewhat unsatisfied after this discussion. The current model has assumed that each piece has a single contrapuntal analysis which every listener infers. But often, our intuitions about the correct contrapuntal analysis are not so clear-cut. As noted in the opening of this chapter, it is often unclear whether two melodic segments form a single stream or not. Moreover, it is often the case that what appears in the score as a single melodic line can be broken down into multiple contrapuntal strands. I believe that the current model can shed interesting light on these matters of ambiguity and hierarchy in contrapuntal structure. However, it seems best to leave this topic for a general discussion of ambiguity, which I present in chapter 8. For now, let us continue with the presentation of the preference rule systems themselves.

5
Pitch Spelling and the Tonal-Pitch-Class Representation

5.1 Pitch-Class, Harmony, and Key

The aspects of music considered so far—meter, grouping, and contrapuntal structure—are crucial ones, and contribute much to the richness and complexity of musical structure. However, an entire dimension of music has so far been neglected. In listening to a piece, we sort the pitch-events into categories called *pitch-classes*; based on their pitch-class identities, notes then give rise to larger entities, harmonies and key sections. These three kinds of representation—pitch-class, harmony, and key—are the subject of the next three chapters.

In the present chapter, I discuss the categorization of notes into pitch-classes. Pitch-classes are categories such that any two pitches one or more octaves apart are members of the same category. These are the categories indicated by letter names in common musical parlance: C, C♯, and so on. (They may also be indicated by integers: by convention, C is 0, C♯ is 1, etc.) Rather than accepting the traditional twelve-category pitch-class system, however, I will argue for a system of "tonal pitch-classes," distinguishing between different spellings of the same pitch-class (for example, A♭ versus G♯). This means that a spelling label must be chosen for each pitch event; a preference rule system is proposed for this purpose. Tonal pitch-classes are represented on a spatial representation known as the "line of fifths," similar to the circle of fifths except stretching infinitely in either direction. The line of fifths is of importance not only for the representation of pitch-classes, but harmonies and keys as well (discussed in chapters 6 and 7). I will begin with a general discussion of this model and my reasons for adopting it.

f	F	d	D	b	B	g♯
b♭	B♭	g	G	e	E	c♯
e♭	E♭	c	C	a	A	f♯
a♭	A♭	f	F	d	D	b
d♭	D♭	b♭	B♭	g	G	e

Figure 5.1
Weber's table of key relationships (1851). Uppercase letters represent major keys; lowercase letters represent minor keys.

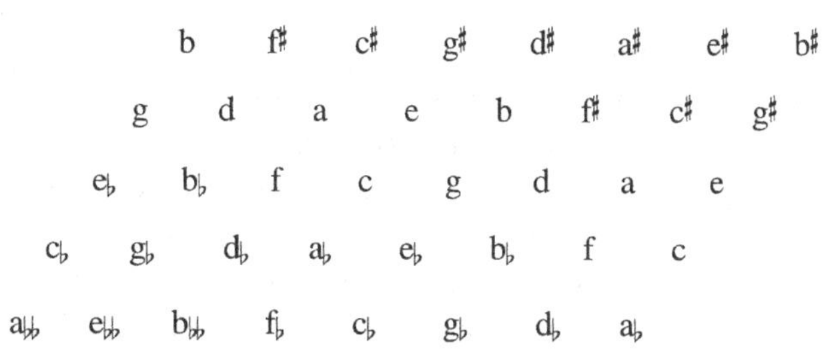

Figure 5.2
Riemann's "Table of Relations" (1915/1992).

5.2 Spatial Representations in Music Theory

The use of spatial models to represent tonal elements—pitches, chords, and keys—has a long history in music theory. The early-nineteenth-century theorist Gottfried Weber (1851) represented keys in a two-dimensional array, with the "circle of fifths" on one axis and alternating major and minor keys on the other (figure 5.1); several later theorists also advocate this model, including Schoenberg (1954/1969) and Lerdahl (1988). Lerdahl also proposes a similar space for chords within each key; both Lerdahl's key and chord spaces are "toroidal," in that they wrap around in both dimensions. Other spatial models have represented pitches. Riemann's influential "table of relations" (1915/1992) is a two-dimensional space with three axes representing perfect fifths, major thirds, and minor thirds (figure 5.2); a major or minor triad emerges as a triangle on this plane. Longuet-Higgins (1962) proposes a model of pitches with perfect fifths on one axis and major thirds on the other. Other work on spatial representations comes from psychology. Shepard (1982) proposes a complex five-dimensional representation of pitches, incorporating octave, fifth, and semitone relations. Krumhansl (1990,

Figure 5.3

46) presents a spatial representation of similarity relations between keys, based on data about the stability of pitches in the context of different keys (discussed further in chapter 7); the model that emerges is quite similar to Weber's two-dimensional model. She also proposes a two-dimensional representation of chords (p. 190), again based on experimental data.

While these models come from a variety of perspectives, their main motivation is the same: to capture intuitions about relations of closeness or similarity between tonal elements. These relations are an important part of our experience of tonal music. Consider the case of keys. C major and G major are closely related, in that modulating between them seems easy and natural; moving (directly) from C major to F♯ major, however, seems much more radical and abrupt. Figure 5.3 gives another example in the area of harmony. The first five chords all feel fairly close to one another; the final chord, F♯7, is far from the others, creating a sense of surprise and tension, and a sense of moving away to another region in the space. This sense of space, and movement in space, is an indispensable part of musical experience; it contributes greatly to the expressive and dramatic power of tonal music.

One feature common to all of the models proposed above is an axis of fifths, such that adjacent elements are a fifth apart. This reflects the traditional assumption that elements related by fifths (pitches, chords, or keys) are particularly closely related. Retaining this assumption, the model I adopt here is simply a one-dimensional space of fifths, as shown in figure 5.4—the "line of fifths."[1] Some discussion is needed as to why this particular space was chosen (as opposed to one of the other spaces discussed above, for example).

The first issue that arises concerns the kind of "fifths axis" that is to be used. One possible representation of fifths is shown in figure 5.5, the well-known "circle of fifths." Essentially, this is a one-dimensional space which wraps around itself, with each pitch-class represented exactly once. By contrast, the line of fifths allows for the representation of dif-

Figure 5.4
The "line of fifths."

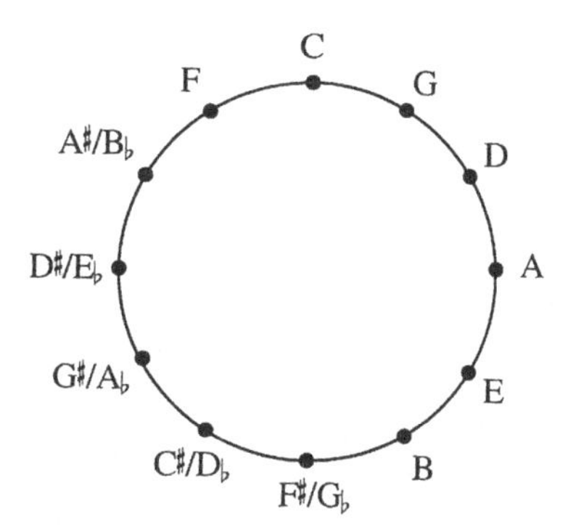

Figure 5.5
The "circle of fifths."

ferent "spellings" of the same pitch: A♭ and G♯ are at different positions in the space. (We can assume, for the moment, that the line extends infinitely in either direction.) We could call categories such as A♭ and G♯ "tonal pitch-classes" (or "TPCs"), as opposed to the twelve "neutral pitch-classes" ("NPCs") represented on the circle of fifths. The question is, which of these models is preferable from a cognitive viewpoint, the circular (NPC) model or the linear (TPC) model? This issue has received surprisingly little attention. Music notation seems to assume a TPC model, as does traditional tonal theory, where distinctions between, say, A♭ and G♯ are usually recognized as real and important. Some of the spatial representations discussed earlier—such as Riemann's and Longuet-Higgins's—also recognize spelling distinctions. On the other hand, some work in music cognition assumes an NPC model of pitch; one example is Krumhansl's key-profile model, discussed at length in chapter 7. (Much music theory also assumes an NPC model of pitch; however, this branch of theory—sometimes known as pitch-class set theory—is mainly concerned with nontonal music.) I will argue here for the tonal-pitch-class view; for a variety of reasons, a linear space of fifths is preferable to a circular one for the modeling of tonal cognition.

The simplest argument for the line of fifths is that it is needed to capture distinctions that are experientially real and important in themselves.

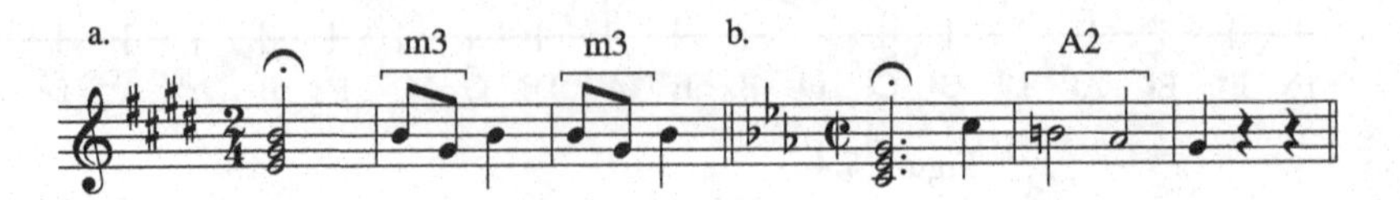

Figure 5.6
From Aldwell & Schachter 1989.

Figure 5.7
"The Star-Spangled Banner," mm. 1–2.

A circle is only capable of representing neutral-pitch-class distinctions; but tonal-pitch-class distinctions are sometimes of great importance. Figure 5.6 offers a nice example, suggested by Aldwell and Schachter. The placid, stable minor third in figure 5.6a has quite a different effect from the restless, yearning augmented second in figure 5.6b. One explanation for this is that the lower note is heard as a G♯ in the first case and an A♭ in the second; A♭ and B are represented differently on the line of fifths (specifically, they are much further apart) than G♯ and B. Tonal-pitch-class distinctions are also experientially relevant in the case of chords and keys. For example, the key of E♭ major seems much closer to C major than the key of D♯ major. Imagine beginning in C major and modulating to E♭ major; then imagine starting in C major, moving first to E major, and then to D♯ major. We feel that we have gone further—and have arrived at a more remote destination—in the second case than in the first. These distinctions are represented quite naturally on the line of fifths, but not on the circle.

So far, the rationale for spatial representations has been that they capture relationships that are of direct experiential importance. However, there is another important motivation for spatial representations: they can play an important role in choosing the correct analysis. Consider figure 5.7, the first six notes of the "Star-Spangled Banner." These could be spelled G-E-C-E-G-C; alternatively, they could be spelled G-E-B♯-F♭-Fx-C. Clearly, the first spelling is preferable to the second, but why? One explanation is that, in choosing spelling labels for pitches, we prefer to locate nearby pitches close together on the line of fifths (figure 5.8). As we will see, a similar principle obtains in harmony, where there is a preference to choose roots that are close together on the line of fifths;

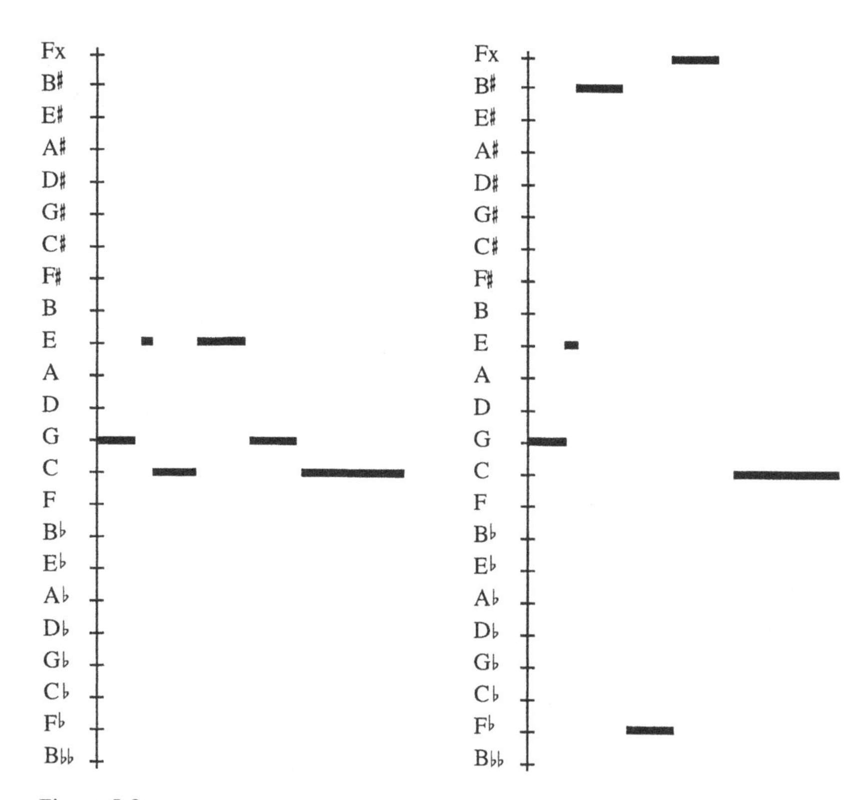

Figure 5.8
Two possible spellings of the excerpt in figure 5.7, represented on the line of fifths.

and in the case of roots, this principle affects *neutral*-pitch-class decisions (C versus A) as well as *tonal* pitch-class ones (C versus B♯). In such cases, the spatial model not only represents important intuitions about closeness between elements, but also helps us to label elements correctly.

A further point should be made about spelling distinctions. The spellings of pitches are not just important in and of themselves; they also influence other levels of representation. Consider figure 5.9; how would the melody note of the last chord be spelled? In the first case, C♯ seems most plausible; in the second case, D♭ is preferable. This could be explained according to the principle presented above: we prefer to spell notes so that they are close together on the line of fifths. In the first case, C♯ is closer to the previous pitches; in the second case, D♭ is closer. (This can be seen intuitively in figure 5.10, though the difference is small.) Notice, however, that the spelling of the last chord in figure 5.9a and

Figure 5.9

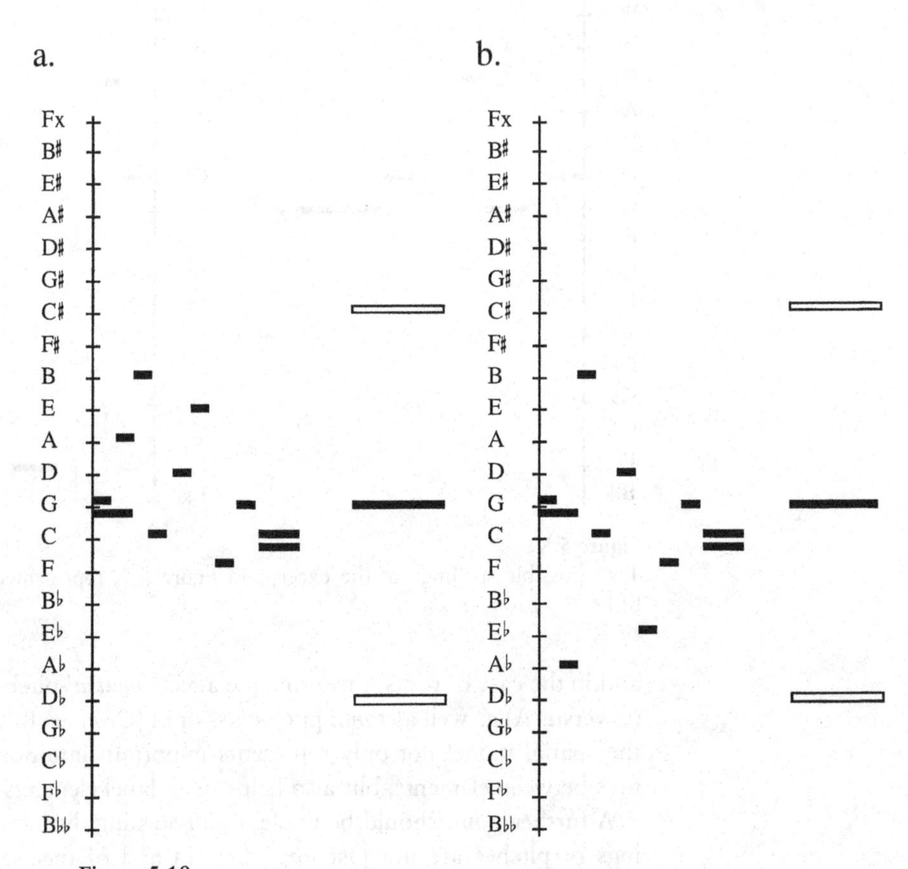

Figure 5.10
Line-of-fifths representations of (a) figure 5.9a, (b) figure 5.9b. White rectangles indicate possible spellings of the final note (C♯ or Db).

5.9b influences the harmonic implications of the chord. A D♭ pitch (combined with G) implies an E♭7 chord, suggesting a move to A♭ major; a C♯ suggests A7, moving to D major (or minor). Again, the point is that spelling distinctions are not only important in themselves; they also help us to attain desired analytical results, in this case in the labeling of harmonies. Spelling distinctions can also have implications for key structure, as I will show in chapter 7.

To summarize the argument so far, there are several motivations for spatial representations, and for the particular representation proposed here. In general, spatial representations capture intuitions about the closeness of tonal elements that are experientially real in and of themselves. There are compelling reasons for adopting a representation that recognizes distinctions of spelling. These distinctions are of direct experiential importance, for pitches, chords, and keys alike. Using the line of fifths also helps us to obtain the correct spelling labels for pitches, and is useful in harmonic analysis as well; the idea here is that there is a preference to choose labels that are close together on the line of fifths. Finally, recognizing spelling distinctions at the level of pitches provides useful input in harmonic and key analysis.

While it may be apparent that the line of fifths has a number of desirable features, it is not clear that it is the *only* model with these features. Since a number of other spatial representations have been proposed, it is worth discussing why the line of fifths was chosen over other alternatives. One might wonder, first of all, why a one-dimensional model was chosen rather than a two-dimensional one. Two-dimensional models of chords and keys have the virtue that they can capture the distinction between major and minor chords and keys, something not possible on the line of fifths. (The current model does capture these distinctions in other ways, as I will explain in later chapters.) One problem with a two-dimensional representation is that each element is represented at infinitely many places. In other words, with such a representation, we would have to choose not only whether a chord was an A♭ or a G♯, but *which* A♭ it was.[2] To my mind, such distinctions do not have a strong basis in musical intuition.

One might also question the decision to use a flat space as opposed to a wraparound one. As mentioned above, one reason for this is that a flat space allows TPC distinctions to be recognized, which are important for a number of reasons. Another reason is that the closeness of elements to previous elements needs to be calculated, and this can be done most easily on a flat space. However, I do not claim to have an airtight argument for why the line of fifths is superior to all other spaces for the problems I address here. This could only be determined by making a serious effort to solve these problems using other spaces, something which I have not attempted.

5.3 Tonal-Pitch-Class Labeling

If we assume that the line of fifths has psychological reality, this has important implications for music cognition. It means that listening to a piece involves choosing tonal-pitch-class labels for each pitch-event, and in so doing, mapping each event on to a point on the line. (Roots and keys must also be mapped on to the line, as I discuss in chapters 6 and 7.) In the remainder of this chapter I propose a preference model for performing this task.

Before continuing, several issues must be considered. One is the psychological reality of TPC distinctions. With each of the other preference rule systems proposed here, I have begun by presenting experimental evidence for the general reality of the kind of structure at issue. In the case of spelling distinctions, however, it must be admitted that such evidence is scarce. The cognitive reality of spelling distinctions simply has not been explored. I argued above that spelling distinctions are experientially important, and that they also have consequences for other aspects of musical structure. While I think there would be general agreement on this among highly trained musicians, the importance of spelling distinctions for listeners of tonal music in general is less clear. (It should be noted that tonal-pitch-class labels inferred in listening are assumed to be *relative*, rather than absolute; I will return to this point.)[3]

Another issue concerns the source of correct tonal-pitch-class analyses. If we assume that listeners infer spelling labels for pitches, how do we know what labels are inferred? One obvious source of evidence here is composers' "orthography," that is, the way they spell notes in scores; this might be taken to indicate the TPC labels they assume. Another source we can use is our own intuitions about how events in a piece should be spelled—for example, in writing down a melody that is dictated to us. We should note, however, that relying on such intuitions—our own or the composers'—is sometimes problematic, since there are cases—"enharmonic changes"—where spelling decisions are clearly based on matters of convenience, rather than on substantive musical factors. In particular, we prefer to avoid remote spellings such as triple flats and triple sharps. But it is usually fairly clear where such practical decisions are being made, and what the "musically correct" spelling would be.

In general, the issue of pitch spelling has been as neglected in artificial intelligence as it has been in psychology. One exception is the algorithm for pitch-spelling proposed by Longuet-Higgins and Steedman (1971). Unlike the model I propose below, the Longuet-Higgins/Steedman model operates only on monophonic music. Moreover, it is only applied after the key of the piece in question has already been determined; the spelling of notes within the scale is then determined by the key signature. (The authors propose a key-finding algorithm which I discuss in chapter 7.) The problem then becomes how to determine the spelling of notes that are "chromatic," or outside the scale. Longuet-Higgins and Steedman consider first the case of chromatic notes that are part of what they call a "chromatic scale," i.e. a sequence of notes forming an ascending or descending stepwise chromatic line. The rule they propose is that the first and last pair of notes in the scale should be separated by a "diatonic semitone": a semitone interval in which the two pitches are on different staff lines (such as G♯ to A rather than A♭ to A).[4] This rule captures an important insight, but it is not sufficient. Consider the sequence G-G♯-A (assuming that the first and last notes are diatonic, thus must be spelled as G and A). The rule here would state that both pairs of notes should be separated by a diatonic semitone, but there is no spelling of the second event that will allow this. As for notes whose spelling is not decided by the first rule, Longuet-Higgins and Steedman propose that they should be spelled so as to be "in the closest possible relation to the notes which have already been heard," where closeness is measured in terms of Longuet-Higgins's two-dimensional representation of pitches (mentioned earlier). This, again, is a valuable idea, and is represented in a somewhat different way in my own algorithm, as will be seen in the next section.

5.4 A Preference Rule System for Tonal-Pitch-Class Labeling

The task of the current model, then, is to determine a tonal-pitch-class label for each event in a piece. The input we assume is the usual one of a note-list or "piano-roll," giving the pitch, on-time, and off-time of each event. Recall from chapter 1 that events are assumed to be categorized into steps of the chromatic scale; pitches are represented as integers, with middle C = 60. We can think of the output of the model as a set of labels (C, F♯, A♭, etc.) attached to each event in the input representation. Another useful way of representing the output is seen in figures 5.8 and 5.10. This is similar to a piano-roll representation, except that the vertical axis now represents the line of fifths, rather than pitch height; we could call this the "tonal-pitch-class representation" of a piece. TPC's, like NPC's, can be labeled with integers, representing their line-of-fifths position. We will adopt the convention that C = 2, with ascending fifths denoted by ascending numbers; thus the C major scale comprises the positive integers 1 through 7.

It is important to note that the possible *tonal*-pitch-class labels for an event are constrained by its *neutral* pitch-class. If an event has an NPC of 0 (to use, once again, the usual convention), it can have a TPC of C, B♯, or D♭♭, but not C♯ or A. Thus the first step to determining the TPC labels of a set of events is to determine their NPC labels. From a computational point of view, this is easy. Given that pitches are labeled with integers, their neutral pitch-class follows automatically. Any pitch whose pitch number is a multiple of 12 has NPC 0; any pitch whose pitch number is a multiple of 12, plus 1, has NPC 1; and so on. To put it differently, the NPC of an event is given by its pitch number modulo 12. (It may be that this is not a very plausible model of how neutral pitch-class is determined perceptually, but I will not explore that here.) The model must then choose a TPC label for the pitch, given the possibilities allowed by its NPC label.

The model consists of three "TPRs" (tonal-pitch-class preference rules). The first rule is a very simple one, but also the most important:

TPR 1 (Pitch Variance Rule). Prefer to label nearby events so that they are close together on the line of fifths.

In many cases, this rule is sufficient to ensure the correct spelling of passages. Return to the case of "The Star-Spangled Banner" (figure 5.8). As noted earlier, the pitch variance rule offers an explanation for why the first spelling is preferred; it locates the events more closely together on the line of fifths. Note that the rule applies to *nearby* events. If one measure contains a B♯ followed by an E♯, there is great pressure to spell the second event as E♯ rather than F; if the two events are widely separated in time, the pressure is much less.

Let us consider how the pitch variance rule might be quantified. Any TPC representation has a "center-of-gravity" (COG), a mean position of all the events of the passage on the line of fifths; a spelling of all the pitch-events is preferred which minimizes the line-of-fifths distance between each event and the center of gravity. In statistical terms, a representation is sought which minimizes the variance among events on the line of fifths. The COG is constantly being updated; the COG at any moment is a weighted average of all previous events, with more recent events affecting it more. This ensures that the pressure for events to be located close together on the line of fifths is greatest for events that are close together in time. One important parameter to be set here is the "decay" value of the COG, determining how quickly the pressure to locate two events close together decays as the events get further apart in time. Roughly speaking, it seems that the pressure is significant for intervals of a few seconds, but decays quickly for longer intervals.

The pitch variance rule is perhaps not what first comes to mind as the main principle behind spelling. Alternatively, one might explain spelling choices in terms of diatonic collections. In "The Star-Spangled Banner," perhaps, we decide from the first few pitches that the key is C major; this implies a certain diatonic collection; this then dictates the spellings of subsequent pitches. This is somewhat more complex than my explanation, since it assumes that key has already been determined and that this "feeds back" to influence spelling. However, there are also cases where the scale-collection explanation simply does not work. Consider figure 5.9; it was argued that C♯ is the preferred spelling of the final melody note in figure 5.9a, D♭ in figure 5.9b. These choices are captured well by the pitch variance rule. In figure 5.9a, the COG of all the events (up to the final one) is 3.21, so that C♯ (9.0) is closer than D♭ (−3.0); in figure 5.9b, the COG is 2.21, making D♭ closer. To explain this in terms of scale collections, however, is problematic. The keys of the first measure in figure 5.9a and b are clearly C major and C minor, respectively; neither C♯ nor D♭ are in either of these scales (regardless of what "minor scale" one chooses). One could also explain the spellings of these pitches in terms of their harmonic and tonal implications. Clearly a C♯-G tritone implies an A7 chord, and anticipates a move to D minor; a D♭-G tritone implies E♭7, moving to A♭ major. One might claim that C♯ is preferred in the first case because the implied key of D minor is somehow closer to C major—or more expected in a C major context—than A♭ major is; these key expectations then govern our interpretation of the pitches. Again, this is a rather complex explanation of something that can be explained very easily using the one rule proposed above. One of the main claims of the current model is that spelling can be accomplished without relying on "top-down" key information.

While the pitch variance rule does not explicitly refer to diatonic collections, such collections do emerge as privileged under this rule. As is well known, a diatonic scale consists of seven adjacent positions on the circle (or line) of fifths; a passage using a diatonic scale collection will therefore receive a relatively good score from the pitch variance rule (as opposed to, for example, a harmonic minor, whole-tone, or octatonic

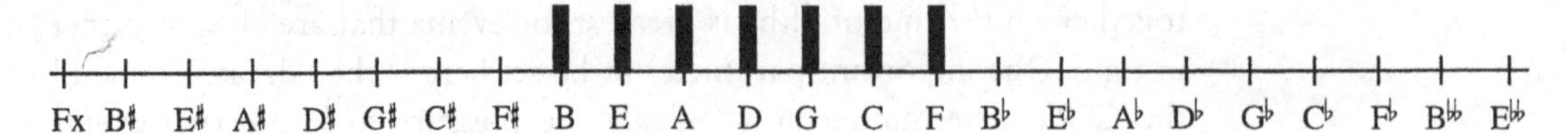

Figure 5.11
The C diatonic scale represented on the line of fifths.

collection), since its pitches permit a very compact spelling on the line of fifths. Note also that the pitch variance rule will naturally find the *correct* spelling of a diatonic scale, since this is the most closely-packed one. For example, imagine a passage whose pitches are all (potentially) within the C major scale, with all seven steps of the scale equally represented. If the C major spelling of the events is chosen, then the mean position of events on the line of fifths will be exactly at D, and all the events of the passage will lie three steps or less from the center of gravity (see figure 5.11). (This seems to be roughly correct in practice as well; for actual musical passages in major, the COG is generally about two steps in the sharp direction from the tonic.)[5] Thus this spelling of the events will be strongly preferred over other possible spellings. If the passage contains a few "chromatic" pitches (but not enough to greatly affect the center of gravity), the pitch variance rule may not express a strong preference as to their spelling, since two alternatives are about equally far from the center of gravity; other rules may then be decisive.

One apparent problem with the pitch variance rule should be mentioned. In "The Star-Spangled Banner," the spelling G-E-C-E-G-C is clearly preferable to G-E-B♯-F♭-Fx-C; but what about Fx-Dx-B♯-Dx-Fx-B♯? Here, the pitches are as "closely-packed" as in the original version; they are simply shifted over by twelve steps. This raises a subtle, but important, point. It is best to regard the TPC representation as relative rather than absolute: what is important is the relative positions of events on the line of fifths, not their absolute positions. For one thing, treating the line-of-fifths space as absolute would assume listeners with absolute pitch. If we treat the space as relative, moreover, this means that a representation of a piece in C major is really no different from the same representation shifted over by twelve steps. This view of the TPC representation is not only cognitively plausible, but musically satisfactory as well. There is no musically important reason for notating a piece in C major rather than B♯ major; it is simply a matter of notational convenience.

While the pitch variance rule alone goes a long way towards achieving good TPC representations, it is not sufficient. Another principle is voice-

Figure 5.12

leading. Given a chromatic note followed by a diatonic one a half-step away, we prefer to spell the first note as being a diatonic semitone away from the second—that is, placing them on different diatonic steps—rather than a chromatic semitone. In figure 5.12a, we prefer to spell the first two melody events as E♭-D rather than D♯-D. The line of fifths provides a way of stating this principle. Note that a diatonic semitone (such as E♭-D) corresponds to a five-step interval on the line of fifths; a chromatic semitone (such as D♯-D) corresponds to a seven-step interval. The voice-leading rule can thus be stated as follows:

TPR 2 (Voice-Leading Rule, first version). Given two events that are adjacent in time and a half-step apart in pitch height, prefer to spell them as being five steps apart on the line of fifths.

Given two events of NPC 3 and 2, then (adjacent in time and in the same octave), it is preferable to spell them as E♭-D rather than D♯-D. This bears a certain similarity to the pitch variance rule, in that it prefers to spell events so that they are close together on the line of fifths (five steps being closer than seven). However, the pitch variance rule applies generally to all nearby pitches, not merely to those adjacent and a half-step apart, and it is clear that a general rule of this kind is not sufficient. Consider figure 5.12a and b; apart from the E♭/D♯, these two passages contain exactly the same TPCs (the correct spelling of these events will be enforced by the pitch variance rule); and yet E♭ is preferred in one case, D♯ in the other. Thus something like the voice-leading rule appears to be necessary.

A problem arises with the voice-leading rule as stated above. Consider figure 5.13a and b. The TPC's (excluding the G♯/A♭) are identical in both cases, thus the pitch variance rule expresses no preference as to the spelling of the G♯/A♭. But the voice-leading rule expresses no preference either. In the first case, spelling the event as G♯ will result in one 7-step

Figure 5.13

Figure 5.14
Beethoven, Sonata op. 31 No. 1, II, mm. 1–4.

gap (G-G♯) and one 5-step gap (G♯-A); spelling it as A♭ will merely reverse the gaps, creating a 5-step gap followed by a 7-step one. Either way, we have one 7-step gap, and hence one violation of the rule. Intuitively, the rule is clear; the chromatic event should be spelled so that it is 7 steps from the previous event, 5 steps from the following event. Again, rather than expressing this in terms of chromatic and diatonic pitches, we will express it in another way. A chromatic pitch is generally one which is remote from the current center of gravity: the mean line-of-fifths position of all the pitches in a passage. We therefore revise the voice-leading rule as follows:

TPR 2 (Voice-Leading Rule, final version). Given two events that are adjacent in time and a half-step apart in pitch height: if the first event is remote from the current center of gravity, it should be spelled so that it is five steps away from the second on the line of fifths.

Here again, it might seem that an appeal to scale collections would be a simpler solution. Why not say, simply, "prefer to spell chromatic notes as five steps away from following diatonic ones"? In fact, however, this traditional rule does not correspond very well to musical practice.

Consider the passage in figure 5.14, from the second movement of Beethoven's Sonata Op. 31 No. 1 (specifically the chromatic scales in mm. 2 and 4). The traditional rule for spelling would have us spell each of the chromatic notes as sharps, since this places them a diatonic semitone away from the following pitch. But notice that Beethoven uses B♭ rather than A♯, violating this rule; the other four chromatic degrees are spelled as the traditional rule would predict. This is in fact the general practice in the spelling of chromatic ornamental tones: $\flat\hat{7}/\sharp\hat{6}$ is generally spelled as $\flat\hat{7}$, regardless of voice-leading context. Similarly, $\sharp\hat{4}/\flat\hat{5}$ is generally spelled as $\sharp\hat{4}$ even in descending lines.[6] The scale-collection approach is of no help here; B♭ is clearly chromatic in the context of C major, and it would seem arbitrary to posit a momentary move to F major in such cases. The current approach offers a solution. An ascending chromatic scale in a C major context, such as that in figure 5.14, presents a conflict for pitches such as D♯/E♭ and A♯/B♭. E♭ is closer to the center of gravity than D♯, and is therefore preferred by the pitch variance rule; but D♯ is preferred by the voice-leading rule. (Recall that for a C major piece, the COG is generally around D, so E♭ is generally slightly closer than D♯.) Similarly with A♯ and B♭. In the latter case, however, perhaps pitch variance favors the flat spelling over the sharp one enough (since it is much closer to the center of gravity) that it is preferred, overruling the voice-leading rule. In this way, the current model offers an explanation of why B♭ is preferred over A♯ in a C major context, even in ascending chromatic lines. The same logic might explain why F♯ is generally preferred over G♭ in a descending chromatic line in C major. Getting these results would depend on exactly how the parameters of the rules are set; it appears that, when a spelling of an event locates it within four steps of the center of gravity, that spelling is generally preferred, regardless of voice-leading.[7]

One final rule is needed to achieve good spelling representations. An example from earlier in the chapter, figure 5.3, provides an illustration. A♯ is clearly preferred over B♭ in the final chord, but why? B♭ is clearly closer to the center of gravity of the passage than A♯. The event in question is neither preceded nor followed by half-step motion, so voice-leading is not a factor. The explanation lies in harmony. With A♯, the pitches of the last chord form an F♯7 chord. As B♭, however, the pitch would be meaningless when combined with F♯ and E. (The pitches could also be spelled G♭-B♭-F♭, to form a G♭7 chord, but this is less preferred by pitch variance.) In this case, then, the spelling of the event is determined by harmonic considerations: a spelling is preferred which permits an acceptable harmony to be formed. We capture this in the following rule:

TPR 3 (Harmonic Feedback Rule). Prefer TPC representations which result in good harmonic representations.

To specify this rule in a rigorous fashion, we would obviously need to define what is meant by a "good" harmonic representation. The following chapter presents a preference rule system for harmonic analysis. For now, the important point is that the TPC representation of a piece forms part of the input to this process: whereas A♯ is considered compatible with a root of F♯, B♭ is not. The harmonic feedback rule then says that, where other factors favor interpreting a group of events in a certain way (for example, analyzing them as a triad or seventh), there will be a preference to spell the pitches accordingly; harmonic considerations thus "feed back" to the TPC representation.

It may be noted that the usual "chordal" spelling of a triad or dominant seventh is the most closely packed one on the line of fifths. Thus one might argue that the effect of harmony on spelling is simply a consequence of the pitch variance rule: since the pitch variance rule applies most strongly to events close together in time, there is naturally very strong pressure for simultaneous events to be compactly spelled. The problem here is that the notes of a harmony are not necessarily simultaneous. Rather, harmonic analysis involves a complex process of grouping notes together and inferring harmonies from them; and it is this grouping that influences how notes are spelled. Consider the variant of figure 5.3 shown in figure 5.15. The first A♯/B♭ and the second A♯/B♭ in the melody are equally close in time to the F♯; but the first one is clearly a B♭, the second one an A♯. The reason is that the meter suggests a change of harmony on the beginning of the third measure (rather than on a weaker beat, such as the last eighth of the second measure). This groups the first

Figure 5.15

A♯/B♭ together with the C-E-G, the second with the F♯-C♯-E; the harmonic feedback rule then prefers B♭ in the first case, A♯ in the second.

5.5 Implementation

The implementations of the TPC-labeling model and harmonic model (described in chapter 6) are closely integrated, and really constitute a single program. Here we will consider the TPC-labeling component; in section 6.4 we will consider the harmonic component and the way the two components are combined. The implementation described in this chapter and the next was largely developed and written by Daniel Sleator.

As usual, the first problem in implementing the preference rule system proposed above is to find a way of evaluating possible TPC analyses. A TPC analysis is simply a labeling of every pitch-event in the piece with a TPC label. One added well-formedness rule is that each note must have the same spelling all the way through; changes of spelling within a note are not permitted. We first divide the piece into very short segments. As with the contrapuntal program (explained in chapter 4), this is done using the metrical structure. We assume that every event begins and ends on a beat; if necessary, we adjust their time-points to the nearest beat. This creates a grid of squares, in which each square is either "black" or "white" (overlapping notes are not permitted).[8] We arbitrarily limit the possible spellings for pitches to a range of four cycles on the line of fifths. As usual, there is a combinatorial explosion of possible analyses. In this case, however, the explosion is worse than usual, because we must consider all possible combinations of different spellings of simultaneous pitches; thus there is a large number of analyses even for a single segment. To solve this problem, we make a heuristic assumption that the spellings of simultaneous pitches are always within a twelve-step "window" on the line of fifths. That is, we assume that A♭ and G♯ will never be present simultaneously—an assumption that seems to be virtually always correct in practice.

To apply the pitch variance rule (TPR 1), we calculate the pitch COG at each segment of the analysis. This involves taking the mean line-of-fifths position of all previous events, given their spelling in the analysis. Events are weighted for their length; they are also weighted for their recency according to an exponential function, as discussed above. (For this purpose, again, we use the short temporal segments described earlier; each part of an event within a segment is treated as a separate event.) Given the COG for a particular segment, we calculate the line-of-fifths distance of each pitch from the COG. Summing these distances for all pitches within the segment provides a measure of their closeness to previous pitches; summing these scores for all segments gives a measure of the closeness of all events in the piece to one another.

For the voice-leading rule (TPR 2), we calculate the distance of each event from the current COG. (Recall that the voice-leading rule applies only to events that are "chromatic"; we define this as four or more steps away from the current COG.) We also consider the inter-onset-interval between the event and the next event a half-step away. We then use some rather tricky reasoning. Suppose that the current COG is right at D; suppose further that we are considering an event of TPC A♭, and it is followed by an event of NPC 9. The second event could be spelled as A or B♭♭; but given the current COG at D, we can assume that A will be the preferable spelling. Thus we know that a spelling of A♭ will result in a voice-leading penalty; a spelling of G♯ will not. In this way we can assign penalties for the voice-leading rule, given only the spelling of the current event, the current COG, and the time-point of the next event a half-step away; we make an assumption about the spelling of the second event without actually knowing it.

Given this way of evaluating analyses, we then search for the best analysis using the usual left-to-right dynamic programming technique. (We have not yet considered the harmonic feedback rule [TPR 3]; we return to this below.) There is a complication here. Dynamic programming assumes that each segment in an analysis only cares about some numerical property of the analysis so far. We keep the "best-so-far" analysis ending with each possible value of that variable, knowing that whatever happens in the next segment, one of these "best-so-far" analyses will yield the best possible score. Here, however, the variable that matters is the COG (center of gravity) of the analysis so far. Ideally, for each COG, we would want to keep the best analysis ending with that COG. But a COG is a real number; the different analyses up to a certain point might lead to an unbounded number of different COGs. We solve this problem in a heuristic fashion by dividing the line of fifths into small ranges or "buckets." For each bucket, we keep the highest-scoring analysis ending up with a COG within that bucket; we also keep the 12-step window for the current segment entailed by that analysis.

The procedure, then, is as follows. We proceed through the piece left to right. At each segment, we consider each possible 12-step window as an analysis of the current segment. We consider adding this segment analysis on to the best-so-far analysis ending with each COG at the previous segment; this makes a "COG-window pair."[9] We calculate the pitch variance scores and voice-leading penalties for the current COG-window pair. Then, for each COG bucket at the *current* segment, we find all COG-window pairs ending up with a new COG in that bucket, and choose the highest-scoring one. (Of course, there may not be *any* window-COG pair leading to a new COG in a given bucket.) This gives us a new set of best-so-far analyses, allowing us to proceed to the next segment.

We have not yet considered how the harmonic feedback rule is implemented (TPR 3). As noted earlier, we want the harmonic and TPC representations to interact. The harmonic analysis should be affected by TPC considerations, but pitch spelling should also be influenced by harmony. In essence, we want to choose the combined TPC-harmonic representation that is preferred by both the TPC and harmonic rules combined. The system must consider all possible combined TPC-harmonic representations, evaluating them by both the TPC and harmonic rules, and choosing the one that scores highest overall. How we do this will be discussed in the next chapter.

5.6 Tests

The model was tested using the Kostka-Payne corpus discussed in section 2.5. The "correct" spellings were assumed to be the ones given in the notation, as shown in Kostka and Payne's workbook (1995b). One

Figure 5.16
Schubert, Originaltaenze Op. 9 (D. 365) No. 14, mm. 9–16.

problem was what to do about "enharmonic changes." In three of the excerpts, the notation features a sudden tonal shift that seems musically illogical. In Schubert's Originaltänze Op. 9, No. 14, for example (figure 5.16), the music shifts from D♭ major to A major, as opposed to the more logical B♭♭ major (the flat submediant key). No doubt this is because B♭♭ major is a difficult key to write and read. In these cases, however, I assumed the musically logical spelling to be the correct one; in the case of the Schubert, for example, I assumed that the entire A major section was spelled in B♭♭ major, with all pitches shifted twelve steps in the flat direction. In a few cases, also, the program spelled the entire excerpt in the wrong "cycle" of the line of fifths; for example, in D♭ minor rather than C♯ minor. In such cases, the program's output was judged correct as long as the *relative* positions of events on the line of fifths were the same as in the original.

Figure 5.17
Schumann, "Sehnsucht," Op. 51 No. 1, mm. 1–5.

The simplest way of evaluating the program's performance was simply to calculate the proportion of notes in the corpus that the algorithm judged correctly. (Recall that the program was not allowed to assign different spellings to different parts of a note.) Out of 8747 notes in the input corpus, the program made 103 spelling errors, a rate of 98.8% correct. As usual, it is difficult to know what to compare this with. It should be kept in mind, however, that a much simpler algorithm could probably achieve quite a high success rate. One very simple algorithm would be to always choose the same spelling for each NPC, using a range of the line of the fifths from (perhaps) C♯ to A♭; it appears that such an algorithm might achieve a success rate of about 84%.[10]

A single problem accounted for a large majority of the program's errors. Recall that, in cases of half-step motion, the program is designed to favor diatonic semitones, in cases where the first event is chromatic (TPR 2). The way we determine whether an event is chromatic is by seeing whether it is remote from the current COG. Frequently, however, an event is close to the current COG, but a sudden tonal shift afterwards causes a shift in the COG, making it chromatic in retrospect; and this is what determines its spelling. An example is the "German sixth," as used in modulation. In figure 5.17, the G♯'s in m. 4 first sound like A♭'s, part of a B♭ dominant seventh. Retrospectively, however, they turn out to be G♯'s, part of a German sixth chord in D minor. Given everything up and including the first half of m. 4, the program interprets the G♯/A♭'s as A♭'s, just as it should. Given what follows, the program should reinterpret the G♯/A♭'s in the context of D minor; in this context both G♯ and A♭ are chromatic, thus the voice-leading rule applies, favoring a G♯ spelling (since this allows a diatonic-semitone resolution to the A's in the second half of the measure).[11] However, the program has no way of doing this. One solution would be to label notes as chromatic based on their closeness to the center of gravity of *subsequent* pitches rather than previous ones, but this is not really possible given the current implementation.

6
Harmonic Structure

6.1 Harmony

The importance of harmony is indicated best, perhaps, by the amount of time devoted to it in undergraduate music education. The typical music student has at least two years of required music theory; and typically, most of this time is spent on harmony. Students spend considerable time and effort learning to analyze music harmonically, and also learning the reverse process—to generate appropriate notes from an explicit or implicit harmonic structure, whether this involves realizing harmonic or figured bass symbols, writing an accompaniment for a melody, or adding upper voices to a bass. In aural training classes, meanwhile, a large proportion of classroom time is normally devoted to identification of harmonies and harmonic structure, either in isolation or in progressions. Clearly, harmony holds a place of central importance in our theoretical conception of common-practice music.

As with all the kinds of structure studied in this book, but particularly in the case of harmony, a nagging question may arise. If we are assuming that harmonic analysis is something done by all listeners familiar with tonal music, why does it takes them so long to learn to do it explicitly? One might argue, first of all, that many undergraduate music students do not really qualify as "experienced listeners" of common-practice music; some, no doubt, have had very little exposure to common-practice music when they enter college. But tonal harmony is ubiquitous in all kinds of music besides common-practice music: children's songs, folk songs, and Christmas carols; music for film, TV, advertising, and computer games; and so on. Moreover, as we will see, there is psychological evidence that even musically untrained listeners possess a basic understanding of tonal harmony. Rather, the important point is this: the fact that people cannot readily perform harmonic analysis explicitly is no argument against the claim that they are performing it unconsciously when they hear music. As discussed in chapter 1, it is one of the basic premises of cognitive science that there are many things going on in our minds of which we are not readily aware. A useful analogy can be drawn with linguistics; it takes students considerable time to learn to perform syntactic analyses of sentences, though it is clear that they are doing this in some form every time they perceive language.[1] This is not to deny that theory and ear-training also involves a component of real "ear-training," teaching students to hear things and make aural distinctions that they were not in any way making before. But a good deal of undergraduate music theory, I would argue, is simply teaching people to do explicitly what they are already doing at an unconscious level.

In this chapter I present a preference rule system for performing harmonic analysis. Before proceeding, I should explain more precisely what I mean by "harmonic analysis." As I conceive of it here, harmonic analysis is the process of dividing a piece into segments and labeling each one with a root. In this sense it is similar to traditional harmonic analysis, or "Roman numeral analysis," as it is taught in basic music theory courses. There is an essential difference here, however. In Roman numeral analysis the segments of a piece are labeled not with roots, but rather with symbols indicating the relationship of each root to the current key: a chord marked "I" is the tonic chord of the current key, and so on. In order to form a Roman numeral analysis, then, one needs not only root information but key information as well. (Once the root and the key are known, the Roman numeral of a chord is essentially determined: if one knows that a chord is C major, and that the current key is C, the relative root of the chord can only be I.) Thus Roman numeral analysis can be broken down into two problems: root-finding and key-finding. My concern here will be with the root-finding problem; I will address the key-finding problem in the following chapter.

Another difference between root analysis and conventional Roman numeral analysis is that the latter gives other information about chords besides their (relative) roots, such as their mode (major or minor), inversion (the relationship of the bass note to the root), and extension (triad, seventh, etc.). While such information is not explicit in the harmonic representation I propose here, it is easily accessible once the harmonic representation is formed; I will return to this point in a later section.

6.2 Experimental and Computational Work on Harmonic Analysis

A good deal of experimental research has focused on harmony. Most of this work is concerned with the effects of harmonic structure on other kinds of intuitions and representations; in addition, however, it indirectly provides an impressive body of evidence for the general psychological reality of harmonic structure. For example, it has been found that chords within a previously established key are judged to follow better than others (Krumhansl 1990, 168–77). Chords are perceived with degrees of relatedness (reflected in judgments of how well one chord follows another) that accord well with tonal theory (Krumhansl, Bharucha & Kessler 1982); the same pair of chords will also seem more closely related to each other if they are within the key of the previous context (Bharucha & Krumhansl 1983). While the focus of these studies was on the ways that the perception of chords is affected by context, they also suggest that chords are being correctly identified. Other research has shown that harmony affects the memory for melodies. Studies by Deutsch have shown that melodies are more easily remembered when they can be encoded using "alphabets" based on tonal chords; we will discuss this in more depth in chapter 12.

Other work has focused on the role of harmony in segmentation. A study by Tan, Aiello, and Bever (1981) shows that melodic fragments implying V-I cadences tend to be heard as segment endings, an effect found for both trained and untrained subjects (this experiment was discussed in section 3.2). This suggests that listeners are sensitive to the closural effect of cadences, which in turn indicates an ability to identify cadences and their component harmonies. This is especially notable since the harmonies were presented in quite a subtle way, with the notes of each harmony presented in succession and with some harmonies incompletely stated. Further evidence for listeners' sensitivity to cadences comes from a study by Rosner and Narmour (1992), in which untrained subjects were asked to indicate the degree of closure implied by various harmonic progressions. Listeners attributed the highest degree of closure to V-I progressions; moreover, they recognized the closural effect of V-I even when the first chord was not in root position (i.e., when a note other than the root was in the bass—for example, V6-I), indicating an understanding of chordal inversion.

Harmony can also affect the perception of key. In a study by Butler (1989), it was found that the same group of pitches arranged in different ways could have different tonal implications. As arranged in figure 6.1a, the pitches F-E-B-C clearly imply C major; as arranged in figure 6.1b, they are much more tonally ambiguous. Both musically trained and

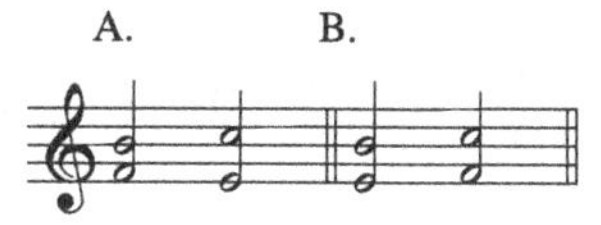

Figure 6.1
The same set of pitches can have different tonal implications when arranged in different ways (Butler 1989).

untrained listeners showed a strong awareness of these distinctions. An obvious explanation for this result is that tonal judgments are affected by harmony: figure 6.1a implies a clear G7-C progression, indicating C major, while figure 6.1b suggests a tonally indecisive E-F progression.[2]

Several attempts have been made to devise computer algorithms which perform harmonic analysis. Particularly notable are the models of Winograd (1968) and Maxwell (1992). Both of these algorithms begin with pitch information, and derive a complete Roman numeral analysis; both root and key information must therefore be determined. I will confine my attention here to the root-finding component of the programs. Examples of the outputs of the two programs are shown in figures 6.2 and 6.3. Both systems essentially analyze the input as a series of "verticals" (where any change in pitch constitutes a new vertical). In Winograd's system (1968, 20), the root of each vertical is determined by looking it up in a table; simple rules are given for labeling two-note verticals. (In Maxwell's model it is not explained how verticals are initially labeled.) There are then heuristics for deciding whether a vertical is a real chord or merely an ornamental event, subordinate to another chord (I will return to these below). Although this approach seems to operate quite well in the examples given, there are many cases where it would not. Very often the notes of a chord are stated in sequence rather than simultaneously, as in an arpeggiation; neither algorithm appears capable of handling this situation. In many other cases, the notes of the chord are not fully stated at all (either simultaneously or in sequence). For example, the pitches D-F may be part of a D minor triad, but might also be B♭ major or even G7; as I shall show, context must be taken into account in interpreting these. (This causes problems in Winograd's example in figure 6.2: the first chord in m. 14 is analyzed as III_6, implying root D, where it should clearly be part of an arpeggiated B♭ 6/4 chord.) Problems arise also with events that are not part of any chord, so-called "ornamental dissonances" such as passing tones and neighbor notes. Both Winograd's and Maxwell's algorithms have rules for interpreting certain verticals as ornamental, but

Figure 6.2
Schubert, Deutsche Taenze Op. 33 No. 7. The analysis shown is the output of Winograd's harmonic analysis program. From Winograd 1968. Reprinted by permission of the *Journal of Music Theory*.

these are not sufficient. For example, Maxwell's algorithm (1992, 340) considers any single note to be ornamental to the previous chord. Figure 6.4 gives a simple example where this will not work; the A is surely not ornamental to the previous chord here. In short, Winograd's and Maxwell's studies leave some important problems unsolved.

Others have attempted to model harmonic perception using a neural-network or "connectionist" approach, notably Bharucha (1987, 1991).[3] Bharucha proposes a three-level model with nodes representing pitches, chords, and keys. Pitch nodes are activated by sounding pitches; pitch nodes stimulate chord nodes, which in turn stimulate key nodes (figure 6.5). For example, the C major chord node is stimulated by the pitch nodes of the pitches it contains: C, E and G. Bharucha's model nicely captures the intuition that chords are inferred from pitches, and keys are in turn inferred from chords. The connectionist approach also offers insight into how harmonic knowledge might be acquired, an important

1 2 3 4 5 6
c: i - - iv - vii° i iv$^6_5$ ii°$^6_5$ V - - i - -
(P)
E^b: vi - - vii° - V$_6$
7 8 9 10 11 12
IV V I - - I ii iii I$_6$ V$^6_5$ - - I IV$_6$ vi ii$_6$ V - -
(P)
13 14 15 16 17 18
f: vii° - - V - - i V - i - -
E^b: ii - - ii - vii°$^6_5$ V - -

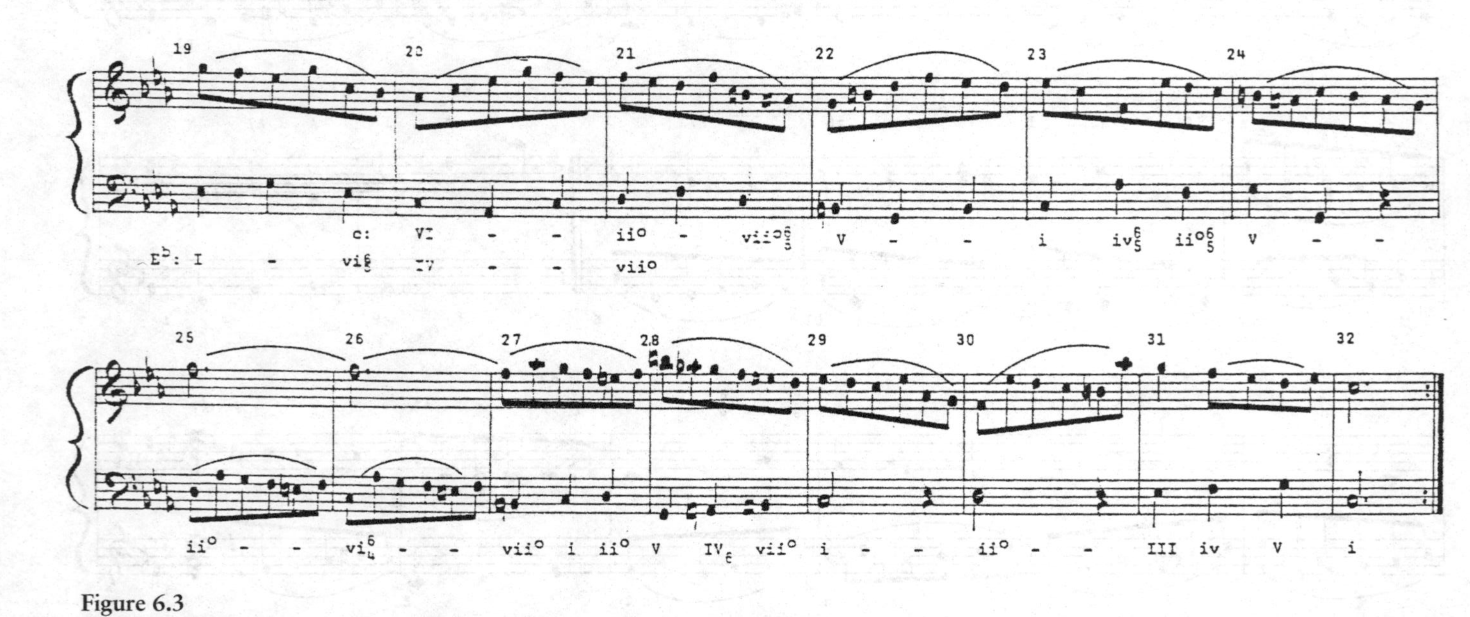

Figure 6.3
Bach, French Suite No. 2 in C Minor, Minuet. The analysis shown is the output of Maxwell's harmonic analysis program. From Maxwell 1992. Reprinted by permission of MIT Press.

Figure 6.4

issue which my own model does not address (Bharucha 1987, 26–7; 1991, 94–5).

Bharucha does not present any examples of his model's output, so it is difficult to evaluate it. However, there is reason to suspect that it might encounter some serious problems. It was noted above that harmonic analysis cannot simply be done by analyzing each vertical sonority one by one; context must also be considered. Bharucha proposes an interesting solution to this problem: a chord node is not merely activated while its pitch nodes are activated; rather, its activation level decays gradually after stimulation (Bharucha 1987, 17–18). This might seem to address some of the failings of the models discussed above, such as the problem of arpeggiations. However, this solution raises other difficulties. In listening to a piece, our experience is not of harmonies decaying gradually; rather, one harmony ends and is immediately replaced by another. Another problem relates to the model's handling of "priming," or expectation. Experiments have shown that, when listeners hear a chord, they are primed to hear closely related chords (for example, they respond more quickly to related chords than to unrelated ones in making judgments of intonation). Bharucha's model attempts to handle this, by allowing the key nodes stimulated by chord nodes to feed back and activate the nodes of related chords (pp. 18–21). The problem here is this: what exactly does the activation of a chord node represent? One would assume that it represents the chord that is actually being perceived at a given moment. But now Bharucha is suggesting that it represents something quite different, the amount that a chord is primed or expected. The phenomenon of priming is indeed an important one; but the degree to which a chord is heard is different from the degree to which it is expected. (I will offer an alternative account of the harmonic priming phenomenon in section 8.8.)

The three models discussed so far all assume that harmonic perception begins with pitch. The theory of Parncutt (1989) challenges this assumption. Parncutt argues that many aspects of musical cognition depend not on pitches as they occur in the score, but rather on "virtual pitches."[4] A musical pitch is made up of a combination of sine tones or pure tones: a fundamental plus many overtones. But the overtones of a pitch may

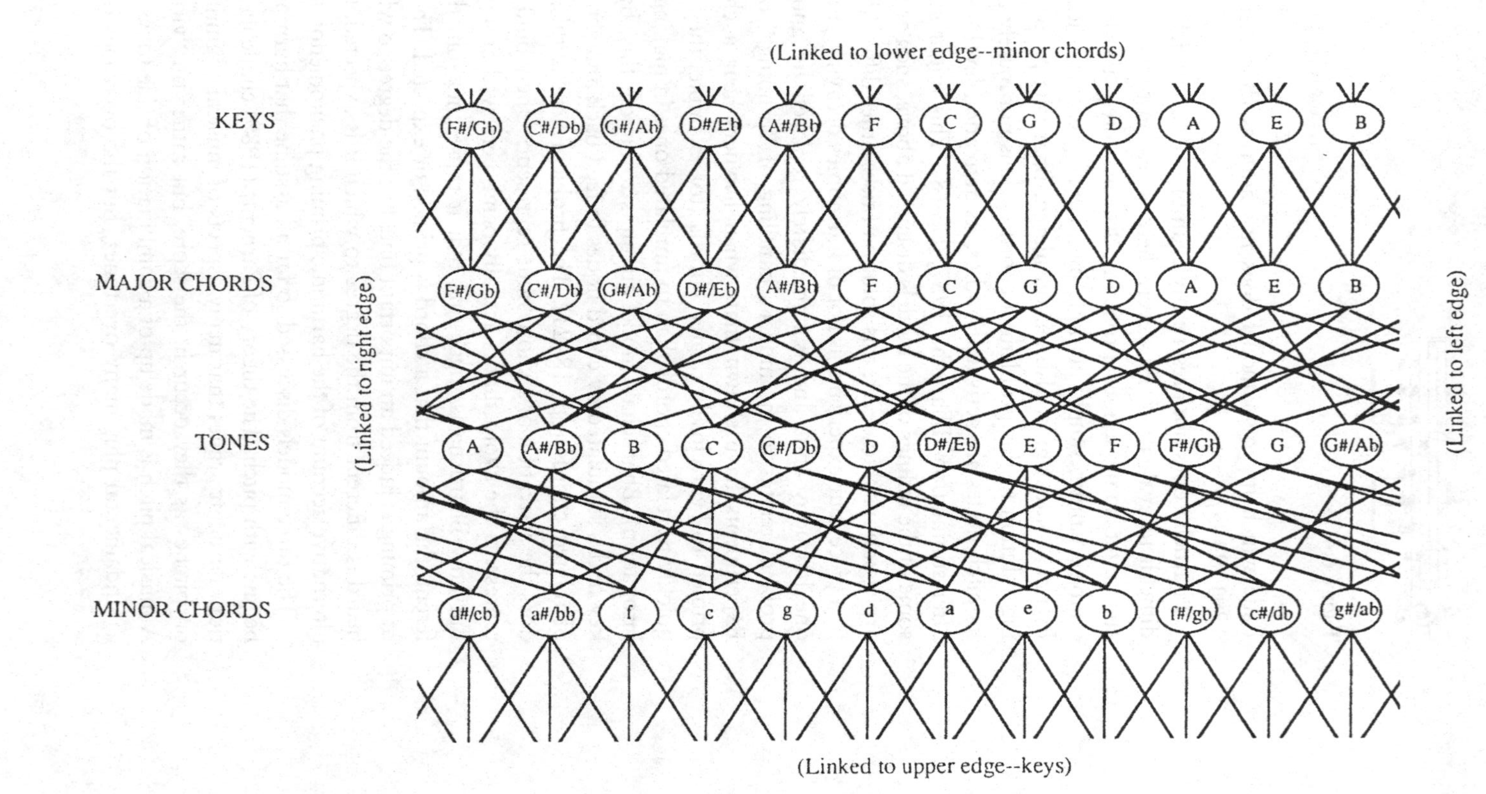

Figure 6.5
Bharucha's connectionist model (1987), showing relationships between pitches, chords, and keys. The ovals represent nodes; the lines represent connections between them. Reprinted by permission of Lawrence Erlbaum Associates, Inc.

also be understood as overtones of different fundamentals. For example, if one plays C4-E4-G4 on the piano, some of the components of E4 and G4 are also overtones of C4; others are overtones of pitches that were not even played, such as C3 and C2. In this way, a set of played pitches may give rise to a set of "virtual pitches" which are quite different in frequency and strength. Parncutt uses virtual pitch theory to make predictions about a number of aspects of musical cognition, such as consonance levels of chords and the number of perceived pitches in a chord; it is also used to predict the roots of chords. The root of a chord, Parncutt proposes, is the virtual pitch which is most strongly reinforced by the pure tone components of the chord (Parncutt 1989, 70, 146–50). The theory's predictions here are quite good for complete chords (such as major and minor triads and sevenths). They are less good for incomplete chords; for example, the root of the dyad C-E♭ is predicted to be E♭ (p. 147), as opposed to C or A♭. In cases where consonance levels or roots of chords are not well explained by his theory, Parncutt suggests that they may have "cultural rather than sensory origins" (p. 141).

The psychoacoustical approach to harmony yields many interesting insights. However, it is rather unsatisfactory that, in cases where the theory does not make the right predictions, Parncutt points to the influence of cultural conditioning. This would appear to make the theory unfalsifiable; moreover, it is certainly incomplete as a theory of root judgments, since some other component will be needed to handle the "cultural" part. It also seems problematic that the model's judgments are based on the actual acoustical composition of sounds; this suggests that harmonic analysis might be affected by *timbral* factors, which seems counterintuitive. But even if Parncutt's theory were completely correct as far as it went, in a certain sense it goes no further than the other studies discussed here in accounting for harmonic perception. It accounts for the fact that certain pitch combinations are judged to have certain roots, and it offers a more principled (though imperfect) explanation for these judgments than other studies we have seen. But as we have noted, there is much more to root analysis than simply going through a piece and choosing roots for a series of isolated sonorities. One must also cope with arpeggiations, implied harmonies, ornamental dissonances, and so on. A psychoacoustical approach does not appear to offer any solution to these problems. This is not to say that psychoacoustics is irrelevant to harmony; clearly it is not (indeed, it might be incorporated into my own approach in a limited way, as I will discuss). But there is much about harmony which—at present, at least—does not seem explicable in psychoacoustical terms.

While there are many valuable ideas in these studies, none of them offers a satisfactory solution to the problem of harmonic analysis. Some of these models also suffer from being highly complex. Maxwell's program (the chord-labeling component alone) has 36 rules; Winograd's program, similarly, has a vast amount of information built into it (as can be seen from his article). Bharucha's and Parncutt's models are more elegant; however, they seem even less adequate than Maxwell's and Winograd's systems in handling the subtleties of harmonic analysis—ornamental dissonances, implied harmonies, and the like. I now propose a rather different approach to these problems.

6.3 A Preference Rule System for Harmonic Analysis

The input required for the model is the usual "piano-roll" representation. The model also requires metrical structure, for reasons which will be made clear below. In addition, the model requires information about the spellings of each note in the input. In the previous chapter I proposed a model for assigning each pitch-event a spelling or "tonal pitch-class" (TPC), thus creating the TPC representation. It was argued there that the TPC representation and the harmonic representation closely interact; spelling can affect harmonic analysis, but harmonic analysis may also affect spelling. For now, however, we will assume that the TPC representation has been completed before harmonic analysis starts. The program's output consists of a division of the piece into segments, called "chord-spans," with each segment labeled with a root. As explained in the previous chapter, under the current framework, roots—like pitches and keys—are given tonal-pitch-class names; that is, we distinguish between different spellings of the same pitch-class (A♭ versus G♯). We can imagine each root as a point on the "line of fifths," shown in figure 6.6. In the following discussion we will use as an example the Gavotte from Bach's French Suite No. 5 in G Major, shown in figure 6.7.

The model has two basic tasks: it must divide the piece into chord-spans, and it must choose a label for each chord-span. For the moment, we will simply take the segmentation of the piece for granted; assume that the Bach Gavotte in figure 6.7 is divided into quarter-note segments.

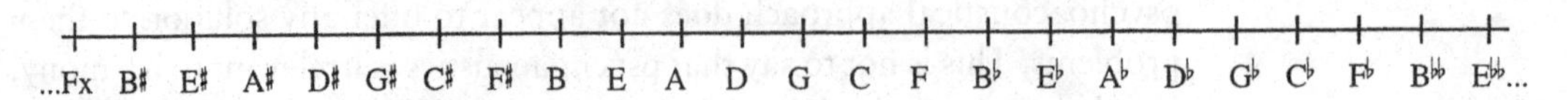

Figure 6.6
The "line of fifths."

Figure 6.7
Bach, French Suite No. 5, Gavotte.

Let us simply consider how correct root labels can be chosen for each segment. One important factor is clearly the pitches that the segment contains. Consider the second chord of the Gavotte: G-B-G. Simply considering the segment out of context, we know its root is unlikely to be D or F; the most likely root is G, since both G and B are chord-tones of G major. E is also a possibility, but even considering this segment in isolation G would seem more likely. The way we capture these intuitions is as follows. Every TPC has a relationship to every root, depending on the interval between them. The TPC G is $\hat{1}$ of G, $\hat{5}$ of C, $\hat{3}$ of E♭, ♭$\hat{3}$ of E, and ♭$\hat{7}$ of A. Certain relationships are more preferred than others; we try to choose roots for each segment so that the relationships created are as preferred as possible. This first HPR (harmonic preference rule) is called the compatibility rule, and is stated as follows:

HPR 1 (Compatibility Rule). In choosing roots for chord-spans, prefer certain TPC-root relationships over others, in the following order: $\hat{1}$, $\hat{5}$, $\hat{3}$, ♭$\hat{3}$, ♭$\hat{7}$, ♭$\hat{5}$, ♭$\hat{9}$, ornamental. (An ornamental relationship is any relationship besides those listed.)[5]

Consider the chord G-B-G. A root of G will result in TPC-root relationships of $\hat{1}$, $\hat{1}$, and $\hat{3}$, while a root of E will result in ♭$\hat{3}$, ♭$\hat{3}$, and $\hat{5}$. The former choice is clearly preferred by the compatibility rule, since on balance, the relationships involved are higher up on the list. Roots which involve "ornamental" relationships—those other than the ones specified—are still less preferred.

Note that the compatibility rule considers TPCs, not neutral pitch-classes. It is for this reason that the program requires spelling labels as input. The importance of this has already been discussed; it allows the model to make the correct root choices in cases like figure 5.4, and also allows for choices of harmony to influence spelling. We will consider later exactly how this interaction is captured by the model.

It may be seen that the TPC-root relationships specified above allow most common tonal chords to be recognized. $\hat{1}$, $\hat{5}$ and $\hat{3}$ form a major triad; $\hat{1}$, $\hat{5}$ and ♭$\hat{3}$ form a minor triad. $\hat{1}$, $\hat{5}$, $\hat{3}$ and ♭$\hat{7}$ form a dominant seventh; substituting ♭$\hat{3}$ for $\hat{3}$ produces a minor seventh. $\hat{1}$, ♭$\hat{3}$, and ♭$\hat{5}$ form a diminished triad; adding ♭$\hat{7}$ produces a half-diminished seventh. The fully diminished seventh is not allowed, but ♭9 allows for an alternative (and well-accepted) interpretation of the diminished seventh as $\hat{3}$-$\hat{5}$-♭$\hat{7}$-♭$\hat{9}$: a dominant seventh with an added ninth and no root. The model also allows certain combinations which are not common tonal chords, such as $\hat{1}$-$\hat{3}$-$\flat\hat{3}$ and other chords which combine the third and flat third (or the fifth and flat fifth). This is a flaw in the model which perhaps should be addressed, although in practice it rarely seems to cause the model to produce the incorrect analysis.

Now consider the second quarter of m. 1 (the first full measure). The E is clearly ornamental (how this is determined will be explained below); the chord-tones of the segment are then F♯-A-F♯ (the A is held over from the previous beat). The correct root here is D, but the compatibility rule alone does not enforce this. A root of D will yield TPC-root relationships of $\hat{3}$-$\hat{5}$-$\hat{3}$, whereas an F♯ root will yield $\hat{1}$-$\flat\hat{3}$-$\hat{1}$; the compatibility rule does not express a clear preference. Consider also the last quarter of m. 1; here the pitches are E-G-E, offering the same two interpretations as the previous case ($\hat{3}$-$\hat{5}$-$\hat{3}$ versus $\hat{1}$-$\flat\hat{3}$-$\hat{1}$). But here, the root is E; in this case, then, the $\hat{1}$-$\flat\hat{3}$-$\hat{1}$ choice is preferable. Clearly, another rule is needed here. The rule I propose is a very simple one: we prefer to give each segment the same root as previous or following segments. In this case, the first beat of m. 1 clearly has root D; there is then strong pressure to assign the second beat the same root as well. Another way of saying this is that we prefer to make chord-spans as long as possible (where a chord-span is any continuous span of music with a single root). This rule—which we could tentatively call the "long-span" rule—also addresses another question: we have been assuming segments of a quarter-note, but why not consider shorter segments such as eighth-note segments? For example, what is to prevent the algorithm from treating the third eighth-note of m. 4 as its own segment and assigning it root D? Here again, the "long-span" rule applies; spans of only one eighth-note in length (that is, an eighth-note root segment with different roots on either side) will generally be avoided, although they may occasionally arise if there is no good alternative.

While it is true that long spans are usually preferred over shorter ones, further consideration shows that this is not really the principle involved. Consider figure 6.8. The first note of the second measure could be part of

Figure 6.8

the previous F segment (as a $\hat{1}$), or it could be part of the following G segment (as a $\flat\hat{7}$). The compatibility rule would prefer the first choice, and the long-span rule stated above expresses no clear preference; why, then, is the second choice preferable? The reason is that we do not simply prefer to make spans as long as possible; rather, we prefer to make spans start on strong beats of the meter. This has the desired effect of favoring longer spans over shorter ones (strong beats are never very close together; thus any very short span will either start on a weak beat itself, or will result in the following span starting on a weak beat). In m. 1 of the Bach, for example, it will mitigate against starting spans on the second and fourth quarter notes, since these are relatively weak beats. However, it has the additional effect of aligning chord-span boundaries with the meter, thus taking care of cases like figure 6.8. We express this as the "strong beat rule":

HPR 2 (Strong Beat Rule). Prefer chord-spans that start on strong beats of the meter.

The strong beat rule raises a complication: it means that the algorithm requires metrical structure as input. The kind of metrical structure I am assuming is that proposed in chapter 2: a framework of levels of evenly spaced beats, with every second or third beat at one level being a beat at the next level up. I explain below how exactly the "strength" of a beat is quantified.

If harmonic changes are preferred at strong beats, one might wonder about the possibility of having a change of harmony at a time-point where there was no beat at all. The current model simply disallows harmonic changes at time-points which are not beats. It seems intuitively reasonable to exclude this possibility. This is also a logical extension of the strong beat rule; if changes of harmony on very weak beats are undesirable, it is naturally highly undesirable to have a harmonic change where there is no beat at all. (Occasionally one might want a change of harmony on an extrametrical note; the model does not allow this.) As we will see, this simplifies the implementation of the model, since it means that only beats must be considered as points of harmonic change.

A further preference rule is nicely illustrated by m. 15 of the Bach Gavotte. Considering just the first half of the measure, let us assume for the moment that the C and A in the right hand and the F♯ and A in the left hand are ornamental dissonances; this leaves us with chord-tones of G and B. The compatibility rule would prefer a root of G for this segment, but E seems a more natural choice; why? This brings us to a consideration discussed in chapter 5: we prefer to choose roots that are close together on the line of fifths. The previous span clearly has root B; we therefore prefer E over G as the root for the following span. The same applies to the first half of m. 18; the chord-tones here, C and E, could have a root of A or C, but because the previous span has root G, C is preferable (in this case, the compatibility rule reinforces this choice). We express this rule as follows:

HPR 3 (Harmonic Variance Rule). Prefer roots that are close to the roots of nearby segments on the line of fifths.

The final rule of the algorithm concerns ornamental dissonances. We have been assuming that certain events are ornamental. This means that, in the process of applying the compatibility rule (that is, looking at the pitches in a segment and their relationship to each root), certain pitches can simply be ignored. But how does the algorithm know which pitches can be ornamental? The key to our approach here is an idea proposed by Bharucha (1984). Bharucha addressed the question of why the same pitches arranged in different orders can have different tonal implications: B-C-D♯-E-F♯-G has very different implications from G-F♯-E-D♯-C-B, the same sequence in reverse. (Bharucha verified this experimentally, by playing subjects each sequence followed by either a C major or B major chord. The first sequence was judged to go better with C major, the second with B major [pp. 497–501].) He hypothesized what he called the "anchoring principle": a pitch may be ornamental if it is closely followed by another pitch a step or half-step away.[6] In the first case, all the pitches may be ornamental except C and G; in the second case, D♯ and B may not be ornamental. It is then the non-ornamental pitches that determine the tonal implications of the passage. The current model applies this same principle to harmonic analysis. The algorithm's first step is to identify what I call "potential ornamental dissonances" ("PODs"). A POD is an event that is closely followed by another pitch a whole-step or half-step away in pitch height. What is measured here is the time interval between the *onset* of each note and the onset of the next stepwise note—what we could call its "stepwise inter-onset interval." For example, the first E in the melody in m. 1 of the Bach Gavotte is a good POD because it is closely followed by F♯; the A in the melody in m. 4 is closely followed by G. However, the G in m. 4 is not closely followed by any pitch in a stepwise fashion; it is not a good POD. (The "goodness" of a POD is thus a matter of more-or-less rather than all-or-nothing.) The algorithm then applies the compatibility rule, considering the relationship between each TPC and a given root. As mentioned earlier, if the relationship between an event's TPC and the chosen root is not one of the "chord-tone" relationships specified in the compatibility rule—$\hat{1}$, $\hat{5}$, $\hat{3}$, $\flat\hat{3}$, $\flat\hat{7}$, $\flat\hat{5}$, or $\flat\hat{9}$—that event is then ornamental. Any pitch may be treated as ornamental, but the algorithm prefers for ornamental events to be good PODs. We express this rule as follows:

HPR 4 (Ornamental Dissonance Rule) (first version). An event is an ornamental dissonance if it does not have a chord-tone relationship to the chosen root. Prefer ornamental dissonances that are closely followed by an event a step or half-step away in pitch height.

The model satisfies this rule in an indirect fashion—not by selecting certain notes as ornamental once the root is chosen (this follows automatically), but by choosing roots so that notes that emerge as ornamental are closely followed in stepwise fashion. However, there is always a preference for considering events as chord-tones rather than ornamental dissonances, even if they are good PODs; this is specified in the compatibility rule. (The least preferred chord-tone relationships listed in the compatibility rule—$\flat\hat{5}$ and $\flat\hat{9}$—are only slightly more preferred than ornamental relationships, reflecting their rather marginal status as chord-tones.)

The "anchoring principle" does a good job of identifying a variety of kinds of ornamental dissonances. It handles ordinary passing tones (such as the eighth-note E in m. 1 of the Bach Gavotte), and neighbor-notes (the D in the left-hand in m. 5), as well as unprepared neighbors and appoggiaturas, notes which are followed but not preceded by stepwise motion (such as the C in m. 4). It also handles "double neighbors," such as the C-A in m. 15: a pair of ornamental tones on either side of a following chord-tone. The C is considered a (fairly) good ornamental dissonance because it is followed (fairly) closely by the B; the fact that there is an A in between is irrelevant. However, not all kinds of ornamental dissonances are captured by this rule. One important exception is escape tones, ornamental notes approached but not followed by stepwise motion (an example is the F♯ at the end of m. 7); another is anticipations, notes (often at the end of a measure) which are followed by another note of the same pitch, such as the G at the end of m. 23. The current version of the model cannot handle such notes, although in principle it should be possible to incorporate them.

Another principle is involved in choosing good ornamental dissonances. Consider the second half of m. 18 of the Bach. Here, every note in both the left and right hands is closely followed stepwise, so according to the rule stated above any of them could be treated as ornamental. The compatibility rule would still prefer an analysis which treated most of the notes as chord-tones; however, several possibilities seem almost equally good. Assigning a root of G to the segment would produce three chord-tones (G-G-B); a root of E would produce four (E-G-G-B). Yet D seems to be the most plausible choice of root. The main reason, I suggest, is that under a D interpretation, the chord-tones (F♯-A-F♯-A) emerge as metrically strong, and the ornamental tones as metrically weak. Of course, ornamental tones are not always metrically weak; this very passage contains several metrically strong dissonances, such as the C on the third beat of m. 17 and the F on the downbeat of m. 18. Other things being equal, however, metrically weak ornamental dissonances are preferred. We add this to the ornamental dissonance rule as follows:

HPR 4 (Ornamental Dissonance Rule) (final version). An event is an ornamental dissonance if it does not have a chord-tone relationship to the chosen root. Prefer ornamental dissonances that are (a) closely followed by an event a step or half-step away in pitch height, and (b) metrically weak.

The four preference rules presented here appear to go a long way towards producing reasonable harmonic analyses. Figure 6.9 shows the analysis of the Bach Gavotte produced by an implementation of the model. (The implementation is discussed further below.) Each chord symbol indicates a chord-span beginning on the note directly beneath, extending to the beginning of the next chord-span. While the program's performance here is generally good, there are several questionable choices. In cases where I consider the program's output to be definitely incorrect, I have indicated my own analysis in brackets. Probably the most egregious error is the omission of the important A harmony in m. 7; this is due to the escape tone F♯ in the right hand (mentioned above), a kind of ornamental tone which the program does not recognize.

6.4 Implementation

As noted earlier, the implementation of the harmonic model described here is really combined with the TPC-labeling model described in the previous chapter to form a single program; this program is largely the

Figure 6.9
Bach, French Suite No. 5, Gavotte, showing the program's harmonic analysis. (In cases where my own analysis differs, it is shown above in brackets.)

work of Daniel Sleator. I will begin by describing the implementation of the harmonic model in isolation.

The harmonic program requires the usual "note-list" input. Again, we assume for now that the TPC representation is complete once the harmonic analysis begins; thus each note also has a TPC label, which is an integer indicating its position on the line of fifths. The program also requires a specification of the metrical structure, consisting of a list of beats, giving the time-point and highest beat level of each beat. Like the contrapuntal and TPC programs already discussed, the harmonic representation requires a division of the piece into short segments, which is provided by the lowest level of the metrical structure. (Onsets and offsets of events are quantized to the nearest beat, so that lowest-level beats form segments in which each pitch is either present or absent.) In addition, however, the harmonic program also requires knowledge of the strength of beat on which each segment begins.

As noted above, the model only allows beats of the metrical structure to be chord-span boundaries. This means that lowest-level beats in the metrical structure can be taken to indicate indivisible segments for the harmonic program; the program only needs to choose a root for each segment. A well-formed analysis, then, is simply a labeling of each segment with a root. What we were earlier calling a "chord-span" emerges as a series of consecutive segments all having the same root. Roots, like TPCs, are positions on the line of fifths, and can therefore be represented as integers.

A given analysis of a piece can be scored by evaluating each segment of the analysis, and summing their scores. For the compatibility rule (HPR 1), each note (or part of a note) in each segment yields a score depending on the relationship of the note to the root given in the analysis. Higher-ranked relationships (such as $\hat{1}$ or $\hat{5}$) receive higher scores. Notes whose relationship to the root are not listed in the compatibility rule at all receive ornamental dissonance penalties (HPR 4); the penalty depends on the inter-onset interval to the next note a whole-step or half-step away in pitch, and the strength of the beat on which the note begins.

Some explanation is needed of how the "strength" of a beat is measured, since this is a bit complex. Recall that a metrical structure consists of several levels of beats. Every level of beats has a certain time interval associated with it, which is the time interval between beats at that level. (Even if time intervals between beats at a given level vary, as they may, we can measure the time interval of a particular beat in a more local way, by calculating the mean of the intervals to the next beat on either side.)

The metrical strength of a beat is then given by the time interval of the highest level containing it. A beat whose highest level has a long time interval is metrically strong; the longer the time interval, the stronger the beat. This provides a way of quantifying beat strength, which can then be used for the ornamental dissonance rule. The same approach is used for the strong beat rule (HPR 2). For each segment whose root differs from the root of the previous segment in the analysis, we assign a penalty based on the strength of the beat at which the segment begins; stronger beats result in lower penalties.

Finally, for the harmonic variance rule (HPR 3), we use an implementation similar to that for the pitch variance rule discussed in chapter 5. For each segment, a center of gravity (COG) is calculated, reflecting the average position of roots of all previous segments on the line of fifths. Segments are weighted by their length; they are also weighted for recency under an exponential curve, so that more recent segments affect the COG more.[7] We then assign a penalty to the current root based on its distance from that center of gravity.

The scores for segments on the compatibility rule and variance rule are weighted according to the length of segments. This means that the length of segments should not have a big effect on the program's analysis; for example, the output for a piece should be the same if the eighth-note or sixteenth-note level is chosen as the lowest level of meter. (Of course, if the eighth-note level is the segment level, then changes of hamony at the sixteenth-note level are not possible.) Note, however, that the program *is* sensitive to absolute time values. The strong beat penalty is based on absolute time between beats; a change of harmony might be tolerable at one tempo, but at a faster tempo the penalty might be prohibitively high. The musical significance of this will be discussed further in the next section.

We now consider how the usual dynamic programming approach could be applied to search for the highest-scoring analysis. For the moment, just consider the compatibility rule, the ornamental dissonance rule, and the strong beat rule. The scores of the first two rules can be calculated in a purely local fashion, and do not depend on context in any way. (The ornamental dissonance penalty depends on the stepwise inter-onset interval, but this does not depend on the *analysis* of neighboring segments.) For the strong beat rule, the score depends on whether the previous segment had the same root as the current one. To find the optimal analysis for this rule, for each segment, we must calculate the "best-so-far" analysis ending at each root. In evaluating the next segment, we can then consider adding each root on to each "best-so-far" analysis at the previous segment (factoring in the penalty from the strong beat rule, if any); we can then choose a new "best-so-far" analysis for each root at the current segment. What makes things complicated is the harmonic variance rule. To apply this rule, we must know how close the current root is to the center of gravity of previous roots. We can implement this rule in the same way that we implemented the pitch variance rule in the previous chapter. We "bucket" the COG into small ranges on the line of fifths, and for each small range, we store the best analysis ending with a COG in that range. To include the strong beat penalty as well, we must keep the best-so-far analysis ending in each *combination* of a particular COG and a particular root at the current segment.

It was argued in chapter 5 that, while spelling distinctions sometimes affect harmony, harmonic structure may also "feed back" to influence spelling. This brings us to an important general issue. Having one preference rule system influence another presents no problem; we simply apply one system first, and use the output as part of the input to the other. But what if there are two preference rule systems that interact, so that the structure generated by each one may influence the other? How can this be modeled at a conceptual level, and how can it be implemented? Lerdahl and Jackendoff's preference-rule-based generative theory has several cases of interactions between components, in that two kinds of structure require input from each other. For example, within the metrical component, *GTTM*'s Metrical PR 9 states "prefer a metrical analysis that minimizes conflict in the time-span reduction"; Time-Span-Reduction PR 1 states "Of the possible choices for head of a time-span T, prefer a choice that is in a relatively strong metrical position." Some have criticized this aspect of *GTTM* as circular (Cohn 1985, 38). In fact, however, there is a straightforward way of formalizing this situation with preference rule systems, though it sometimes creates computational problems.

The way to think about two interacting preference rule systems is as follows. Each preference rule system has rules which are internal to the system and which do not rely on input from the other system. However, there also must be a rule (or more than one rule) which monitors the interaction between the two systems. In the present case, that rule is the compatibility rule (HPR 1), which yields a score based on how compatible the TPCs of a segment are with the root chosen for that segment. Now, suppose we treat the two systems as, essentially, a single system, generating a combined TPC-harmonic representation: a segmentation of the piece labeled with roots, as well as a labeling of each note with a TPC. As usual, we can break this problem down into an "evaluation" problem and a "search" problem. To evaluate a particular analysis, we score the TPC component of the analysis by the TPC rules, score the harmonic component by the harmonic rules, and evaluate the compatibility of the two with the compatibility rule; we then judge the analysis overall by summing all of these rule scores. One point to note is that it becomes arbitrary whether we think of the compatibility rule as a harmonic rule or a TPC rule; it is really both. (Since we are assuming that both the harmonic and TPC representations might be adjusted to satisfy the compatibility rule, the harmonic feedback rule proposed in chapter 5 [TPR 3] becomes superfluous—though it is useful to maintain it as a reminder of the influence of harmony on pitch-spelling.) Another point is that the relative weights of the TPC and harmonic rules now become important. For example, if the harmonic rule scores are large relative to the TPC scores, the TPC scores will carry very little weight, and the combined analysis will be chosen almost entirely on the basis of the harmonic rules.

The search problem is harder. At each segment, we consider each possible TPC analysis combined with each harmonic analysis; each pair creates an overall segment analysis. (Recall that a TPC analysis of a segment is simply a 12-step window on the line of fifths.) The best analysis of the current segment now depends on the previous root, the previous harmonic COG, and the previous TPC COG. Thus we must store the best analysis ending in each combination of these three things, and consider combining it with each possible analysis of the current segment. Not surprisingly, this leads to a tremendous amount of computation. We "prune" the search by throwing out all the "best-so-far" analyses of a given segment whose score is more than a certain value below the highest-scoring analysis. This is a heuristic solution, since it means that the system is not guaranteed to find the highest-scoring analysis overall, but we have found a "pruning value" which maintains a reasonable speed for the program while rarely affecting the outcome.

6.5 Some Subtle Features of the Model

Figure 6.10, the melody "Yankee Doodle," illustrates an important virtue which sets the current model apart from those discussed earlier: its handling of unaccompanied melodies. As noted above, it is generally assumed that unaccompanied melodies have an implied harmonic structure, though for the most part the harmonies implied are not fully stated.

Figure 6.10
The unaccompanied melody "Yankee Doodle," showing the program's analysis.

The current model is designed to handle unaccompanied melodies as easily as fully harmonized textures. In most cases, the harmonies chosen by the program in figure 6.10 are not uniquely identified by the pitches present. For example, the F♯ at the end of m. 2 might easily imply an F♯ or B harmony in another context. The harmonic variance rule is crucial here; given previous harmonies of G and D, D is much closer on the line of fifths than F♯ or B would be. Similarly, in m. 5, the variance rule favors a choice of C over E (which might be favored if the measure were considered in isolation). The strong beat rule also plays an important role in selecting good places for harmonic change, and preserving a reasonably slow harmonic rhythm. Without it, for example, the compatibility rule would favor treating the third note of the melody as a separate chord-span with root D or A.

As well as its handling of implied harmony, the model captures some other subtle features of tonal harmony which deserve mention. In the first place, the model is sensitive to the voicing of chords, in particular to the doubling of pitch-classes in chords. As mentioned above, for each analysis of a segment, the compatibility rule assigns a score which is the sum of individual scores for each pitch in the segment. Specifically, the score for a pitch is given by the length of the note multiplied by a "compatibility value," reflecting how compatible that pitch is with the given root; pitch-root relationships ranked higher in the compatibility rule yield higher compatibility values. In both of the chords of figure 6.11a, both F and A are possible roots. In the first chord, however, the presence of four A's will yield an especially high score for an A analysis, since A is $\hat{1}$ of A; analyzing A as $\hat{3}$ of F is less preferred. In the second chord, the prepon-

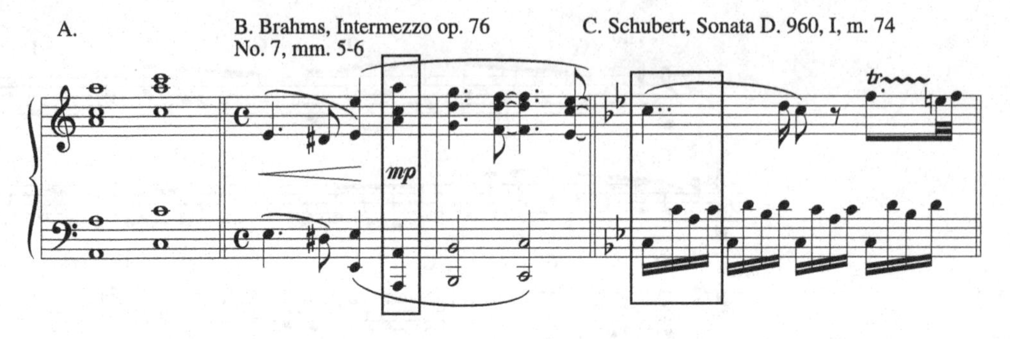

Figure 6.11
The doubling of pitches in a chord can affect its harmonic implications.

derance of C's (which could be either $\hat{5}$ of F or $\flat\hat{3}$ of A) will favor an F analysis. This accurately reflects a subtle aspect of harmonic perception: the doubling of pitches in a chord *does* affect its harmonic implications. The following two excerpts show the importance of doubling in real musical situations. While both of the harmonies marked with boxes contain only A and C, the first implies a root of A, while the second more strongly suggests F. (The larger context reinforces these differing implications, but they seem clear even when the chords are heard in isolation.) Another feature of the model worth mentioning is its handling of combined ornamental dissonances. Note that in the Bach Gavotte (figure 6.9), the program is generally reluctant to posit harmonic changes on weak eighth-note beats. One exception is the fourth eighth-note beat of m. 5. The three pitches in this chord segment—F♯, D, F♯—are all followed stepwise, and could well be considered ornamental. However, since each ornamental dissonance carries a small penalty and these are summed, the penalty for three ornamental tones in combination can be quite high; normally, in such cases, the program prefers to consider them as chord-tones, despite the high penalty from the strong beat rule. Again, this seems intuitively right; when three or more ornamental tones are combined, there is a strong tendency to hear them as a harmony.

A final emergent feature of the model is its sensitivity to tempo. Consider the passage in figure 6.12a, from a Haydn quartet. At a tempo of quarter = 120 (somewhat slow for this piece), the program assigns a separate harmony to each of the dyads in the upper voices. However, suppose the passage were performed four times as fast; it can then be rewritten (approximately) as shown in figure 6.12b. Now, the program finds only a single D harmony and a single A harmony, treating many of

Figure 6.12
Haydn, String Quartet Op. 64 No. 5, I, mm. 1–4. (A) The program's analysis of the excerpt, at a tempo of quarter = 120; (B) The same excerpt four times as fast, again with the program's analysis.

the notes as ornamental. The program's behavior here is due to the fact, mentioned above, that the rules are sensitive to absolute time values. In figure 6.12b, having a harmonic change on each dyad results in very high penalties from the strong beat rule, since some of these dyads occur on very weak beats; in figure 6.12a, the corresponding beats are stronger, and are thus more acceptable as points of harmonic change. (See the discussion in the previous section of how the strong beat rule is quantified.) Similarly, the penalties for the ornamental dissonance rule—which depend on the stepwise inter-onset-interval for ornamental notes—are calculated in terms of absolute time, and thus exert stronger pressure against ornamental tones at slower tempos. While one might dispute the program's exact analysis of figure 6.12a, the model's general sensitivity to tempo seems justified; positing a harmonic change on each dyad in figure 6.12a is surely more plausible than in figure 6.12b.

6.6 Tests

The program was tested on the corpus of excerpts from the Kostka-Payne workbook (1995b), already described in section 2.5 with regard to the meter program. (See the discussion there of how the corpus was selected.) The workbook is accompanied by an instructors' manual (Kostka 1995), containing Roman numeral analyses of the excerpts done by the author, with key symbols and Roman numeral symbols. While Roman numeral analysis does not give roots explicitly, a root analysis is implied; for example, if a section is notated as being in C major, a I chord implies a root of C, a V chord a root of G, and so on.

The excerpts were analyzed by the program, and the program's root analyses were compared with Kostka's harmonic analyses.[8] The pro-

Figure 6.13
Haydn, Sonata Op. 30, I, mm. 85–9, showing the program's harmonic analysis. Where Kostka's analysis differs, it is shown above in brackets.

gram's success was judged by the number of measures it analyzed correctly. In some cases, the program analyzed part of a measure correctly; in this case a fractional score was awarded. In a small number of cases (about 2%), a label was given in Kostka's analysis with no implicit root, such as "Ger + 6" (for "German sixth"); these were excluded from the count. Out of 525.1 measures, the program analyzed 439.8 correctly, a rate of 83.7% correct.

The program's errors fall in a variety of categories. One error, which is hardly an error, is that the program's treatment of the diminished seventh—as a dominant seventh on the fifth degree, with a flat ninth and no root—differs from Kostka's, who treats it as a diminished seventh on the seventh degree. (The program's interpretation is a well-established one, reflected in some theory textbooks; see Piston 1987, 328.) Other errors are due to the program's fairly limited vocabulary of chords. For example, it cannot handle the major seventh, major ninth, and sixth degrees when treated as chord tones. In such cases the program must find some other way of interpreting the problematic note. Sometimes it treats the note as an ornamental dissonance; this is often a reasonable interpretation, and often leads to the correct choice of root. In other cases, however, errors arise. An example is seen in figure 6.13, from a Haydn sonata; consider the A major seventh in m. 88. The seventh degree of the chord, G♯, cannot easily be treated as ornamental; thus the program chooses to analyze it as a C♯ minor chord, with the A's being ornamental. (The D major seventh in the following measure causes a similar problem.)

A few errors are due to the fact, discussed earlier, that the program is unable to handle certain kinds of ornamental dissonances, such as escape tones. The inability to handle pedal tones is also an occasional problem.

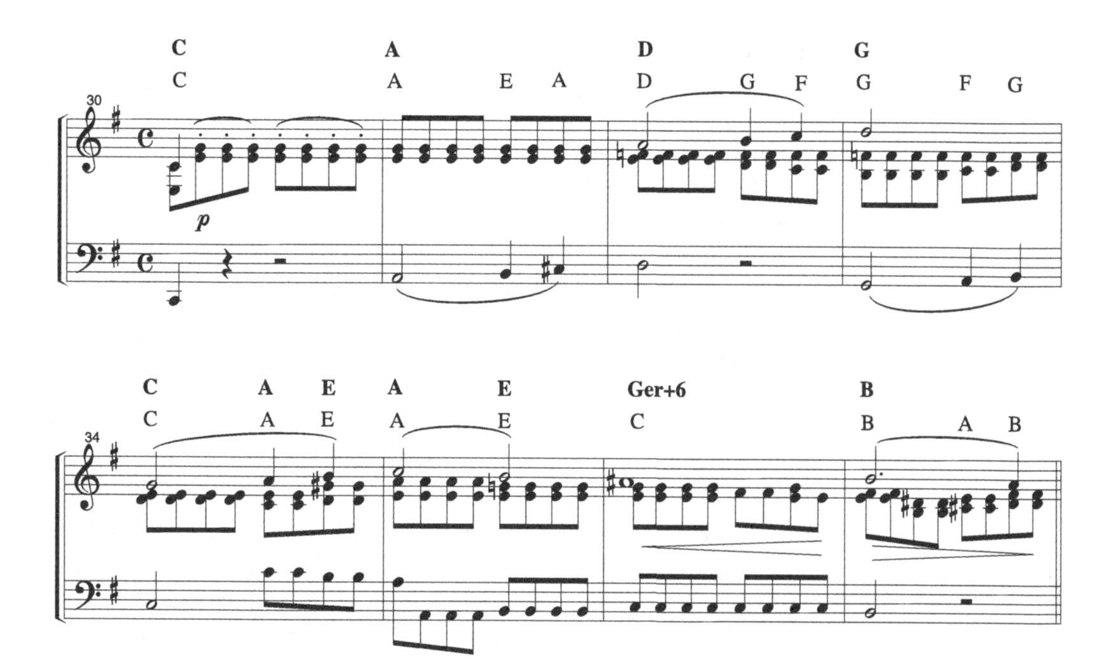

Figure 6.14
Haydn, String Quartet Op. 74 No. 3, II, mm. 30–7, showing Kostka's analysis (above) and the program's (below).

Finally, the program's harmonic rhythm is often somewhat too fast. Consider its analysis of figure 6.14, from a Haydn string quartet. The program's analysis is not ridiculous, but in several cases a note is analyzed as a chord-tone when it might better be treated as ornamental, as in Kostka's analysis. Slightly altering the program's parameters—in particular, increasing the weight of the strong beat penalty—might improve its performance somewhat, at least with respect to the Kostka-Payne corpus.[9]

6.7 Other Aspects of Harmonic Structure

The program described above is concerned solely with one aspect of harmonic information, namely root analysis. This is certainly an essential part of harmonic analysis, and we have seen that it is far from trivial. However, there is much else to know about tonal harmony of which the program is quite ignorant. In the first place, the program has no knowledge of the functions of harmonies relative to the current key; recovering this information would depend on determining the key, a problem we will address in the next chapter. Second, the program labels harmonies in terms of roots only, and omits other important information about chords such as their mode (major, minor or diminished), extension (triad or seventh) and inversion (which note of the chord is in the bass). We should note, however, that these labels depend mainly on which chord-tones of the root are present, and this is determined by the program in the course of applying the compatibility rule. For example, if a harmony contains $\hat{1}$, $\flat\hat{3}$, and $\hat{5}$, it is a minor triad; if it also contains $\flat\hat{7}$, it is a minor seventh; and so on. Thus while the program does not make information about mode, inversion and extension absolutely explicit, it does much of the work needed to find this information.[10]

Finally, the program has no knowledge of the conventional progressions of tonal music. A tonal piece does not simply string together harmonies at random; rather, much of tonal harmony can be explained in terms of a few fundamental patterns, in particular motion on the line (or, more conventionally, the circle) of fifths. Also of vital importance are cadences, the conventional harmonic gestures (in particular V-I) that indicate closure in common-practice music. I would argue that the recognition of such patterns should not be regarded as part of harmonic analysis itself. Rather, such higher-level "schemata" are presumably identified by some kind of pattern-recognition system which uses harmonic structure as input. (See chapter 12 for further discussion.) There is a real question, however, as to whether more specific knowledge of conventional progressions would be useful to the program in a top-down fashion in arriving at the correct root analysis. Perhaps there is a perceptual preference for certain patterns such as cadences, so that we assign a bonus to any analysis that features them. Such a preference might allow the program to correctly identify the ii°-V (F♯-B) half-cadence that is missed in mm. 11–12 of the Bach Gavotte, as well as the V-I cadence in mm. 7–8.[11] Attempting to incorporate such patterns into the program would be an interesting further step. The current model does suggest, however, that there are limits on how much such information is needed; one can go a long way toward correct harmonic analyses without knowledge of conventional progressions.

7
Key Structure

7.1 Key

In the previous two chapters, we considered the labeling of pitch-classes and the identification of chords. No one could deny that these are important kinds of musical information. In themselves, however, they are of limited use. Knowing that a pitch is a C, for example, tells us little. What is important is knowing that it is (for example) the fifth degree of the current scale. Similarly, the fact that something is an F major chord says little about its function or significance in the current context; however, if we know that it is a IV chord—that is, the chord based on the fourth degree of the current key—then we know much more. This brings us to the fundamental significance of key: key provides the framework whereby pitches and chords are understood.[1]

Key is important in other ways as well. The key structure of a piece—the sequence of keys through which it passes, and the relationships among them—can contribute greatly to its expressive effect. As Charles Rosen (1971, 26) has shown, a modulation can act as a kind of "large-scale dissonance," a conflict demanding a resolution; it is largely this that allows tonal music to convey a sense of long-range motion and drama.

In this chapter I propose a system for inferring the key of a piece and sections within a piece. The model I present builds on an earlier proposal, the key-profile algorithm of Krumhansl and Schmuckler (described most extensively in Krumhansl 1990). I will point to some important weaknesses of the Krumhansl-Schmuckler (hereafter K-S) algorithm, and propose modifications. With these modifications, I will suggest that the key-profile model can provide a highly effective approach to key-finding.

First, however, let us review some other important research relating to the perception of key.

7.2 Psychological and Computational Work on Key

The perception and mental representation of key has been the subject of considerable study in music cognition. Some of this work has explored the way the perception of other musical elements is affected by the key context. It has been found, for example, that the perceived stability or appropriateness of pitches and chords is greatly affected by context: pitches and chords within a contextually established key are judged to "fit" or "complete" the context better than others (Krumhansl 1990; Brown, Butler & Jones 1994). Other studies have shown that the perception of melody is affected by tonal factors as well. Melodic patterns that project a strong sense of key are more easily remembered than others (Cuddy, Cohen & Mewhort 1981); melodies are more easily recognized if they are presented in a suitable tonal context (Cuddy, Cohen & Miller 1979). These studies, along with others by Dowling (1978) and Deutsch (1980), suggest that the scales associated with keys play an important role in musical encoding, a subject we will discuss further in chapter 12.

Other studies have explored the perception of modulation: change from one key to another. Thompson and Cuddy (1992) found that both trained and untrained listeners were sensitive to changes in key, and that the perceived distances of modulations corresponded well with music-theoretical ideas about key distance. Another study by Cook (1987) explored listeners' ability to detect whether a piece began and ended in the same key. While listeners were indeed sensitive to this for short pieces, their sensitivity declined greatly for longer pieces. This is a notable finding, given the importance accorded to large-scale key relationships in music theory. We should note, however, that this finding relates only to listeners' perception of relationships between keys, not their judgments of the current key. Still another important aspect of key is the distinction between major and minor keys. A variety of studies have shown this to be an important factor in the emotional connotations of musical excerpts (Dowling & Harwood 1986, 207–10); indeed, even very young children appear to be sensitive to these associations. In a study by Kastner and Crowder (1990), subjects were played pairs of melodies that were identical except that one was major and the other minor, and were shown cartoon faces reflecting various emotions; for each melody, subjects were told to point to a face which went with the melody. Subjects as young as three years old showed a reliable ability to associate major melodies with happy faces and minor melodies with sad ones.

A number of computational models of "key-finding" have been proposed. Moreover—uniquely, among the kinds of musical structure considered in this book—some of these models have been systematically tested, allowing comparison with the current model. These tests will be discussed later.

Perhaps the earliest attempt at a computational key-finding system was the algorithm of Longuet-Higgins and Steedman (1971). Designed for monophonic pieces only, the algorithm processes a piece left to right; at each pitch that it encounters, it eliminates all keys whose scales do not contain that pitch. When it is left with only one key, this is the preferred key. If it makes a step which causes all keys to be eliminated, it undoes that step; it then looks at the first note of the piece, and if the first note is the tonic (or, failing that, the dominant) of one of the remaining eligible keys, it chooses that key. If it reaches the end of the melody with more than one eligible key remaining, it again performs the "first-note" test to choose the correct key. Aside from the fact that this algorithm is limited to monophonic music, there are some major limitations to it. For example, it cannot handle chromatic notes; any note outside a scale will cause the key of that scale to be eliminated from consideration. This surely does not occur perceptually; the chromatic notes in figure 7.1a (F♯ and D♯) do not prevent a strong implication of C major. Consider also figure 7.1b; while the correct key here is clearly C major, the algorithm would be undecided between C major and G major (since neither one is eliminated), and would be forced to rely on the "first-note" rule, which yields the incorrect result in this case. Despite these problems, the basic idea behind the Longuet-Higgins/Steedman algorithm is an interesting and valuable one; we will reconsider this model later in the chapter.

A similar algorithm was developed by Holtzmann (1977), building on the work of Longuet-Higgins and Steedman. Like the earlier model, Holtzmann's algorithm works only on melodies; however, rather than treating pitches one at a time and eliminating keys, it searches for certain features—the tonic triad, the tonic and dominant, or the tonic and mediant—at certain structural points, namely the first and last few pitches of the melody. The key whose features are most strongly present at these points is the preferred one. The model's reliance on the last few notes of the melody seems problematic. For one thing, it does not seem true to perception; with an extended melody, we are surely capable of

A

B

Figure 7.1
(A) Greek folk melody. (B) George M. Cohan, "You're a Grand Old Flag."

identifying the key well before the melody ends. One could, of course, run the model only on the first section of a melody; but in deciding where the first section ends, we are no doubt biased towards choosing a note that we have already identified as the tonic. (This problem arises in Holtzmann's own tests, as we will see later on.)

Both the Longuet-Higgins/Steedman and Holtzmann algorithms fail to address an important aspect of key structure, namely, modulation. In listening to a piece of music, we not only infer a key at the beginning of the piece, but can also detect changes of key within the piece. Vos and Van Geenen (1996) present a proposal for monophonic key-finding which attempts to handle modulation. In this model, a melody is processed from left to right; for each pitch, points are added to each key whose scale contains the pitch or whose I, IV or V7 chords contain it. (It is unclear why the algorithm considers chord membership. The same effect could be obtained by simply adjusting the weights added to pitches for different scale degrees; for example, the members of the I triad are always the $\hat{1}$, $\hat{3}$ and $\hat{5}$ scale degrees.) There is a "primacy" period of the first five notes where special conditions obtain. The first pitch gives an extra weight to the key of which it is the tonic. (There are other rather complex rules regarding the primacy period which I will not explain here.) The algorithm's key judgments at any moment are based on choosing the key with the highest score. However, only the last 40 notes of the melody are considered in each key judgment; in this way, the algorithm is capable of modulation.

The performance of these three monophonic key-finding systems —Longuet-Higgins and Steedman's, Holtzmann's, and Vos and Van Geenen's—will be discussed further below. Attempts have also been made to model key-finding in polyphonic music. Two important efforts here have been the harmonic analysis systems of Winograd (1968) and Maxwell (1992), discussed in the previous chapter. As Winograd and Maxwell conceive of it, harmonic analysis involves labeling chords with Roman numerals, indicating the function of each chord relative to the key; this involves both determining the root of the chord and the key. The root-finding aspect of these systems has already been discussed (see section 6.2); here we consider the key-finding aspect once root labels have been determined. Winograd's program starts at the ending of a piece, and works backwards, generating a series of key sections or "tonalities." The possible analyses for a piece are somewhat constrained by the rules of the "grammar" that Winograd is using: for each chord, a tonality must be chosen (if possible) such that the chord has some function in the tonality. Moreover, according to the grammar, a well-formed tonality must contain either a dominant → tonic progression or a dominant-preparation → dominant. Given these rules, the program first searches through the piece to find all possible tonalities in the piece; then only these tonalities need be considered in the labeling of each chord. The set of possible tonalities is then further reduced through the use of heuristics, which Winograd calls "semantic rules"; for example, tonalities are preferred which yield common progressions such as II-V-I. Maxwell's handling of key is similar to Winograd's; it is based on a search for potential cadences or "p-cadences." Some p-cadences are better than others (i.e. V7-I cadences are best; V6-I cadences are somewhat less good; and so on). Based on a point system, the p-cadences are scored, and the best ones are taken to be indicative of the key of the previous section.

Winograd and Maxwell each provide a few examples of their programs' output, which suggest that their key-finding systems are quite successful (see figures 6.2 and 6.3). Since their models are highly complex, it is difficult to simulate them or to predict their results in other cases. One oddity is that, in both Maxwell's and Winograd's systems, the key signature is included in the input to the program. (Maxwell's system actually uses this information directly in determining the main key of the piece; Winograd's system does not.) This suggests that Maxwell and Winograd were more interested in modeling the process of harmonic analysis as it is done explicitly with a score, rather than as it occurs in listening. The fact that Winograd's program begins at the end of the piece, rather than the beginning, is another indication of this. This attitude is also reflected in Maxwell's commentary; in discussing the problem of judging the main key of a piece, Maxwell states: "A human analyst glances at the key signature and the final cadence (if necessary) and makes a quick judgment. Our method will be the same" (p. 342).

Finally, mention should be made of two very different approaches to key-finding. One is the connectionist model of Bharucha (1987, 1991). As described in section 6.2, this model features three levels of nodes, representing pitches, chords, and keys. Pitch nodes activate the chord nodes of major and minor triads that contain them; chord nodes, in turn, activate nodes for the keys in which they have diatonic functions. As noted in our earlier discussion, Bharucha presents no examples of the model's output on actual pieces, nor can its behavior be easily predicted; thus it is difficult to evaluate it.

Bharucha's model and Winograd's and Maxwell's models, though very different in perspective, have an important feature in common: they all base their judgments of key primarily on harmonic information. As noted in section 6.2, the experiments of Butler (1989)—showing that the same pitches arranged in different ways can have different tonal implications—suggest that, indeed, harmony may play a role in key-finding. I will return to this issue in a later section.

The model of Leman (1995) stands apart from the others discussed here in that it analyzes sound input directly, making judgments of key without first extracting pitch information. (In this respect it is somewhat similar to Parncutt's system for analyzing harmony, discussed in section 6.2.) The model begins by performing subharmonic analysis of the sound input; this indicates how strongly the harmonics of each pitch are present in the sound. For any chord, a vector can be produced indicating how strongly it supports each pitch-class as a virtual root (here again, Parncutt's earlier model comes to mind). Similar vectors for keys are then generated, by combining the vectors for primary chords in each key: I, IV, and V. Keys whose vectors correlate well with that of a given chord or sonority in a piece are assumed to be good candidates for the key at that point. The integration of these judgments over time means that the model's key choice for a given segment may be influenced by both previous and subsequent segments. In tests on two pieces in which the model's judgments were compared to those of experts, the model scored in the neighborhood of 75–80%. Leman acknowledges some significant weaknesses with the model, such as the inadequate weight given to the leading-tone, and problems in distinguishing between parallel major and minor keys. In addition, an objection that was raised regarding Parncutt's model of harmony (see section 6.2) is relevant again here. Since sound input is used directly as input to the model, a given piece might be represented quite differently depending on the instrumentation and timbral variation used in performance; this raises the possibility that timbral factors might affect the model's judgment of key. Alternatively, if the model's judgments are *not* affected by the timbre of notes, and depend only on the fundamentals—i.e., the pitches themselves—one wonders why this information is not used instead. (As argued in chapter 1, there is considerable evidence that listeners do recover pitch information in listening.) Despite these problems, Leman's model appears to achieve quite a high level of performance.

7.3 The Krumhansl-Schmuckler Key-Finding Algorithm

We now turn to a discussion of the Krumhansl-Schmuckler algorithm. I will begin by presenting the model in its original form; I will then discuss some problems that arise, and propose solutions, leading to a modified version of the model.

The Krumhansl-Schmuckler key-finding algorithm is based on "key-profiles." A key-profile is a vector of twelve values, representing the stability of the twelve pitch-classes relative to a given key (Krumhansl 1990, 79–80). The key-profiles were based on data from experiments by Krumhansl and Kessler (1982) in which subjects were asked to rate how well each pitch-class "fit with" a prior context establishing a key, such as a cadence or scale. A high value in the key-profile means that the corresponding pitch-class was judged to fit well with that key. Each of the 24 major and minor keys has its own key-profile. The key-profiles for C major and C minor are shown in figure 7.2; other profiles are generated simply by shifting the values around by the appropriate number of steps. For example, whereas the C major vector has a value of 6.35 for C and a value of 2.23 for C♯, C♯ major would have a value of 6.35 for C♯ and a value of 2.23 for D.[2] As Krumhansl notes (1990, 29), the key-profiles reflect well-established musical principles. In both major and minor, the tonic position (C in the case of C major/minor) has the highest value, followed by the other two degrees of the tonic triad (G and E in C major, G and E♭ in C minor); the other four degrees of the diatonic scale are next (D, F, A and B in C major; D, F, A♭ and B♭ in C minor—assuming the natural minor scale), followed by the five chromatic scale steps.

The algorithm judges the key of a piece by correlating each key-profile with the "input vector" of the piece. The input vector is, again, a twelve-valued vector, with each value representing the total duration of a pitch-class in the piece. Consider figure 7.3, the first measure of "Yankee

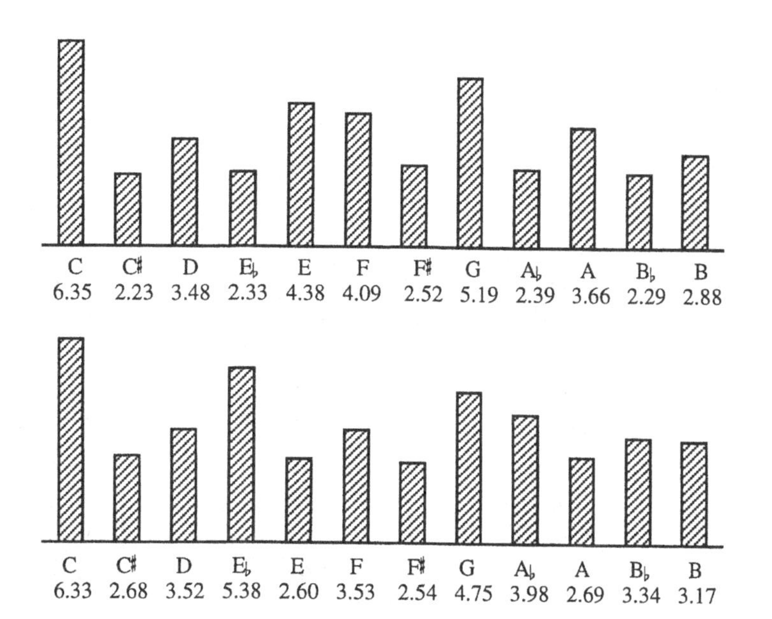

Figure 7.2
The Krumhansl-Kessler key-profiles (1982), for C major (above) and C minor (below).

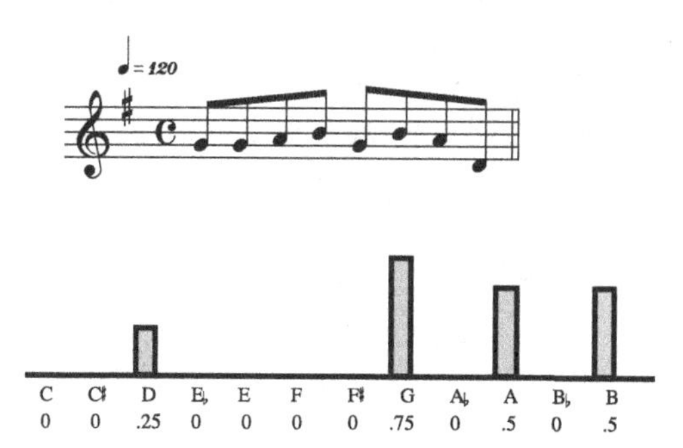

Figure 7.3
Measure 1 of "Yankee Doodle," with input vector showing total duration of each pitch class.

Table 7.1
Key-profile scores for the first measure of "Yankee Doodle" (figure 7.3)

Key	Score	Key	Score
C major	0.274	C minor	−0.013
C♯ major	−0.559	C♯ minor	−0.332
D major	0.543	D minor	0.149
E♭ major	−0.130	E♭ minor	−0.398
E major	−0.001	E minor	0.447
F major	0.003	F minor	−0.431
F♯ major	−0.381	F♯ minor	0.012
G major	**0.777**	G minor	0.443
A♭ major	−0.487	A♭ minor	−0.106
A major	0.177	A minor	0.251
B♭ major	−0.146	B♭ minor	−0.513
B major	−0.069	B minor	0.491

Doodle"; assume a tempo of quarter note = 120. Pitch-class G has a total duration of 0.75 (seconds); A has a duration of 0.5; B has a duration of 0.5; D has a duration of 0.25; the other eight pitch-classes have durations of 0, since they do not occur at all in the excerpt. The input vector for this excerpt is shown in figure 7.3. The correlation value, r, between the input vector and a given key-profile vector is then given by

$$r = \frac{\sum(x - \bar{x})(y - \bar{y})}{(\sum(x - \bar{x})^2 \sum(y - \bar{y})^2)^{1/2}}$$

where x = input vector values; $\bar{x}$ = the average of the input vector values; y = the key-profile values for a given key; and $\bar{y}$ = the average key-profile value for that key.

To find the key for a given piece, the correlations must be calculated between each key-profile and the input vector; the key-profile yielding the highest correlation gives the preferred key.

Table 7.1 shows the results of the algorithm for the first measure of "Yankee Doodle." G major is the preferred key, as it should be. It can be seen that all the pitches in the excerpt are in the G major scale (as well as several other scales); moreover, the first and third degree of the G major tonic triad are strongly represented, so it is not surprising that G major receives the highest score.

The first modification I wish to propose is a simplification in the way the key-profile scores are calculated. The formula used in the K-S algorithm is the standard one for finding the correlation between two vectors (Howell 1997). In essence, this formula takes the product of the corresponding values in the two vectors (the input vector and the key-profile vector, in this case) and sums these products. To use Krumhansl's metaphor, this amounts to a kind of "template-matching": if the peaks in the key-profile vector for a given key coincide with the peaks in the input vector, this number will be large. The correlation formula also normalizes both vectors for their mean and variance. However, if our only goal is to find the algorithm's preferred key for a given passage of music, these normalizations are not really necessary. We can obtain essentially the same result by defining the key-profile score simply as the sum of the products of the key-profile and input-vector values, or $\sum xy$—sometimes known as the "scalar product." The algorithm then becomes to calculate this score for all 24 keys, and choose the key with the highest score.[3]

7.4 Improving the Algorithm's Performance

Krumhansl (1990, 81–106) reports several tests that were done of the K-S algorithm. First, the algorithm was tested on the first four notes of each of the 48 preludes of Bach's *Well-Tempered Clavier*. (In cases where the fourth note was simultaneous with one or more other notes, all the notes of the chord were included.) The algorithm chose the correct key on 44 of the 48 preludes, a rate of 91.7%. Similar tests were done on the first four notes of Shostakovich's and Chopin's preludes, yielding somewhat lower correct rates: 70.8% and 45.8%, respectively. In another test, the algorithm was tested on the fugue subjects of the 48 fugues of the *Well-Tempered Clavier*, as well as on the subjects of Shostakovich's 24 fugues. For each fugue, the algorithm was given a series of note sequences starting from the beginning of the piece: first the first note, then the first two notes, then the first three notes, and so on. At the point where the algorithm first chose the correct key, the test for that piece was terminated. On 44 of the Bach fugue subjects and 22 of the Shostakovich fugue subjects, the algorithm eventually found the correct key. As Krumhansl acknowledges (p. 93), this test is somewhat problematic, since it is unclear how stable the algorithm's choice was; it might choose the correct key after four notes, but it might have shifted to a different key if it was given another note. Finally, the algorithm was tested on each measure of Bach's Prelude No. 2 from Book II of the *Well-Tempered Clavier*, and its judgments were compared to the judgments of two experts as to the key of each measure. In this case, however, a different version of the algorithm was applied, using Fourier analysis to map each input vector onto a four-dimensional space of keys. Moreover, the algorithm's judgment for each measure was based on a weighted sum of the pitch durations in the current measure and also previous and subsequent measures, in order to reflect the effect of context on key judgments.

While all of these tests are of interest, only the first group of tests provide clear data as to the algorithm's judgments of the preferred key for an isolated segment of music. The algorithm's performance here was mixed, performing much better on the Bach preludes than on the Shostakovich and Chopin preludes; however, one could argue that judging the key after four notes is unrealistic in the latter two cases, given their more complex tonal language. In any case, further tests seem warranted.

An easy and informal way of testing the algorithm is by giving it a piece, having it judge the key for many small segments of the piece in isolation—measures, say—and comparing the results to our own judgments. This was done using a computer implementation of the algorithm, exactly as it is specified in Krumhansl 1990. (For this test, then, the original formula was used, rather than the modified formula proposed above.) In deciding what we think is the correct key for each measure, it is important to stress that each measure is to be regarded *in isolation*, without considering its context, since this is what the algorithm is doing. This is not of course how we naturally listen to music, but considering the tonal implications of a small segment of music taken out of context is, I think, not difficult to do. (Note that the current test differs from Krumhansl's test of the Bach Prelude No. 2, where both the experts and the algorithm were taking the context of each measure into account in judging its key.) Figure 7.4 shows the first half of the Courante of Bach's Suite for Violoncello No. 3. The algorithm's preferred key is shown above each measure (the top row of symbols, labeled "K-S"). In a number of cases, the algorithm's choice is clearly correct: m. 1, for example. In some cases, the key is somewhat unclear, and the algorithm chooses one of several plausible choices (in measure 17 it chooses G major, although E minor would certainly be possible). In a number of cases, however, the algorithm's choice is clearly wrong; these cases are indicated with an exclamation mark. In m. 4, the algorithm chooses G major, although the measure contains an F—this pitch is not present in a G major scale and, as a $\flat\hat{7}$ scale degree, is indeed highly destabilizing to the key; however, all the notes of the measure are present in the C major scale. Similar errors occur in mm. 14 and 16. In a number of other cases (mm. 8, 22, 29, 30,

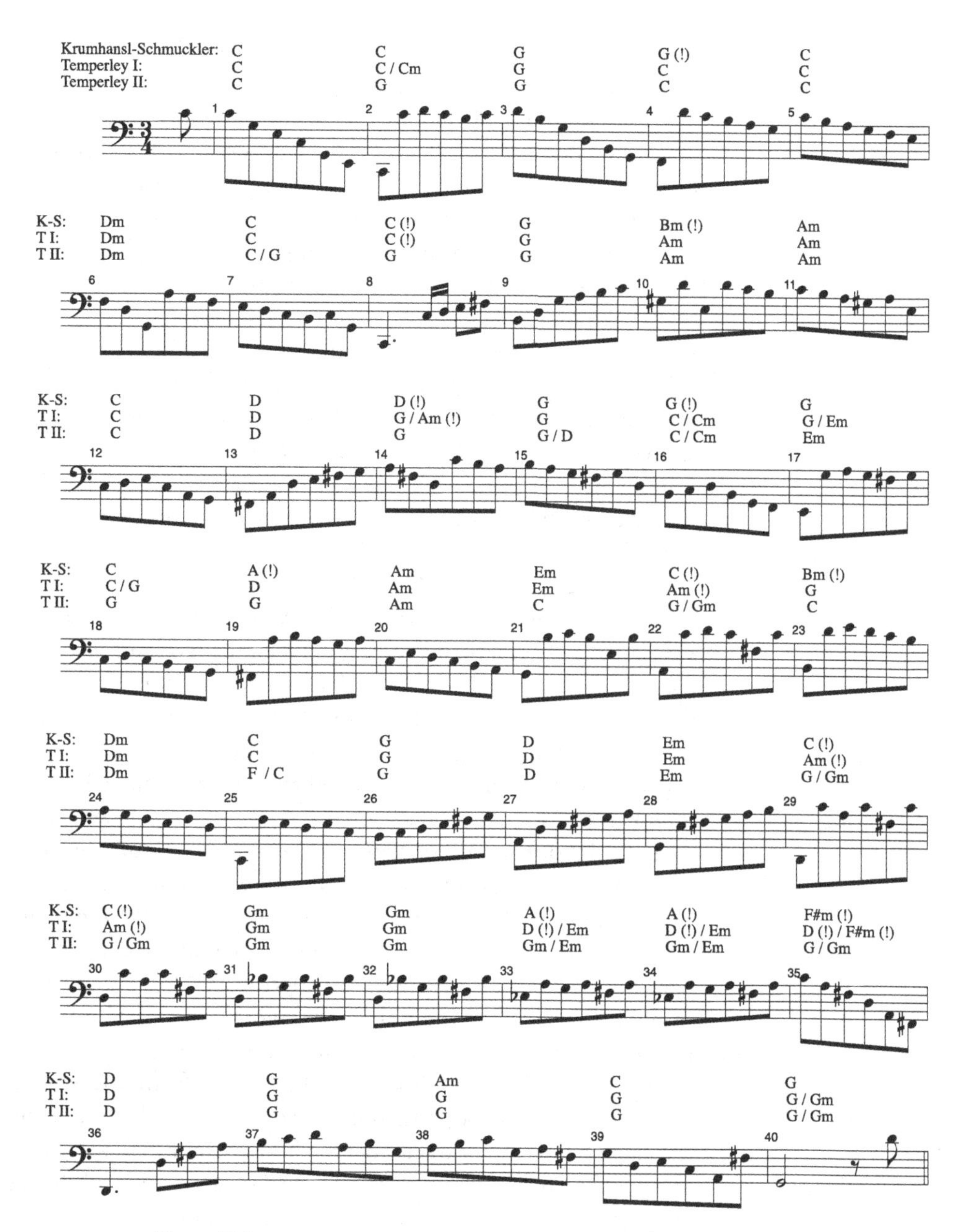

Figure 7.4

Bach, Suite for Violoncello No. 3, Courante, mm. 1–40, showing judgments for each measure from three different key-finding algorithms. Minor keys are marked with "m"; all other keys are major.

33, 34, and 35), the algorithm chooses a key despite the presence of a pitch outside the scale of that key, and despite the existence of another key that contains all the pitches of the segment. This suggests that the algorithm does not distinguish strongly enough between diatonic and chromatic scale degrees. The algorithm also sometimes chooses minor keys when the lowered seventh or raised sixth degree of the key is present (mm. 10 and 38, for example), though these are much less common than the raised seventh and lowered sixth degree, that is to say the "harmonic minor" scale. Altogether, the algorithm makes incorrect judgments on 13 of the 40 measures, a correct rate of 67.5%. The discrepancy between this result and the algorithm's performance on the opening four-note segments of the Bach preludes is worth noting. Inspection of the preludes shows that a great number of them begin by outlining or elaborating a tonic triad. The Courante would seem to provide a wider variety of melodic and harmonic situations, although of course it, too, is a highly limited sample.

Inspecting the key-profile values themselves (figure 7.2), it becomes clear why some of these errors occur. While diatonic degrees have higher values than chromatic degrees, the difference is slight; in C major, compare the values for B (2.88) and F♯ (2.52). In minor, we find that the flattened seventh degree (B♭ in the case of C minor) has a higher value than the leading-tone (B). This seems counterintuitive; as mentioned earlier, the flat seventh is quite destabilizing to the tonic, while the leading-tone is often a strong indicator of a new tonic (consider the way G♯ in measure 10 of the Courante points towards A minor). In major, too, it seems odd that the leading-tone has the lowest value of the seven diatonic degrees. A related problem should be mentioned: the dominant seventh, which is usually taken as strongly implicative of the corresponding tonic key, is not so judged by the K-S algorithm. Rather, the G dominant seventh (for example) most strongly favors G major, B minor, D minor, D major, G minor, and F major, in descending order of preference, with C major seventh on the list (C minor is even further down). Finally, it seems likely that some of the minor values are too high; in general there is a slight tendency for the model to choose minor keys more often than it should. (The total score for the minor triad, 16.46, is slightly higher than that for the major triad, 15.92, which seems odd.)

A revised version of the profiles is shown in figure 7.5, which attempts to solve these problems. These values were arrived at by a mixture of theoretical reasoning and trial and error, using a variety of different

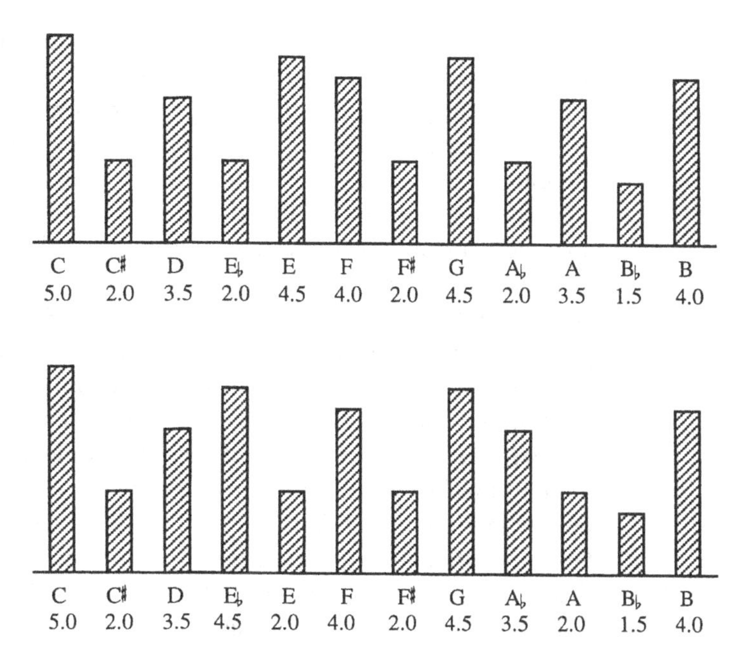

Figure 7.5
A revised version of the key-profiles, shown for C major (above) and C minor (below).

pieces for testing. (An attempt was made to keep all the values in the same range as those in the K-S algorithm, to permit easy comparison.) The basic primacy of the diatonic scale is still reflected; all diatonic steps have higher values than chromatic ones. In the case of minor, I assume the harmonic minor scale, so that $\flat\hat{6}$ and $\hat{7}$ are within the scale and $\hat{6}$ and $\flat\hat{7}$ are chromatic. All the chromatic degrees have a value of 2.0, with the exception of $\flat\hat{7}$, which has a value of 1.5. The unusually low value for $\flat\hat{7}$ proved necessary, in part, to achieve the right judgment for the dominant seventh, but it appears to lead to good results in general. All the diatonic degrees have a value of at least 3.5; $\hat{2}$ and $\hat{6}$ in major, and $\hat{2}$ and $\flat\hat{6}$ in minor, have exactly this value. $\hat{4}$ and $\hat{7}$ are given slightly higher values (4.0), reflecting their importance (more on this below). The triadic degrees—$\hat{1}$, $\hat{3}$, and $\hat{5}$ in major, $\hat{1}$, $\flat\hat{3}$, and $\hat{5}$ in minor—receive the highest values; the value for $\hat{1}$ is highest of all. The same values are used for both the major and minor tonic triads.

These revised key-profiles improve performance considerably on the Bach Courante. (I now use my modified version of the key-profile formula, rather than the original version.) The results of the algorithm are shown as "Temperley I" in figure 7.4; questionable choices are again marked with exclamation marks. In a few cases, two keys receive exactly equal scores; in such cases both keys are shown, for example "C / G." The algorithm now makes only 6.5 errors instead of 13. (If the algorithm chooses two keys and one of them seems incorrect, this is counted as half an error). There is another problem with the key-profile approach, however, which cannot be solved merely by tweaking the key-profile values. Consider m. 29; the pitches here, D-C-A-C-F♯-C, outline a D dominant seventh. Given a simple D dominant seventh, with four notes of equal length played once, my algorithm (unlike the K-S algorithm) chooses G major and G minor equally as the preferred key. In m. 29, however, my algorithm chooses A minor. The reason is clear: there are three C's and one A, all members of the A minor triad, giving a large score to this key which swamps the effects of the other two pitches. Yet perceptually, the repetitions of the C do not appear to strongly tilt the key implications of the measure towards A minor, or indeed to affect them very much at all.

This raises a fundamental question about the key-profile approach. Even if we accept the basic premise of "template-matching," there are several ways this could be done. A different approach to template-matching is found in Longuet-Higgins and Steedman's earlier key-finding algorithm (1971). As discussed above, this algorithm processes a (monophonic) piece from left to right; at each pitch that it encounters, it eliminates all keys whose scales do not contain that pitch. When it is left with only one key, this is the preferred key. (Here we will consider just this simplified version of the algorithm, ignoring the "first-note" rule discussed earlier.) We could think of the Longuet-Higgins/Steedman model as implying a very simple key-profile model. In this model, each key has a "flat" key-profile, where all the pitch-classes in the corresponding diatonic scale have a value of one, and all chromatic pitch-classes have a value of zero. The input vector is also flat: a pitch-class has a value of one if it is present anywhere in the passage, zero if it is not. Choosing the correct key is then a matter of correlating the input vector with the key-profile vectors—which in this case simply amounts to counting the number of pitch-classes scoring one in the input vector that also score one in each key-profile vector. We might call this a "flat-input/flat-key" profile model, as opposed to the "weighted-input/weighted-key" profile model proposed by Krumhansl and Schmuckler. Note that the Longuet-Higgins/Steedman model handles a case such as m. 29 of the Bach better than the K-S model. All four pitch-classes—D-F♯-A-C—

are present in the G major and G (harmonic) minor scales, and no others, thus these keys (and these keys alone) receive a winning score of 4. However, the Longuet-Higgins/Steedman approach also encounters problems. In particular, the algorithm has no way of judging passages in which all the pitches present are in more than one scale. Consider measure 1 of the Bach; this C major arpeggio clearly implies C major, yet all of these pitches are also present in G major and F major (and several minor scales as well), thus the Longuet-Higgins/Steedman algorithm (or rather the simplified version of it presented here) would have no basis for choosing between them. In this case, the K-S algorithm is clearly superior, since it makes important distinctions between diatonic scale degrees.

This suggests that the best approach to key-finding may be a combination of the Krumhansl-Schmuckler and Longuet-Higgins/Steedman approaches: a "flat-input/weighted-key" approach. That is, the input vector simply consists of "1" values for present pitch-classes and "0" values for absent ones; the key-profile values, on the other hand, are individually weighted, in the manner of Krumhansl's profiles (and my revised version). (Judging pitch-classes simply as "present" or "not present," without considering their frequency of occurrence at all, might not work so well for longer passages; but my model does not do this for longer passages, as we will see below.) The output of this version for the Bach Courante is shown in figure 7.4 as "Temperley II." The algorithm's choice is at least reasonable on all 40 measures. In its judgments for small pitch-sets, then, the current algorithm seems to represent an improvement over the original K-S model.

Some connections should be noted between the values I propose and other theoretical work. Lerdahl's theory of tonal pitch space (1988) is based on a "basic space" consisting of several levels, corresponding to the chromatic scale, diatonic scale, tonic triad, tonic and fifth, and tonic, as shown in figure 7.6. Lerdahl notes the similarity between his space and

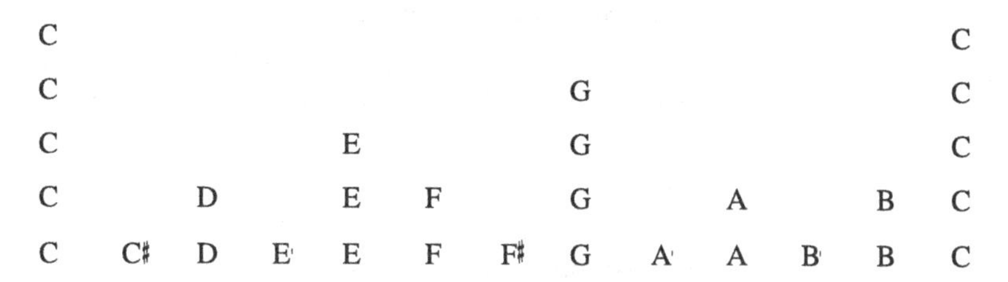

Figure 7.6
Lerdahl's "basic space" (1988), configured for C major.

the Krumhansl key-profiles (p. 338); both reflect peaks for diatonic pitch-classes and higher peaks for triadic ones. My own key-profile values correspond more closely to Lerdahl's space than Krumhansl's do, in that pitch-classes at each level generally have the same value—the exceptions being the higher values for $\hat{4}$ and $\hat{7}$ and the lower value for $\flat\hat{7}$. (My profiles also assign equal values to the third and fifth diatonic degrees, thus omitting the "fifths" level.)

The higher values for $\hat{4}$ and $\hat{7}$ bring to mind another theoretical proposal, the "rare-interval" approach to key-finding. Browne (1981) and Butler (1989) have noted that certain small pitch-class sets may have particular relevance for key-finding, due to their "rarity": the fact that they are only present in a small number of scales. A major second such as C-D is present in five different major scales; a tritone such as F-B is present in only two. Similarly, F-G-C is present in five major and two harmonic minor scales; F-G-B is only present only in C major and minor. While this is clearly true, it is not obvious that such considerations should be explicitly reflected in key-profiles. This point requires some discussion. Let us assume a "flat-key/flat-input" profile model, in which the key of a passage is simply given by the scale which contains the largest number of the pitch-classes present. In such a case, both F-G-B and F-G-C receive a score of 3 for C major (let us consider only major keys for the moment). In the case of F-G-B, however, C major is the only key receiving this score, and will thus be the clear favorite, whereas F-G-C will also receive scores of 3 from four other keys, and therefore should be ambiguous. In this sense the importance of certain "rare" pitch-sets would be an emergent feature of the system, even though $\hat{4}$ and $\hat{7}$ are treated no differently from other scale degrees in the key-profile itself. Nonetheless, my trial-and-error tests suggest that it *is* necessary to give special weight to $\hat{4}$ and $\hat{7}$ in the key-profile (relative to the other scale degrees outside the tonic triad, $\hat{2}$ and $\hat{6}$). In particular, it is very difficult to achieve correct results on the dominant seventh (and related pitch-sets) unless this is done. Perhaps the extra weight required by $\hat{4}$ and $\hat{7}$ in the key-profile somehow relates to their "rarity"—their special capacity for distinguishing each scale from the others. This is a question deserving further study.

One final modification is needed to the key-profiles themselves. Both the original K-S algorithm and the modifications I have proposed up to now assume a "neutral" model of pitch-class, in which pitch-events are simply sorted into twelve categories. In music notation and tonal theory,

Figure 7.7
Bach, Suite for Violoncello No. 3, Courante, mm. 65–6.

however, further distinctions are normally made between different spellings of the same pitch, for example A♭ and G♯—what I call "tonal pitch-classes" (see section 5.2). These distinctions are often applied to chords and keys as well. I argued in chapter 5 that the spelling labels of pitch-events are an important part of tonal perception, and can be inferred from context without relying on top-down key information, using preference rules.

If it is possible to infer spelling labels without using key information, this raises the possibility that spelling might be used as input to key determination. Tests of the original K-S algorithm showed a number of cases where this might be useful. Consider figure 7.7—m. 65 from the Bach Courante discussed earlier, containing the pitches G♯-F-E-D-C-B. The K-S algorithm chooses F minor for this measure. If the first pitch event were spelled as A♭, this would at least be a possibility. If the first pitch is G♯, however, F minor is quite out of the question; A minor is much more likely. It seems plausible that the spelling of the first event as G♯ could be determined from context—for example, by its voice-leading connection to the A in the next measure; the key-profile model could then distinguish between the different tonal implications of G♯ and A♭. A further example of the role of TPC distinctions in key-finding is shown in figure 7.8. Excerpts A and B show two passages discussed in chapter 5; it was argued there that, though the NPC content of the two passages is the same, voice-leading makes us hear E♭ in the first case, D♯ in the second. Now consider following each of these excerpts with the two possible continuations in excerpts C and D. Excerpt C follows excerpt A relatively smoothly, but excerpt D following excerpt A produces a real jolt. Similarly, excerpt D follows excerpt B much more naturally than excerpt C does. The reason, I submit, is that the differing TPC content of excerpts A and B affects their tonal implications. The fact that excerpt A contains an E♭ makes it somewhat compatible with a move to E♭ major (though we would not say that excerpt A implies E♭ major as the preferred key), but much less compatible with a move to E major. In excerpt B, conversely, the presence of D♯ supports a move to E major much more than

Figure 7.8
This example demonstrates the influence of TPC labels on tonal implication. Excerpts A and B are identical in terms of NPC content, and are very similar harmonically (except that the G and C harmonies are reversed in the two cases). Because of voice-leading, we tend to hear the E♭/D♯ as E♭ in excerpt A, D♯ in excerpt B; this in turn affects the tonal implications of the passages. Excerpt A is much more compatible with a continuation in E♭ major (excerpt C) than in E major (excerpt D); with excerpt B, the reverse is true.

E♭ major. It can be seen how a key-profile model which distinguished between E♭ and D♯ could capture these distinctions.

What would such a TPC key-profile look like? A straightforward proposal is shown in figure 7.9. The "line of fifths" is represented on the horizontal axis. TPCs that are diatonic relative to the given key have the same values as in the NPC profile proposed earlier (figure 7.5). All other TPCs close to the tonic (within five steps to the left or six steps to the right) are given a value of 2.0, except for $\flat\hat{7}$ (this is the same as for chromatic NPC's in my original profile); all other TPCs are given a value of 1.5. For example, E is $\hat{3}$ of C major, and thus is given a value of 4.5 (as in my NPC profile); F♭ is chromatic relative to C major, thus its value is 1.5. Profiles were constructed for minor keys on the same principles: chromatic pitches within the range of $\flat\hat{2}$ to $\sharp\hat{4}$ were given a value of 2.0, while all others were given a value of 1.5.

As well as providing a possible improvement in performance in cases like figure 7.7, taking spelling information as input has another benefit: it

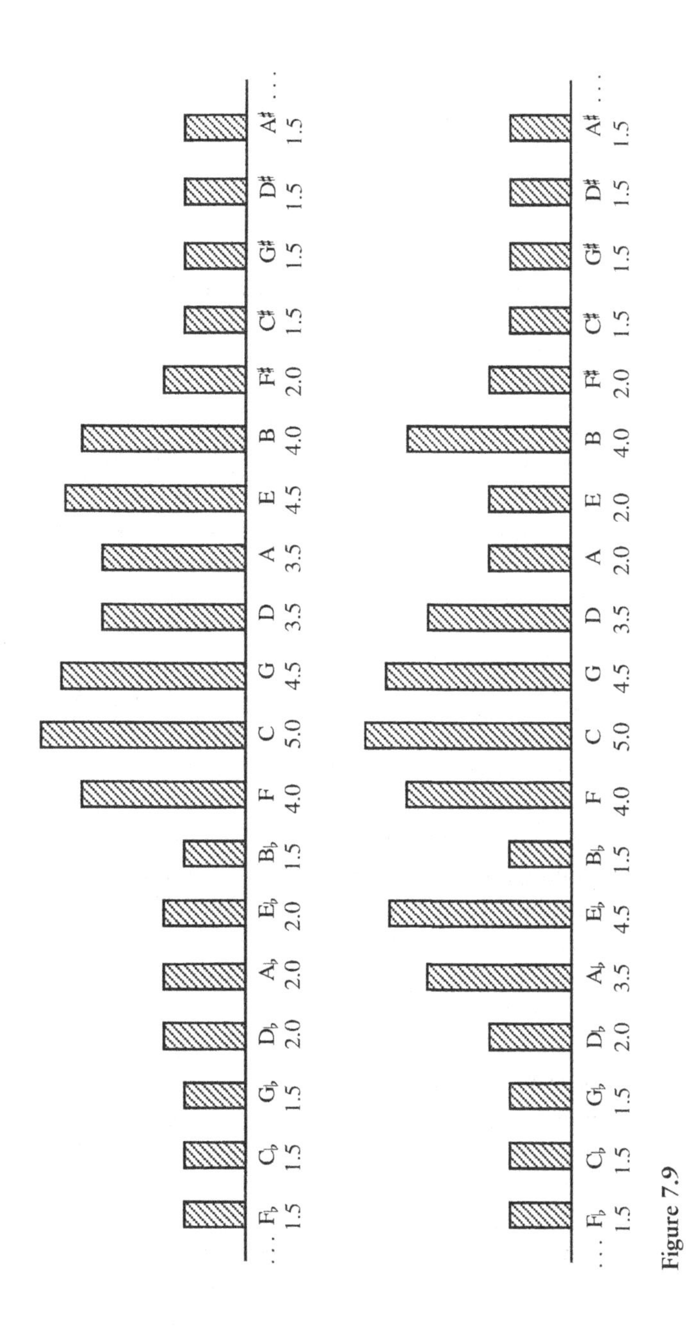

Figure 7.9
Key-profiles recognizing TPC distinctions, for C major (above) and C minor (below).

allows the model to distinguish between keys which have the same NPC label but differ in TPC label, so-called "enharmonically related" keys. The original Krumhansl-Schmuckler model would obviously have no way of choosing between, for example, A♭ major and G♯ major as the key for a passage. With the TPC model proposed above, however, different enharmonically related keys have different key-profiles; thus the model should be able to choose intelligently between G♯ and A♭ major, depending on the TPC content of the input.

7.5 Modulation

In one important sense, the K-S algorithm is not so much flawed as incomplete. The algorithm produces a single key judgment for a passage of music it is given. However, a vital part of tonal music is the shifts in key from one section to another. This is a complex matter, because there are different levels of key. Each piece generally has one main key, which begins and ends the piece, and in relation to which (in music theory anyway) intermediate keys are understood. An extended piece will generally contain modulations to several secondary keys—for example, the Bach Courante moves to G major around m. 9, then (in the second half, not shown in figure 7.4) to A minor, and then back to C major; and there may be even briefer tonal motions as well, so-called tonicizations (for example, the momentary move to A minor in mm. 10–11 of the Courante). One might propose the key-profile system as a way of determining the global level of key. I believe, however, that this is not the most sensible use of the key-profile model. What the key-profile system does well is determine the keys of sections of pieces. It is probably true that most pieces spend more time in their main tonic keys than in other keys, in which case a key-profile model might often work. It seems to me, however, that the global key of a piece really depends on other factors: in particular, the key of the beginning and ending sections. (One global key algorithm which would succeed in the vast majority of cases would simply be to choose the key of the first—or last—section of the piece.)[4]

The key-profile algorithm could easily be run on individual sections of a piece, once these sections were determined. Ideally, however, the division of the piece into key sections would be determined by the algorithm also; presumably, the same information that allows us to infer the correct key also allows us to infer when the key is changing. One possibility is that the algorithm could simply examine many small segments of a piece in isolation; changes of key would then emerge at places where one segment's key differed from that of the previous one. However, this is not very satisfactory. Consider the Bach Courante. The preferred key of m. 3, considered in isolation, is probably G major; heard in context, however, it is clearly outlining a V chord, part of a C major section. Once we begin to get a series of segments that clearly imply G major, though, we sense a definite shift in key. Intuitively, key has inertia: we prefer to remain in the key we are in, unless there is strong and persistent evidence to the contrary. A simple way of modeling this suggests itself: we apply the key-profile algorithm to each segment in isolation, but also impose a penalty for choosing a key for one segment which differs from that of the previous one. Generating a key analysis for a piece thus involves optimizing the key-profile match for each segment while minimizing the number of key changes. In some cases, this might lead the model to choose a key for a segment that is not the best choice for that segment in isolation (in m. 3 of the Bach, for example). However, if the scores for a segment or a series of segments favor another key strongly enough, then it will be worth switching.

The reader may have noted that this now begins to look a lot like a preference rule system—a very simple system, involving just two "KPRs" (key preference rules):

KPR 1 (Key-Profile Rule). For each segment, prefer a key which is compatible with the pitches in the segment, according to the (modified) key-profile formula.

KPR 2 (Modulation Rule). Prefer to minimize the number of key changes from one segment to the next.

7.6 Implementation

The algorithm described above was computationally implemented. As usual, the input required for the program is a "note list," with a pitch, on-time, and off-time for each note. The program also requires information about the spelling of each note: thus each note statement is supplemented with another number giving the TPC, or line-of-fifths position, of the note. (We adopt the convention that C = 2 on the line of fifths, as discussed in chapter 5.) In addition, the input to the program must contain a segmentation of the piece into low-level segments. The program searches for the highest-scoring key analysis of the piece it is given. A key analysis is simply a labeling of each segment of the piece with a key. A key is a TPC—that is, a point on the line of fifths—along with a specification of mode, major or minor. Each segment in an analysis yields a numerical score based on (1) how well its pitches fit the key chosen for that segment (according to the modified key-profile formula), and (2) whether the current segment has the same key as the previous one (if not, a "change penalty" is applied). The score for an analysis is simply the sum of the key-profile scores and change penalties for each segment.

The program's search for the highest-scoring analysis involves the usual dynamic programming approach, which in this case is applied in a very simple way. At each segment, we store the best analysis of the piece so far ending in that key.[5] Moving on to the next segment, we then consider each possible key added on to each "best-so-far" analysis at the previous segment (factoring in the change penalty, if necessary); this allows us to generate a new set of "best-so-far" analyses. At the end of the piece, the highest-scoring analysis at the final segment can be traced back to yield the best analysis for the entire piece.

Aside from the key-profiles themselves, there are two main numerical parameters in the program. One is the change penalty, the penalty for choosing a key for one segment different from that of the previous segment. By choosing a higher value, one can push the program towards less frequent changes and longer key sections; choosing a lower value has the opposite effect. Another, related, parameter is the length of segments. The length of segments matters for two reasons. First, with shorter segments, the total key-profile scores will be higher, and thus will carry more weight relative to the change penalty. Since it seemed desirable for the relative weight of these scores to be unaffected by the length of segments, the program cancels this effect by multiplying the key-profile scores by the length of segments. The length of segments also matters in a more subtle way. Recall that the input vector within each segment is always flat; this means that if a pitch-class occurs and is then repeated within the same segment, the repetitions will have no effect on the key analysis, but if it repeats in a different segment, this will have an effect. With longer segments, then, the algorithm is less affected by the frequency of occurrence of pitch-classes, and more by which pitch-classes are present or absent. Segments of about one measure—typically between one and two seconds—proved to yield the best results in this regard.

We should note, in passing, that the key program described here in combination with the harmony program described in the previous chapter allows for the possibility of performing "harmonic analysis" in the conventional sense (or "Roman numeral analysis"), showing the harmonic function of each chord relative to the key. As noted in chapter 6, once the root of a chord and the current key are known, its Roman numeral label follows automatically: a C major chord in a G major section must be IV. Other information, such as inversion and mode, can be determined in a simple "lookup" fashion by examining which pitches are in the chord. The key program just described was modified to take as input the output of the harmonic program, giving pitch and meter information (recall that metrical information is needed as input to the harmonic program) as well as a list of chord-spans (with their time-points and roots) and the chosen spelling of each note (recall that this is needed by the key program in any case). The key program generates a key analysis in the way described above; it then uses this analysis to assign each chord a Roman numeral symbol. Figure 7.10 shows the output of the program for an entire short piece—the Gavotte from Bach's French Suite No. 5 in G Major. (The score for this piece can be seen in figure 6.7.) The key

```
            1            2               3               4
G: I  .  |V6 .  vi .  |iii6 .  IV ii6 |V  V43 .  .  |I  .  .  vi |

   5               6               7              8
D: V42 .  ii42 V64 |V42 .  I  V65 |I  .  .  .  |.  .  .  .  |

   9               10             11                   12
G: V6 .  .  V42 |I6 .  .  .  |IV iii7 ii V43 |Em: V  .  .  .  |

13           14          15             16
.  .  .  .  |V7 .  .  .  |iv7 i6 .  V7 |i  .  .  .  |

   17           18               19              20
G: I6 .  ii7 .  |v42 ii ii42 .  |V42 I  .  .  |V64 .  V7 .  |

             21           22              23
.  .  .  .  |.  .  .  .  |vi ii6 V  .  |I  .  ]|
```

Figure 7.10
The output of a Roman-numeral-analysis program, incorporating the harmonic and key programs described in chapters 6 and 7. The analysis shown is of the Gavotte from Bach's French Suite No. 5 (the score is shown in figure 6.7). Roman numeral symbols indicate harmonies; letters indicate keys (keys are major unless indicated with an "m" for minor); vertical lines indicate barlines. Measure numbers have been added above for convenience.

analysis seems to be exactly correct; there are a few errors in the chord labels, due to errors in the root analysis (see section 6.3 for discussion).

7.7 Tests

The model described above was subjected to two formal tests. First, it was tested on the 48 fugue subjects from Bach's *Well-Tempered Clavier*; then, on a series of excerpts from the Kostka-Payne theory textbook. Some general comments are needed on how the tests were done.[6]

The program requires a list of segments, with each segment having a start-time and end-time. (The piece must be exhaustively partitioned into non-overlapping segments.) It seemed logical to have segments correspond to measures, or some other level of metrical unit (such as half-measure or two-measure units). It also seemed important to have strict criteria for the length of segments, since this may influence the analysis. The following rule was used: given the tempo chosen, the segments used corresponded to the fastest level of metrical unit above one second. (How the tempi were chosen will be explained below.) In nearly all cases, this resulted in a segment length of between 1.0 and 2.0 seconds.[7]

As required by the program, the input files also contained information about the spelling of each note. The spellings used were exactly those given by the composers' scores, with one qualification. Occasionally, the spelling of notes in a passage seems determined more by notational convenience than by musical logic: for example, in one excerpt from the Kostka-Payne corpus, Schubert modulates from D♭ major to A major, instead of the more logical B♭♭ major (see figure 5.16). There were three such cases in the Kostka-Payne corpus. In such cases, the musically logical spelling was used rather than the notated one, and the implied modulation resulting from this musically logical spelling (e.g. a modulation to B♭♭ major rather than A major) was taken to be the correct one.[8] In both tests, all extrametrical notes—trills, grace notes, and the like—were excluded, due to the difficulty of deciding objectively on the realization of these.

The key-profiles themselves were not in any way modified to improve the program's score on these tests. (As noted earlier, the key-profile values were set based on theoretical considerations and tests on other pieces.) However, the change penalty value *was* modified; different values were tried, and the value was used on each test which seemed to yield the best performance. (Both the Bach and Kostka-Payne corpora involve modulations, as I will explain.) For the Bach fugues, a penalty of 6.0 was used; for the Kostka-Payne corpus, the penalty was 12.0. As discussed earlier, key structure is generally thought to be hierarchical; a piece may have one level of large-scale key changes and another level of tonicizations. It seemed fair to adjust the program's change penalty to allow it to maximally match the level of key change in each test corpus.

The set of fugue subjects from the *Well-Tempered Clavier* was an obvious choice as a test corpus, since it has been used in several other key-finding studies, and therefore provided a basis for comparison. Longuet-Higgins and Steedman, Holtzmann, and Vos and Van Geenen all tested their systems on the corpus (Holtzmann uses only Book I, the first 24 fugues). Krumhansl also tested the K-S algorithm on the fugue subjects; however, as noted earlier, her test was problematic, in that she stopped the algorithm when it had reached the correct key, without giving any measure of how stable that decision was. The other algorithms either self-terminated at some point, or ran to the end of the subject and then made a decision.

Since it is not always clear where the fugue subjects end, decisions have to be made about this. Vos and Van Geenen rely on the analyses of Keller (1976) to determine where the subjects end; I used this source as well. (It is not clear how Longuet-Higgins and Steedman and Holtzmann made these decisions.) Another problem concerns modulation; many of the fugue subjects do not modulate, but several clearly do. Here again, Vos and Van Geenen rely on Keller, who identifies modulations in six of the fugue subjects.[9] Longuet-Higgins and Steedman's and Holtzmann's systems are not capable of modulation; they simply sought to identify the correct main key.

Longuet-Higgins and Steedman's algorithm found the correct main key in all 48 cases. Holtzmann's found the correct key in 23 out of the 24 cases in Book I. Vos and Van Geenen's program detected the correct key as one of its chosen keys on 47 out of 48 cases; it also detected modulations on two of the six cases noted by Keller. However, it also found modulations (and hence multiple keys) in 10 other cases in which there was no modulation.

The current program was run on all 48 fugue subjects. For the tempi, I used those suggested in Keller (1976). Segments were determined in the manner described earlier; the program chose one key for each segment. The current system, of course, has the option of modulating, and was allowed to modulate wherever it chose. Its performance on the corpus is shown in table 7.2. On 42 out of the 48 fugues, the system chose a single opening key (without a tie) that was the correct one. In two cases there were ties; in both cases, the correct key was among the two chosen. If we award the program half a point for the two ties, this yields a score of 43 out of 48. Out of the six modulating fugues, it modulated on two of them (moving to the dominant in both cases, as is correct). In one modulating theme (Book I No. 10), the program chose the second key as the main key. There were thus three nonmodulating themes where the program chose a single, incorrect, key.

It is rather difficult to compare the performance of the various programs on this test. Longuet-Higgins and Steedman's system, and then Holtzmann's, perform best in terms of finding the main key of subjects (although their inability to handle modulation is of course a limitation). (Compared to Vos and Van Geenen's system, the current program seems slightly better, producing a perfectly correct analysis in 39 out of 48 cases, versus 34 out of 48 cases for the Vos and Van Geenen program.) While the success of the Longuet-Higgins/Steedman and Holtzmann programs is impressive, we should note that they are rather limited, in that they can only handle monophonic passages at the beginning of pieces. When the Longuet-Higgins/Steedman algorithm is unable to choose a key by eliminating those whose scales do not contain all pitches present, it chooses the key whose tonic or dominant pitch is the first note of the theme (this rule is needed on 22 out of the 48 Bach fugue subjects). Holtzmann's approach also relies heavily on the first and last notes of the theme. Clearly, this approach is only useful at the beginnings of pieces; it is of no help in determining the keys of internal sections, since there is no obvious "first note." (Relying on the last notes of the theme—as Holtzmann does—is even more problematic, as it requires knowing where the theme ends.) Still, it is of course possible that special factors operate in key-finding at the beginning of pieces; and it does seem plausible, in some of these cases, that some kind of "primacy factor" is involved. (A "first-note" rule of this kind could possibly be incorporated into the current algorithm as a preference rule, but I will not address this here.)

Next, an attempt was made to give the program a more general test. Since part of the purpose was to test the algorithm's success in judging modulations, it was necessary to have pieces where such key changes were explicitly marked. (Musical scores usually indicate the main key of the piece in the key signature, or in other ways, but they do not generally indicate changes of key.) A suitable corpus of data was found in the workbook and instructor's manual accompanying Stefan Kostka and

Table 7.2
The key-finding program's performance on the 48 fugue subjects of Bach's *Well-Tempered Clavier*

Fugue number	Opening key	Modulating? (No unless marked yes)	Program's opening key (if incorrect or tie)	Modulation found? (No unless marked yes)
Book I				
1	C major			
2	C minor			
3	C♯ major			
4	C♯ minor			
5	D major		G major	
6	D minor			
7	E♭ major	yes		
8	D♯ minor			
9	E major			
10	E minor	yes	B minor	
11	F major			
12	F minor			
13	F♯ major			
14	F♯ minor			
15	G major			
16	G minor			
17	A♭ major			
18	G♯ minor	yes		yes (correct)
19	A major	yes		
20	A minor			
21	B♭ major			
22	B♭ minor			
23	B major			
24	B minor	yes		yes (correct)
Book II				
1	C major			
2	C minor			
3	C♯ major			
4	C♯ minor			
5	D major		G major	
6	D minor			
7	E♭ major		E♭ major/ A♭ major	
8	D♯ minor			
9	E major			

Table 7.2 (continued)

Fugue number	Opening key	Modulating? (No unless marked yes)	Program's opening key (if incorrect or tie)	Modulation found? (No unless marked yes)
10	E minor			
11	F major			
12	F minor			
13	F♯ major		B major	
14	F♯ minor			
15	G major			
16	G minor		G minor/ B♭ major	
17	A♭ major			
18	G♯ minor			
19	A major			
20	A minor	yes		
21	B♭ major			
22	B♭ minor			
23	B major			
24	B minor			

Dorothy Payne's textbook *Tonal Harmony* (Kostka & Payne 1995a, 1995b; Kostka 1995); this is the same corpus that was used in testing of the meter, pitch-spelling, and harmony programs (see section 2.5).

The program's analyses of the 46 excerpts from the corpus were compared to the analyses of the excerpts in Kostka's instructor's manual. For each segment in which the program's key choice was the same as Kostka's, one point was given. One problem was what to do if a segment was notated (by Kostka) as being partly in one key and partly in another. A related problem concerned "pivot chords." The Kostka analyses contained many such chords, which were notated as being in two keys simultaneously. (As noted earlier, the program itself does not allow changes of key within a segment; nor does it allow multiple keys for a segment, except in rare cases of exact ties.) The solution adopted was this. When a segment contained either a pivot chord or a change of key—we could call such segments "bi-tonal"—then 1/2 point was given if the key chosen by the program was one of the keys given by Kostka; otherwise, zero points were given. In cases where the program produced an exact tie, in which one of the keys chosen was the correct one, half a point was given for the segment.

Out of 896 segments, the program attained a score of 783.5 correct: a rate of 87.4%. The program found 40 modulations, exactly the same number as occurred in Kostka's analyses. It is useful to divide the sample according to where the excerpts occur in the workbook. Like most theory texts, Kostka and Payne's begins with basic chords such as major and minor triads and dominant sevenths, and then moves on to chromatic chords—augmented sixths, Neapolitans, and the like. Thus we can divide the examples into two groups: those in the chapters relating to diatonic chords (chapters 1–20), and those in the later chromatic chapters (chapters 21–26). Viewed in this way, the program scored a rate of 94.7% on the earlier chapters, 79.8% on the later ones, demonstrating a better ability for more diatonic passages. (A possible reason for this will be discussed below.)

To my knowledge, no earlier key-finding system has been subjected to a general test of this kind, so it is difficult to draw comparisons.[10] While the program's performance on the *Well-Tempered Clavier* and Kostka-Payne tests certainly seems promising, it is not perfect, and it is instructive to consider the errors it made. In a number of cases in the Kostka-Payne corpus, the program's rate of modulation was wrong: it either modulated too rarely, missing a move to a secondary key, or it modulated too often, changing key where Kostka does not. An obvious *ad hoc* solution is to modify the change penalty. In almost all such cases, a virtually perfect analysis was obtained simply by making the change penalty higher or lower. Still, it is clearly a flaw in the program that this parameter has to be adjusted for different pieces. Possibly the program could be made to adjust the change penalty on its own, but it is not clear how this might be done.

In some cases, harmony proved to be a factor. Consider figure 7.11, part of an excerpt from a Schumann song. Kostka analyzes this passage as being in B♭ major, but the program finds it to be in F major. The problem is the French sixth chords, G♭-B♭-C-E. (A French sixth, or "Fr6," can be thought of as an inverted dominant seventh—C7, in this case—with a flattened fifth.) The French sixths are followed by F major chords, as is customary; by convention, this progression would normally be interpreted as Fr6-V in B♭ major (or minor). However, the progression involves two pitches outside of B♭ major (G♭ and E), so it is not surprising that B♭ major is not chosen by the current model. The Fr6-V progression is then repeated in other keys, causing further problems for

Figure 7.11
Schumann, "Die beide Grenadiere," mm. 25–8.

the model. In order to get such a passage right, the model would presumably have to know something about harmony: specifically, the conventional tonal implications of Fr6-V progressions. Similar problems arise in excerpts involving other chromatic chords, such as Neapolitan chords and other augmented sixth chords.

The current tests suggest, then, that harmonic information is a factor in key-finding. This is not surprising, in light of the findings of Butler (1989), discussed earlier. In practice, however, it does not appear that such situations arise with great frequency; it is relatively rare that harmonic information is necessary. Most of the cases where it is needed involve chromatic chords whose tonal implications contradict the normal tonal implications of their pitch-classes.

Another area where the program might be improved is the key-profile values themselves. These could undoubtedly be refined, although it is not clear how much gain in performance would result. This could be done computationally, using a "hill-climbing" technique to arrive at optimal values. It could also be done by taking actual tallies of pitch-classes in pieces (relative to the key), and using these as the basis for the key-profile values.[11]

Finally, we must return once more to the issue of spelling distinctions. As noted earlier, the correct spelling of each note was given to the program as input. I argued in chapter 5 that the spellings of notes could be determined in a way that did not depend on key information. As reported in chapter 5, a program for choosing TPC labels had a high degree of success on the Kostka-Payne corpus (correctly identifying the spelling of 98.8% of the notes). Nevertheless, one might question the decision to provide the correct spelling of notes as input in the current tests. Perhaps, one might argue, we really use key information in some cases to determine the spellings of notes; by using the correct spelling as input, then, we may be partly begging the question of how key-finding is accomplished. To address this concern, the same tests done above were also performed with an "NPC" version of the model. This model uses only NPC information as input; it applies an NPC version of the key-profiles (exactly as shown in figure 7.5), and outputs NPC judgments of key. (Thus this model is not expected to make TPC distinctions in its key judgments; as long as the NPC is correct, the answer is judged to be correct.) On the *Well-Tempered-Clavier* fugue subjects, this model's performance was exactly the same as the TPC version, with one difference: on Book I, Fugue 14, the NPC program found a tie between two keys (one of which was correct), instead of finding a single (correct) key. On the Kostka-Payne test, the NPC program received a score of 83.8% of segments correct, as opposed to the 87.4% score achieved by the TPC program. In some cases, the reasons for the superior performance of the TPC model are quite subtle. Consider figure 7.12 (Chopin's Mazurka Op. 67 No. 2); while the TPC version of the program correctly analyzed this excerpt as being in G minor, the NPC version mistakenly identified the key as B♭ major. This excerpt contains many F♯'s; the TPC version considers F♯ to be less compatible with B♭ major than G♭ would be. That G♭ (♭$\hat{6}$) is more compatible with B♭ major than F♯ (♯$\hat{5}$) may seem, at best, a subtle distinction, but taking these distinctions into account allowed the program to modestly improve its performance.

In short, the advantage of the TPC model over the NPC one is noticeable, but not large. Whether the use of correct TPC information as input is "cheating" is perhaps debatable, but in any case it does not appear to make a tremendous difference to the performance of the model.[12]

7.8 An Alternative Approach to Modulation

The above algorithm adopts a simple approach to modulation which proves to be highly effective: impose a penalty for changing keys, which is balanced with the key-profile scores. Another possible way of incorporating modulation into the key-profile approach has been proposed by Huron and Parncutt (1993). Huron and Parncutt suggest that the key at each moment in a piece is determined by an input vector of all the pitch-events so far in the piece, weighted according to their recency. An exponential curve is used for this purpose. If the half-life of the curve is one second, then events two seconds ago will weigh half as much in the input

Figure 7.12
Chopin, Mazurka Op. 67 No. 2, mm. 1–16.

vector as events one second ago.[13] (Huron and Parncutt's algorithm also involves weighting each pitch-event according to its psychoacoustical salience; I will not consider this aspect of their model here.) Huron and Parncutt compared the results of this model with experimental data from studies by Krumhansl and Kessler (1982) and Brown (1988) in which subjects judged the key of musical sequences. In Krumhansl and Kessler's study, subjects heard chord progressions, and judged the stability of different pitches at different points in the progression; these judgments were compared with key-profiles to determine the perceived key. In Brown's study, subjects heard monophonic pitch sequences, and indicated the key directly at the end; the same pitches were presented in different orderings, to determine the effect of order on key judgments. The Huron-Parncutt model performed well at predicting the data from Krumhansl and Kessler's study; it fared less well with Brown's data.

An implementation of the Huron-Parncutt algorithm was devised, in order to test it on the Kostka-Payne corpus. A number of decisions had to be made. It seemed sensible to use the modified version of the key-profile values, since these clearly perform better than Krumhansl's original ones. Exactly the same input format was used as in my test; the same segments were used in this test as well. For each segment, a local input vector was calculated. These input vectors were flat, as in my algorithm; each value was either one if the corresponding pitch-class was present in the segment or zero if it was not. For each segment, a "global input vector" was then generated; this consisted of the sum of all the local input vectors for all the segments up to and including that segment, with each local vector weighted according to the exponential decay function. The key chosen for each segment depended on the best key-profile match for that segment's global input vector. The use of segments here requires some explanation. Unlike my algorithm, the Huron-Parncutt does not actually require any segmentation of the input; it could conceivably make very fine-grained key judgments. For example, the piece could be divided into very narrow time-slices, say a tenth of a second, and a new key judgment could be made after each time-slice, with each prior segment weighted under the exponential curve. (In this way, the duration of events would also be taken into account.) In the case of my algorithm, however, using segments and calculating flat input vectors for each segment was found to work better than counting each note and duration individually; as noted earlier, when notes are counted individually, a repeated note can have too strong an effect. It seemed likely that the same would be true for the Huron-Parncutt algorithm.

The algorithm was tested with various different half-life values, to find the one yielding the best performance. This proved to be a value of 4 seconds. With this value, the algorithm scored correctly on 629 out of 896 segments, a rate of 70.2%.[14] Inspection of the results suggests that there are two reasons why this system performs less well than the preference rule system proposed earlier. One reason is its inability to backtrack. Very often, the segment where a key change occurs is not obviously in the new key; it is only, perhaps, a few seconds later that one realizes that a modulation has occurred, and what the new key is. Another problem with the algorithm is that it has no real defense against rapid modulation; with a half-life of 4 seconds, the algorithm produced 165 modulations (Kostka's analyses contained 40 modulations). Raising the half-life value reduced the number of modulations, but also reduced the level of performance.

It might be possible to improve the decay model. For example, the input vectors could be weighted with subsequent pitches as well as previous ones, perhaps allowing the system to handle modulations more effectively. However, I will not pursue this further here.

I have argued here that the key-profile model can provide a successful solution to the key-finding problem. While I have proposed some modifications to Krumhansl and Schmuckler's original model, the basic idea behind it proves to be a very useful and powerful one. However, it is important to bear in mind a point made in chapter 1: the fact that a model performs well at a task performed by humans does not prove that humans do it the same way. This is particularly worth emphasizing with regard to the role of harmony in key-finding. As noted earlier, several proposals for key-finding—Winograd's, Maxwell's, and Bharucha's—have relied mainly on harmonic information. While it appears that a key-profile model can perform key-finding pretty well without harmony, it might also prove to be the case that a harmony-based model can perform the task well without key-profile information. To put it another way, it may be that key information is often contained in musical stimuli in more than one way. If this proves to be the case, then other evidence will have to be considered—experimental psychological data, for example—in deciding which factors truly are "key" in how key-finding is actually done.

II Extensions and Implications

8
Revision, Ambiguity, and Expectation

8.1 Diachronic Processing and Ambiguity

In previous chapters, I presented a series of preference rule systems for deriving basic kinds of musical structure in common-practice music. While each of the systems requires refinement and improvement, they achieve considerable success in generating the desired analyses, and many of the problems that do arise seem susceptible to solution within the preference rule framework. This in turn suggests—returning to the main point of this whole enterprise—that the preference rule approach is worth taking seriously as a hypothesis about music cognition.

At this point, however, the reader may be wondering how much the analyses presented in previous chapters really correspond to the analyses formed in listeners' minds. The general reality of harmony, meter, and the like has been amply demonstrated by experimental research. Even so, the kinds of representations discussed so far fail to do justice to those we experience in listening, in at least two important ways. First of all, we have said very little about the *diachronic* (or "across-time") aspect of musical listening. Preference rule systems consider all possible analyses of a piece, and then output a complete analysis as the preferred one; but this is clearly not how music is actually perceived. Secondly, it seems too rigid to assume, as I have in previous chapters, that we entertain only a single preferred analysis, even at a given moment. Some moments in music are clearly ambiguous, offering two or perhaps several analyses that all seem plausible and perceptually valid. These two aspects of music—diachronic processing and ambiguity—are essential to musical experience, and accommodating them is an important challenge for any model of music cognition. In this chapter I will try to show that preference rule systems can meet this challenge.

8.2 Modeling the Diachronic Processing of Music

So far, we have viewed the listening process mainly in terms of its end product: a final "preferred" analysis of a piece. Our concern has simply been with devising models that predict this final analysis as accurately as possible.[1] But a fundamental aspect of musical listening is the way it occurs across time. As we listen to a piece, it is unfolded to us gradually. We begin analyzing it as soon as we hear it (inferring structures such as meter, harmony, and so on), and build our analysis up gradually as we go along, perhaps revising our initial analysis of one part on the basis of what happens afterwards. Does the current theory shed any interesting light on this complex diachronic process?

At this point we must think back to the "dynamic programming" approach to implementation presented in chapter 1. It was argued there that this approach allows the huge number of possible analyses of a piece to be searched in an efficient manner. The program processes a piece from left to right; at each segment, for each possible analysis of that segment, it computes the best analysis of the piece so far ending with that segment analysis. (I will speak of "the program" here, although what follows applies equally well to each of the five programs proposed in earlier chapters.) In this way, the program is guaranteed to find the optimal analysis of the entire piece, without having to actually consider an exponentially huge number of analyses. Up to now, we have regarded the dynamic programming approach as an aspect of implementation, a way of allowing the program to find the optimal final analysis. It may be seen, however, that this strategy accomplishes something else as well: It allows the program to build up its analysis of the piece in a "left-to-right" manner. At each moment, the program is keeping a set of "best-so-far" analyses of the portion of the piece that has been processed so far; one of these analyses will be the highest-scoring one, and this is the program's preferred analysis at that moment. This procedure has another feature as well, which is of particular importance. Assume that the system has just processed a certain segment S_n; it has chosen a certain best-so-far analysis as the best one, and this entails a certain analysis for segment S_n. Now the system moves on to segment S_{n+1}, and computes a new set of best-so-far analyses. It may be, however, that the best-so-far analysis chosen at S_{n+1} entails a different analysis for segment S_n than the one initially chosen. In this way, the program in effect goes back to an earlier segment and revises its initial analysis. I will argue that this feature of the model is of considerable musical interest.

Essentially, then, what the program is doing is analyzing a series of ever-growing portions of the piece: segment 1, then segments 1–2, then segments 1–3, and so on. At each step, it finds the highest-scoring analysis of everything heard so far. The dynamic programming approach is simply a very efficient way of accomplishing this, which does not require consideration of a huge number of analyses. In the following discussions, however, it is perhaps simplest to imagine that the system is building up a larger and larger portion of the piece, always considering all possible analyses of whatever portion has been heard.

The problem of applying preference rule models to the process of real-time listening was first explored by Jackendoff (1991), with regard to the rule systems of *GTTM*. Jackendoff discusses three possible models of analyzing music in a left-to-right manner—what he calls "musical parsing." In one, a "serial single-choice" model, the system always maintains only one analysis. If this analysis proves to be untenable, the parser backtracks to some previous point and tries another analysis. As Jackendoff notes, such a parser may end up generating a number of different analyses for the same passage, one after another, until it finds an acceptable one. Another model is the "serial indeterministic" model, which does not generate any analysis of a passage until it is over. Jackendoff points out that this approach is simply unworkable; there is no way of choosing an analysis unless possible analyses have been generated for consideration. (It is also counterintuitive, he suggests, in that it predicts that we would not be aware of *any* interpretation until the passage is over, which is clearly incorrect.) In other words, while each of these models might seem at first to avoid the generation of a number of alternative analyses, in fact they do not. This leads to the final possibility—a "parallel multiple-analysis" model; this model generates many analyses simultaneously, as long as there is reasonable evidence for them (totally implausible analyses may be discontinued). There is a "selection function" which chooses whichever analysis is most satisfactory at any given moment; this is the analysis that is actually present in our musical experience. Jackendoff suggests that since many analyses must be generated in any case, it seems most plausible to assume that they are generated simultaneously (avoiding the constant backtracking of the serial single-choice model).

Figure 8.1
Bach, "Ich bin's, ich sollte buessen," mm. 1–4.

Jackendoff's "parallel multiple-analysis" model is essentially similar to the processing model advocated here. Like Jackendoff's, the model I propose maintains a number of analyses, always choosing one analysis as the most preferred at any given moment. The two models are also similar in that they abandon analyses which are implausible, though they do this in rather different ways. Under Jackendoff's scheme, analyses would be abandoned when they fell below some "threshold of plausibility"—presumably a numerical score. Under the dynamic programming scheme, the model in effect abandons analyses which it *knows* could never be the preferred ones, no matter what happens in the future. But both the current approach and Jackendoff's fulfill the same purpose, namely keeping the number of analyses manageably small.

Jackendoff also offers an interesting analysis of the opening phrase of a Bach chorale (figure 8.1). He shows how the first few beats of the piece (up to the third beat of m. 1) are metrically ambiguous, and could plausibly be interpreted in different ways. Jackendoff provides five hypothetical continuations of the passage, each of which reinforces one of the possible hearings of the opening (figure 8.2). It is not until m. 2 that the notated meter is clearly established—the main factors being the suspension and long melody note on the downbeat of m. 2, along with the long A♭ in the left hand on the third beat of m. 1, which makes both of these events seem like strong beats. This thought experiment is compelling, for two reasons. First, it demonstrates the reality of revision. It is not obvious that our interpretation of the first few beats would be affected by what happens afterwards; yet apparently it is. I find that the different continuations in figure 8.2 *do* make me interpret the opening chords in different ways; in figure 8.2c, the first chord seems strong, while in figure 8.2b, the first chord seems weak and the second chord strong. Secondly, the fact

Figure 8.2
From Jackendoff 1991. Hypothetical continuations of the opening of the Bach excerpt in figure 8.1, with Jackendoff's metrical analyses shown below the score. Reprinted by permission of the Regents of the University of California.

that the process of inferring the meter in these various passages is relatively effortless—lacking any sense of backtracking—suggests that the different interpretations are, indeed, present all along; a particular continuation does not cause us to generate a particular analysis of the opening chords, but merely causes us to choose among analyses that are already generated. (We should not give too much weight to such introspective evidence, however; it is possible, for example, that the system is backtracking, but is just doing it so quickly and effortlessly that we do not notice.)

8.3 Examples of Revision

While capturing the left-to-right nature of listening is important for its own sake, a particularly interesting implication of this phenomenon is what I will call *revision*: the modification of an initial analysis in light of subsequent events. We might also think of revision as a kind of musical "garden-path" effect. The term "garden-path" is used in linguistics to refer to sentences like "the old man the boats." When you hear the first three words, you assume that "man" is a noun; but given the rest of the sentence, you must revise your initial interpretation—treating "man" as a verb instead—in order to make any sense out of it. In this section I will explore some examples of revision in meter and harmony; in the following section I will focus on revision in key structure, where it plays a particularly important role. In some cases I will discuss the output of the implementations used in previous chapters. The programs are not normally set up to display their "provisional" analyses of pieces (analyses only up to a certain point in the piece); however, it can easily be determined what their provisional analysis at a certain point would be, simply by submitting an input representation of the piece only up to that point.

Jackendoff offers an example of metrical revision in his analysis of the Bach chorale. Though he argues that the opening chords of the chorale are metrically somewhat ambiguous, he also suggests that a certain analysis is most salient at each moment. (What is at issue here is the level above the quarter note, what we would call level 3; the tactus and lower levels are taken for granted.) Up to the fourth chord, a duple meter interpretation with the first chord strong is most preferable (the structure shown in figure 8.2a); this is due to the preference for strong beats near the beginning of groups (favoring the first chord as metrically strong), and the preference for duple over triple meter. Once the long A♭ in the left hand is heard (on the fourth beat), this chord is favored as metrically strong; maintaining the strong beat on the first chord requires a triple meter interpretation, as shown in figure 8.2c. However, given the long event on the sixth quarter-note beat, there is great pressure to locate a strong beat there; maintaining the strength of the fourth quarter-note beat leads to the correct metrical structure, corresponding to the notation in figure 8.1.

While Jackendoff's real-time analysis of this piece is not implausible, it is difficult to examine our own intuitions when the interpretations change so rapidly (at least according to his analysis) and the preferred analysis at any moment is not strongly favored over others. Another case where the experience of revision is clearer is shown in figure 8.3: the famously deceptive opening of the final movement of Beethoven's Sonata Op. 14

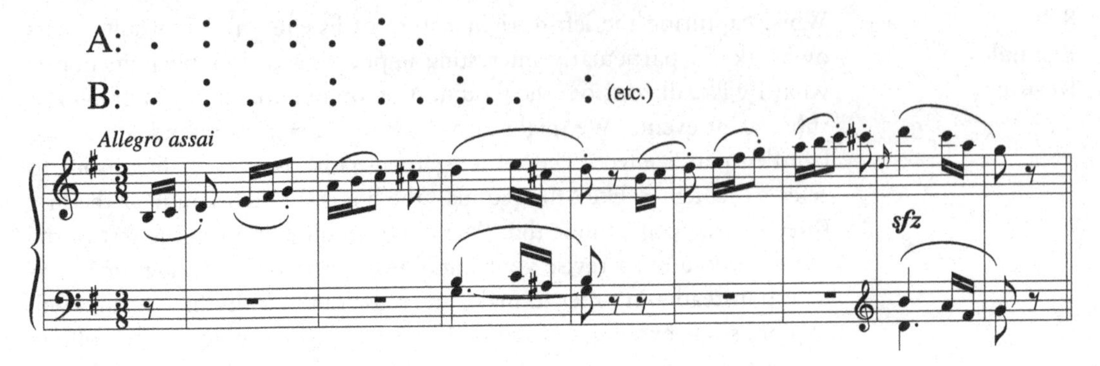

Figure 8.3
Beethoven, Sonata Op. 14 No. 2, III, mm. 1–8.

Figure 8.4
Mozart, Sonata for Violin and Piano K. 526, I, mm. 1–6.

No. 2. Given only the first ten notes, there is no reason why we would entertain the triple meter indicated by the notation; rather, we would more likely assume a duple interpretation, with the eighth-notes metrically strong (structure A). It is only the following two measures—particularly the long chords on the downbeats of the third and fourth measures—that establish the correct meter (structure B). (Remember that the length of an event is defined by its "registral inter-onset-interval"; see section 2.3.) The main factor favoring a duple interpretation of the opening is parallelism—a factor discussed in chapter 2 (MPR 9), but not incorporated into the computer model proposed there. Another deceptive opening is shown in figure 8.4, from Mozart's violin sonata K. 526. The two long bass notes make it fairly clear that the eighth-note beats are grouped triply rather than duply. However, the much greater length of the second left-hand note favors this as metrically strong (MPR 2), leading to structure A. It is only the long notes in m. 5 that prove this to be erroneous. The metrical program captures this well, producing structure A if given only the first three measures, but producing structure B if given the first six measures. We should note that there are other factors in this passage of which the program is ignorant, but which might well influence human listeners. For example, the repeated two-eighth-note motive in mm. 1–2 gives a hint of a quarter-note level of meter (again due to parallelism), which adds to the charming complexity of the passage. (To my ears, the particular V-I cadence in m. 4 also suggests a strong-weak metrical pattern, reinforcing the notated meter even before the long notes in m. 5; this is a kind of stylistic cue to which the program is oblivious.)

Revision is of particular importance at higher levels of meter, due to the frequent irregularities at these levels. Figure 8.5, from Beethoven's Sonata Op. 10 No. 1, offers an example. The metrical level at issue is the two-measure level: are odd-numbered measures strong, or even-numbered measures? (We could describe these as the "odd-strong" and "even-strong" interpretations, respectively.) The opening clearly demands an odd-strong hearing; the big chords in mm. 1 and 5 favor strong beats at these measures, due to the event rule (MPR 1). The odd-strong hearing persists right up to m. 22, where the return of the theme at an even-numbered measure causes a shift to an even-strong hearing. (The big chord at m. 22 is probably enough to force this interpretation, although the factor of parallelism also exerts pressure for analyzing mm. 22–3 in the same way as mm. 1–2.) Once m. 22 is heard, however, the chords in mm. 18 and 20 also seem strong in retrospect, relative to the "empty" downbeats on mm. 17 and 19. Once again, parallelism is a factor, given the strong motivic parallel between mm. 21–2 and the previous two two-measure groups—though even considered in isolation, mm. 17–20 clearly favor an even-strong hearing, which allows them to be easily tipped in that direction once m. 22 is heard.[2]

Harmony is also an area where revision can occur. It is important to bear in mind here that what we mean by harmony is root structure, not Roman numeral analysis. An example of harmonic revision would be interpreting a chord first as an A minor chord, and then later as a C major chord; not, for example, interpreting a C major chord first as I of C and then as IV of G. (The latter would be a case of tonal revision; what

Figure 8.5
Beethoven, Sonata Op. 10 No. 1, I, mm. 1–23.

is being revised there is the key interpretation, not the harmonies themselves.) Figure 8.6 gives an illustration of harmonic revision—the opening of "Aus meinen Thraenen spriessen," the second song in Schumann's song cycle *Dichterliebe*. Hearing just the first chord, the natural choice of root is A major. This is accounted for by the compatibility rule (HPR 1), which favors interpreting A and C♯ as $\hat{1}$ and $\hat{3}$ of A rather than $\flat\hat{3}$ and $\hat{5}$ of F♯ (or any other interpretation). When the bass descends through G♯ to an F♯ minor triad, this gives the entire first three chords the sense of a single F♯ minor harmony; the third chord clearly has a root of F♯ minor, and the strong beat rule (HPR 2) prefers to avoid a harmonic change by including the previous chords in the same harmony. The unambiguous harmonic moves to D major and A major in the following measure, however, favor an A major interpretation of the first chord; this is due to the harmonic variance rule (HPR 3), which prefers an analysis whereby roots of nearby segments are close together on the line of fifths. Another

Figure 8.6
Schumann, *Dichterliebe*: the final two measures of "Am wunderschoenen Monat Mai" and the first two measures of "Aus meinen Thraenen spriessen." The figure shows how the harmonic interpretation of the first chord of the second song (marked by a box) depends on the context. Each bracket indicates a context in which the chord can be heard; the symbol beneath indicates how the chord is interpreted under that context.

complicating factor is the ending of the previous song in the cycle, "Am wunderschoenen Monat Mai." Since this song ends (remarkably) on a dominant seventh, it seems clear that it should follow smoothly into the second song with little pause. Given the C♯7 chord at the end of the first song, the harmonic variance rule exerts pressure on the first chord of "Thraenen" to be interpreted as F♯ minor rather than A major. To summarize the changing interpretations of the first chord: heard completely in isolation, it implies A major; in the context of the first three chords, it is heard as F♯ minor; in the context of the entire first phrase of the second song, it is heard as A major; but when preceded by the end of the previous song, it once again implies F♯ minor.[3]

The harmonic program proposed in chapter 6 has partial success in capturing these changing interpretations. When given just the single chord, the first three chords, or the entire first phrase of the second song, the program analyzes the first chord as A major. (Thus it fails to capture the F♯ minor analysis of the first chord in context 2.) However, when it is given the ending of the first song plus the first phrase of the second, it analyzes the first chord as F♯ minor.

8.4 Revision in Tonal Analysis

While revision manifests itself in all aspects of musical structure, it plays a particularly important role in key structure. A number of authors have observed cases where a segment of music first appears to be in one key, but then demands reinterpretation in another key in light of what follows.[4] For testing of the key program's handling of revision, a special version of the program was designed; this program produces what I call a "running" analysis, showing its provisional analysis for each segment of the piece. (Recall that the key program requires the input to be divided into segments; high-level beats of the metrical structure are used for this purpose.) A sample of the program's output is shown in figure 8.7. The segments of the piece are listed vertically; for each segment, the program's provisional analysis of the piece up to and including that segment is shown horizontally. The diagonal edge of the chart indicates the program's initial analysis for each segment at the moment it is heard. When the choice of key for a segment in a particular provisional analysis is identical to the choice for that segment in the previous analysis, only a hyphen is shown. If a key name is shown rather than a hyphen, this means that the program revised its initial key choice for a segment and chose something else instead. The program's final analysis is indicated by the final (lowest) key choice for each segment; it is also listed along the bottom for convenience.

The running analysis in figure 8.7 is for the Gavotte from Bach's French Suite No. 5; this score for this piece is shown in figure 6.7. While we would not think of this as a particularly ambiguous piece in tonal terms, the program does find several "garden-path" effects, where a segment first analyzed one way is later reinterpreted. For example, consider mm. 5 and 6. At m. 5, the program was still considering the first five measures to be in G major. Given m. 6, however, the program decided that m. 5 would be better interpreted as being in D. Other garden-path effects are found at m. 9, mm. 12–13, and mm. 17–20. While I am not certain that these revisions exactly capture my moment-to-moment hearing of the piece, they seem generally plausible. Notice that two of the provisional key sections in the running analysis—the move to B major in mm. 12–13 and the move to C in mm. 17–20—are completely obliterated in the final analysis.

The current approach also sheds light on another phenomenon. It is widely agreed that modulations typically involve "pivot chords," chords that are compatible with both the previous key and the following one. In the Bach Gavotte, for example, the D major chord in m. 9 (the first half of the measure, anyway) could either be considered as I of D or V of

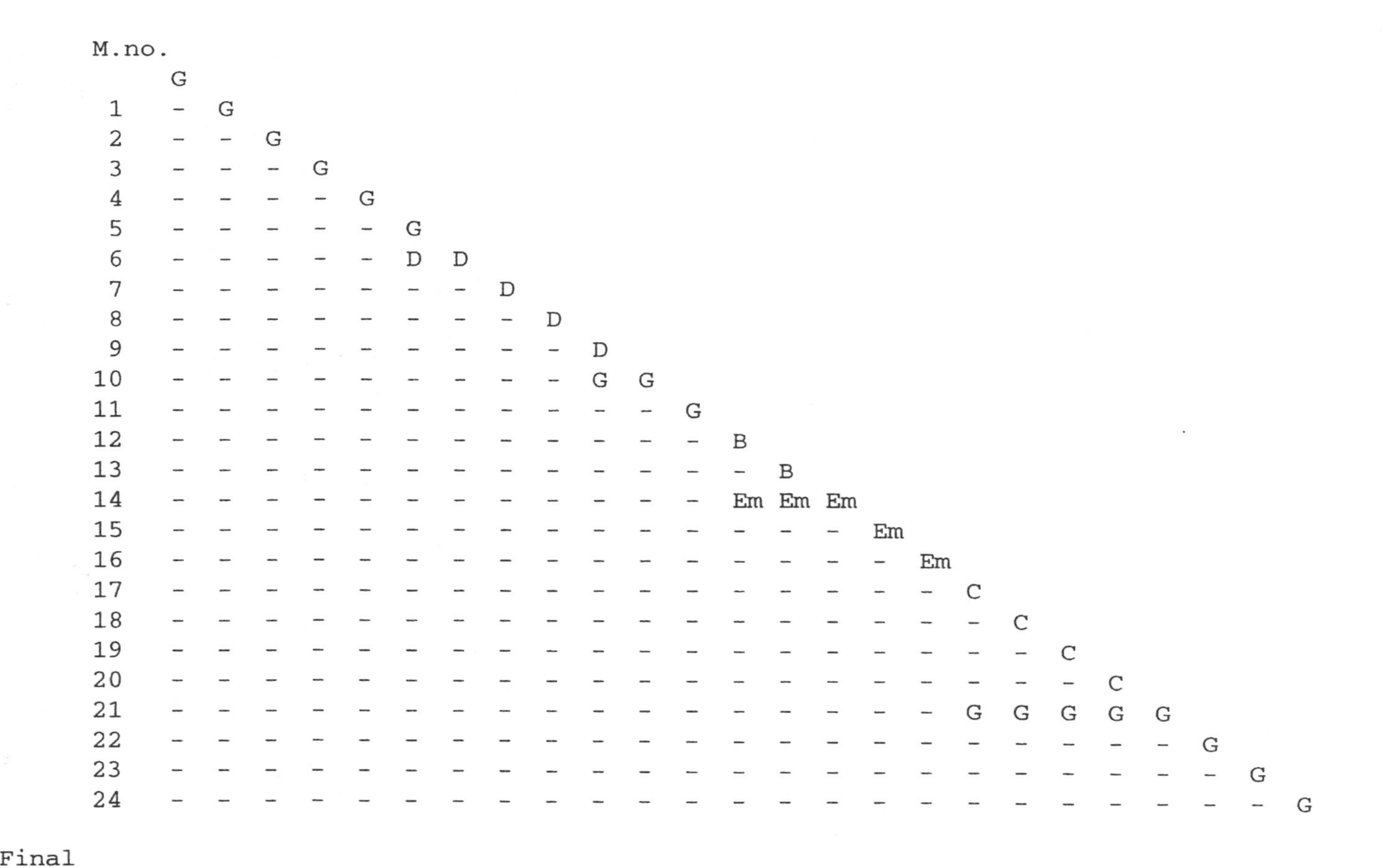

Figure 8.7
A "running analysis" of the Gavotte from Bach's French Suite No. 5 in G major. (The score in shown in figure 6.7.) Segments, corresponding to measures, are listed vertically; the program's provisional analysis at each segment is listed horizontally. (The first segment contains just the first half measure of the piece.) In cases where the program's choice for a segment is the same as the choice for that segment in the previous analysis, only a hyphen is shown. The final choice for each segment is shown at the bottom of the figure.

M.no.																
1	Gm															
2	-	Gm														
3	-	-	Gm													
4	Bb	Bb	Bb	Bb												
5	-	-	-	-	Bb											
6	-	-	-	-	-	Bb										
7	-	-	-	-	-	-	Bb									
8	Gm	Gm	Gm	Gm	Gm	Gm	Gm	Gm								
9	-	-	-	-	-	-	-	-	Gm							
10	-	-	-	-	-	-	-	-	-	Gm						
11	-	-	-	-	-	-	-	-	-	-	Gm					
12	Bb	Bb	Bb	Bb	Bb	Bb	Bb	Bb	Bb	Bb	Bb	Bb				
13	-	-	-	-	-	-	-	-	-	-	-	-	Bb			
14	-	-	-	-	-	-	-	-	-	-	-	-	-	Bb		
15	Gm	Gm	Gm	Gm	Gm	Gm	Gm	Gm	Gm	Gm	Gm	Gm	Gm	Gm	Gm	
16	-	-	-	-	-	-	-	-	-	-	-	-	-	-	-	Gm
Final analysis:	Gm	Gm	Gm	Gm	Gm	Gm	Gm	Gm	Gm	Gm	Gm	Gm	Gm	Gm	Gm	Gm

Figure 8.8
A "running analysis" of Chopin's Mazurka Op. 67 No. 2, mm. 1–16. (The score is shown in figure 7.12.)

G. This would imply that key sections generally overlap by at least one chord. The current approach holds out another possibility, however, which is that pivot chords are essentially a diachronic (across-time) phenomenon. It is not that a chord is understood as being in two keys at one time, but rather it is first interpreted in one way, based on the previous context, and then in another way, in light of the following context. There is nothing in the program that actively searches for, or prefers, such diachronic pivot effects, but they often do seem to emerge at points of modulation.

Chopin's Mazurka Op. 67 No. 2 yields an interesting and highly unusual running analysis. (The score for mm. 1–16 is shown in figure 7.12, the analysis in figure 8.8.) Essentially, the program changes its mind several times as to the main key of the piece. The first three measures are analyzed as being in G minor; in m. 4, however, the program decides, not implausibly, that the first four measures have been in B♭ major. At m. 8, it decides that G minor is a preferable choice for the

Figure 8.9
A recomposition of mm. 14–16 of the Chopin Mazurka excerpt in figure 7.12.

Figure 8.10
Beethoven, Sonata Op. 13, I, mm. 134–6

entire first 8 measures. At m. 12—parallel to the point in the first half where it switched to B♭ major—it again switches to this key. And finally, at m. 15, it reverts to G minor once again. The model's uncertainty as to the main key of the piece seems true to our experience of it. Clearly, if the last two measures had cadenced in B♭ major rather than G minor (as in the recomposition shown in figure 8.9), the model would have chosen B♭ major for the entire section; I suspect a human listener might easily have done so as well.[5]

Many cases of tonal revision involve "enharmonic modulation." This is a modulation involving some kind of reinterpretation of spelling. A typical case involves a diminished seventh which is first interpreted as a dominant of some previous chord, but then reinterpreted as the dominant of a following tonic. For example, in figure 8.10, the diminished seventh on the third beat of m. 135 is interpreted first as vii°4/3/G (with E♭ in the melody)—parallel to m. 134—and then as vii°4/2/Em (with D♯ in the melody). In such cases, then, we have both a revision of spelling and a revision of tonal structure. (A similar example, involving a German sixth chord, was discussed in section 5.6.) Such cases are not easily handled under the current system. The reason is that the key program requires a complete TPC representation in order for it to operate. However, it is not difficult to see how such phenomena could be handled in principle. Suppose the TPC representation was being generated in a left-to-right manner, and at every moment, whatever portion of the representation was complete was fed to the key system for analysis. Immediately after the fifth eighth-note of m. 135 (before the bass descends to B), the TPC representation would presumably spell this chord with an E♭, and the key system would interpret the key as G minor. Once the remainder of m. 135 was heard, the TPC representation would realize that the chord on the third beat was really to be spelled with D♯ (this would be enforced mainly by the harmonic feedback rule—a D♯ spelling allows a B7 chord to be formed); the key system, in turn, would then see that E minor was a more plausible key choice. At present, however, such interactions are beyond the reach of the implemented model.

8.5 Synchronic Ambiguity

Revision could be regarded as a kind of ambiguity. An event, or group of events, is interpreted first one way, then another. There are also cases, however, where a segment of music seems to offer two equally plausible interpretations, even from a particular vantage point in time. We might think of revision as diachronic (across-time) ambiguity, while the latter phenomenon is synchronic (at-one-moment-in-time) ambiguity. (The term "ambiguity" alone usually refers to synchronic ambiguity, and I will sometimes use it that way here.)

Lerdahl and Jackendoff (1983, 9, 42) recognized in *GTTM* that the preference rule approach was well-suited to the description of ambiguity. Informally speaking, an ambiguous situation is one in which, on balance, the preference rules do not express a strong preference for one analysis over another. This can also be nicely handled by the quantitative approach to preference rules assumed here. Recall that the system evaluates analyses by assigning them numerical scores, according to how well they satisfy the preference rules (again, this applies to any of the models presented earlier). At any moment, the system has a set of "best-so-far" analyses, the analysis with the highest score being the preferred one. In some cases, there may be a single analysis whose score is far above all others; in other cases, one or more analyses may be roughly equal in score. The latter situation represents synchronic ambiguity. To actually test the current models in this regard would be quite complex. One would first have to produce a set of excerpts that are agreed to be ambiguous in some respect. It would then have to be decided how exactly the pro-

Figure 8.11
Brahms, Sonata for Violin and Piano Op. 100, I, mm. 31–4.

gram's scores are to be interpreted; how close do the scores of the most favored analyses have to be in order for us to consider that the program judged the passage as ambiguous? I will not attempt such a rigorous test here, but will informally explore some important kinds of ambiguity and the current model's potential for handling them.

The preference rule approach accounts for a number of kinds of situations where ambiguity has long been recognized. An example in the case of metrical structure is the hemiola, a switching back and forth between duple and triple grouping at a high metrical level. Figure 8.11 shows a typical case, from the first movement of Brahms's Sonata for Violin and Piano Op. 100. The first two measures (along with the previous context, not shown here), strongly establish the metrical structure indicated by the notation, with every third quarter-note beat strong. Measures 33 and 34, however, suggest an alternative structure, with every second quarter-note strong. (The main factors here are the onsets of long events in the violin on every second quarter-note beat, as well as the repeated two-quarter-note motive in the piano part.) An alternation between the 3/4 and 3/2 structures follows, leading back to a strongly established 3/4 in m. 43.[6] The effect of mm. 33–34 is one of real ambiguity; we hear the 3/2 meter suggested by the hemiola, but also retain the 3/4 meter persisting throughout, with the implied metrical accent on the downbeat of m. 34. Whether the meter program's output here constitutes recognition of the ambiguity is, again, difficult to say, but at least it recognizes the main factors in favor of both interpretations.

Another important area of ambiguity is harmony. In doing harmonic analysis, one frequently encounters passages which could just as well be

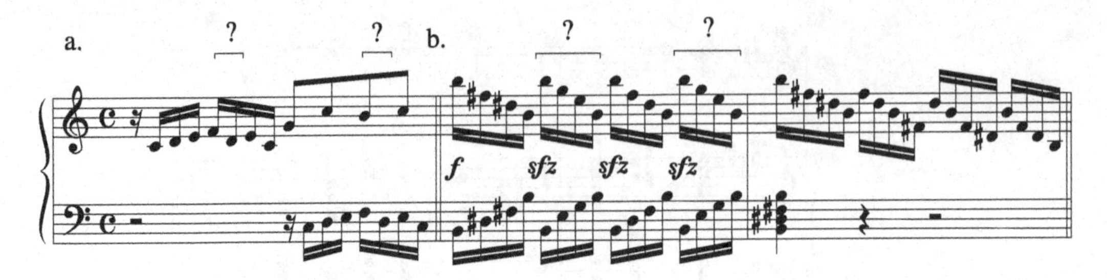

Figure 8.12
Two examples of harmonic ambiguity. (a) Bach, Two-Part Invention No. 1, m. 1; (b) Beethoven, Sonata Op. 53, I, mm. 28–9.

analyzed in more than one way. Perhaps the most common source of uncertainty is whether something constitutes a chord, or is just an ornamental event. This is reflected in conventional harmonic terminology in the concept of a "passing chord"—an entity somewhere between an ornamental event and a true harmony. Consider the opening of Bach's C Major Two-Part Invention, shown in figure 8.12a. Does the F-D in the first half of the measure represent a double-neighbor ornamental figure, part of the opening I chord, or is it an implied V7 harmony (or—less plausibly—ii)? By the current model, the reason for treating the D-F as a harmony is that the compatibility rule favors treating notes as chord-tones rather than ornamental tones. (The two notes are reasonably acceptable ornamental dissonances—the D more so than the F—since they are closely followed stepwise, but a chord-tone interpretation is still preferable.) The pressure for treating them as ornamental comes from the fact that the strong beat rule resists changes of harmony on weak beats, particularly the change back to I that would follow the V chord-span. Thus the model finds factors in favor of both interpretations. (A similar situation arises in the second half of the measure; here, the presence of yet a third tone in the segment—the eighth-note B—adds further pressure for a chordal interpretation over an ornamental one, since three ornamental tones carry a higher cost than two.) A similar case concerns 6/4 chords. In figure 8.12b, the second and fourth beats of the first measure could be regarded as i6/4 chords (in E minor); alternatively, the notes of the 6/4's—the E's and G's in the right and left hands—could be treated as ornamental to the notes of the following V chords. The same issue arises with the cadential 6/4—sometimes called either a I 6/4 or a V 6/4, to reflect these two interpretations. Again, pressure for the chordal interpretation comes from the compatibility rule and the ornamental dissonance rule; pressure for the ornamental interpretation comes from the strong beat rule. Rather than insisting on one interpretation or another of figure 8.12b, it might be most satisfactory to say that we hear it both ways at once: there is the sense of root motion to i, but also the sense of an underlying V harmony. The current model offers an account of how and why this ambiguous perception might come about.

Synchronic ambiguities of a similar kind often arise in key analysis. In many cases, a brief move occurs to the pitch collection of another key, and it is unclear whether to consider it a modulation or not. The term "tonicization" is sometimes used for this phenomenon; frequently it simply involves a V-I harmonic progression (or "secondary dominant") in the foreign key. In such cases, the key-profile rule favors a change of key, while the modulation rule discourages it. Examples of this are ubiquitous in common-practice music; see, for example, the brief move to C major in mm. 17–18 of the Bach Gavotte in figure 6.7.

With regard to both key and harmony, one might offer a different account of the phenomena I have described here. Rather than reflecting ambiguity, they might be seen to reflect the *hierarchical* nature of these representations. In terms of harmonic structure, there can be more than one level of harmony present in a piece, with surface harmonies elaborating more prolonged structural ones. With key, also, it is generally accepted that each piece has a main key; within this main key, there may be moves to secondary keys, and perhaps even briefer tonicizations within and between those. The hierarchical nature of key and harmony is an important issue that I have hardly considered here.[7] However, the above examples suggest that, to some extent, the perception of multi-level tonal structures might be accounted for by the general principles of harmonic and key perception presented earlier.

An area where the importance of ambiguity has been less widely acknowledged is in spelling. I noted earlier that diachronic ambiguities of spelling arise quite frequently, and are often accompanied by tonal revisions as well. For example, a diminished seventh may be interpreted first one way, then another, or what first seems to be a dominant seventh may be reinterpreted as a German sixth. In general, the spelling of a chord is determined by the subsequent context; by convention (and also I think in perception), our ultimate interpretation of a diminished seventh chord, for example, depends on what chord it resolves to. There are also cases, however, where even once the following context is known, the correct spelling of an event is unclear. These ambiguities cannot easily be represented in music notation, of course, since a single spelling has to be

Figure 8.13
Brahms, Intermezzo Op. 116 No. 6, mm. 1–8.

chosen for each event; but they are important nonetheless. Several examples are found in Brahms's Intermezzo Op. 116 No. 6 (figure 8.13). Consider the D/Cx in the alto line in m. 1. This note is $\flat\hat{7}$ of the current key, and would normally be spelled as $\flat\hat{7}$ (D) rather than $\sharp\hat{6}$ (see section 5.4); D also forms a Bm7 chord, which makes it preferred by the harmonic feedback rule (TPR 3). Yet Brahms notates the pitch as Cx, which does indeed seem like a plausible choice; perhaps this is due to the parallelism between Cx-D♯ and the previous B♯-C♯. The E/Dx in the corresponding position in m. 5 seems less ambiguous; E seems to be strongly favored here, maybe because the pitch variance rule (TPR 1) gives an even stronger preference for E versus Dx (in an E major context) than for D versus Cx. The G♭/F♯'s in mm. 19–20 present another striking ambiguity (figure 8.14). Here, the voice-leading rule (TPR 2) exerts pressure for a G♭ spelling, since the G♭/F♯'s resolve to F's; the harmonic feedback rule would argue for F♯, allowing a D7 chord to be formed. In m. 21 the conflict is settled; in this case even voice-leading favors the F♯ intepretation. (Other examples of spelling ambiguity are discussed in Temperley 2000.)

Perhaps the aspect of structure where ambiguity is most common and pervasive is grouping. As noted in chapter 3, it is frequently difficult to

Figure 8.14
Brahms, Intermezzo Op. 116 No. 6, mm. 18–22.

decide between several possible grouping interpretations of a passage. Figure 3.6 gave one case of ambiguous grouping: do the group boundaries in mm. 3–4 occur before the first sixteenth-note of the measure, or afterwards? Or do they occur both before and after the note, resulting in a series of grouping overlaps? Or does the vagueness of our intuitions here suggest that there are really no grouping boundaries at all, and the entire passage is a single group? Figure 7.4 shows an even more problematic example, though the piece is monophonic. With the exception of the three clear breaks—m. 8, m. 36, and at the double bar—there is no clear phrase boundary anywhere. As discussed in section 3.6, these problems become even more acute in complex polyphonic textures, where it is often impossible to find a single satisfactory segmentation. Indeed, London (1997) has argued that ambiguity in grouping is so pervasive as to undermine any effort to produce a single preferred grouping analysis for a piece. While I would not go quite that far, I would certainly concede that ambiguity poses a serious complication in the modeling of grouping structure.

8.6 Ambiguity in Contrapuntal Structure

A particularly interesting kind of ambiguity is found in contrapuntal structure. As noted in section 4.1, there are frequent moments of uncertainty in contrapuntal analysis—for example, where it is not clear whether two inner-voice lines should be joined into a single line. Moreover, even when it is clear that a certain set of notes constitutes a line, one often feels that that is not the whole story. Consider figure 8.15, the opening of a Bach fugue. There is no question that all the notes of the first two measures constitute a single contrapuntal voice, which then continues beneath the second voice entering in m. 3. At the same time,

Figure 8.15
Bach, *Well-Tempered Clavier* Book I, Fugue No. 2 in C Minor, mm. 1–3.

one could also break this voice down into two contrapuntal lines—a lower-register, gently descending line G-A♭-G-F-G-A♭-G-F-E♭, and an upper-register line consisting of repetitions of the C-B-C-(D) motive.

The view of counterpoint I am proposing here is due primarily to the music theorist Heinrich Schenker (1935/1979). Schenkerian analysis is concerned, in part, with teasing apart the contrapuntal threads which form the fabric of most tonal pieces. Schenker is not concerned with superficial lines such as the parts of a string quartet or the melody and accompaniment lines in a sonata; these of course are obvious, and no analytical work is needed to reveal them. Rather, the focus of Schenkerian analysis is on the more subtle contrapuntal lines that may partition and overlap the contrapuntal voices of the surface. For example, the analysis of the Bach fugue subject discussed above, segmenting the notes of the fugue subject into two contrapuntal lines, is essentially that proposed by Schenker; Schenker's actual analysis is shown in figure 8.16. (Schenkerian analysis is also hierarchical, in that it analyzes certain events as being ornamental to others in a recursive fashion; this accounts for the omission of several notes in his analysis and the unusual rhythmic notation and other symbols. This aspect of Schenker's thought is beyond our scope here.)

Schenker's analyses of contrapuntal strands within melodies can be quite perceptually compelling. The streams within the melody in figure 8.15 do not weaken the sense of the entire fugue subject as a coherent melodic line; they might best be regarded as "sub-streams" within the primary stream of the fugue subject itself. It is for this reason that I consider such structures to represent a kind of ambiguity. Does the current model shed any light on these sub-streams? In its normal state, the counterpoint program does not identify the streams proposed by Schenker; rather, it identifies the entire fugue subject as a single stream. However, it can be seen that Schenker's contrapuntal analysis is quite a plausible one according to the rules proposed earlier. The pitches within each stream are quite proximate to each other (indeed, more so than in

Figure 8.16
Schenker, analysis of Bach, *Well-Tempered Clavier* Book I, Fugue No. 2 in C Minor, mm. 1–3, from *Das Meisterwerk in der Musik* (1926/1974). Reprinted by permission of Cambridge University Press.

Figure 8.17
Bach, *Well-Tempered Clavier* Book I, Fugue No. 2 in C Minor, mm. 1–3, showing the program's contrapuntal analysis (with modified parameters).

the "one-stream" analysis), thus favoring this analysis by CPR 1; however, penalties are incurred due to the addition of a second stream (CPR 2) and the numerous rests within streams (CPR 3). By raising the pitch proximity rule penalty (CPR 1) from 1 point to 4 (per chromatic step), and lowering the white square rule penalty (CPR 3) from 20 points to 10 (per second), the model can be made to produce the analysis in figure 8.17: essentially the same as Schenker's. By adjusting the parameters, then, we can make Schenker's analysis the preferred one. With the normal parameters, it is not preferred, but it is perhaps one analysis among several reasonably high-scoring ones. As suggested above, we can imagine that our cognitive preference rule system for contrapuntal analysis considers a number of different analyses in its search for the best one; perhaps our ephemeral experience of an analysis such as figure 8.17 indicates that such an analysis is being considered as part of this search process.

The possibility of sub-streams arises in many situations. Consider the opening of the Mozart sonata discussed in chapter 4 (figure 4.1). It was suggested there that the left-hand part constitutes a single melodic line. However, it might better be treated as three melodic lines, as shown in figure 8.18. Among other things, this allows us to make better sense of the F-A chord in the left hand on the downbeat of m. 5 (conveniently

Figure 8.18
Mozart, Sonata K. 332, I, mm. 1–5: A "sub-stream" analysis of the left hand.

Figure 8.19
Bach, *Well-Tempered Clavier* Book I, Fugue No. 3 in C♯ Major, mm. 1–3. The dotted lines show the analysis produced by the contrapuntal program.

ignored in my initial discussion of this passage): this chord does not represent the sudden addition of a second line, but rather, the continuation of two (or perhaps three) lines already present. Again, by adjusting the program's parameters, it was possible to produce something quite similar to this (although I was not able to produce *exactly* this analysis). In this case, however, positing sub-streams is more problematic. As noted in section 4.2, one of the main sources of psychological evidence for streams is that the notes of the stream emerge as a clearly identifiable unit. In Dowling's experiments with interleaved melodies, for example, the fact that a certain group of notes could be easily recognized under certain conditions indicated that it was being perceived as a stream. But can we really say that the individual lines of the analysis in figure 8.18 constitute easily recognizable units? The middle line, taken in isolation, would not easily be recognized as part of the Mozart sonata; indeed, it hardly seems like a coherent melodic line at all, partly due to its syncopated rhythm. One might argue that the sub-streams that are inferred here are rhythmically transformed versions of the actual "surface" sub-streams, perhaps with each line reduced to a single sustained note in each measure. (Indeed, this is roughly how they would probably be represented in a Schenkerian analysis.) All of this requires further study; the point for now is that we should be cautious about positing sub-streams such as those in figure 8.18 as psychologically real entitites.

Contrapuntal structure may also involve diachronic ambiguities. Consider figure 8.19, the opening of a Bach fugue (discussed earlier in section

4.6). If given only the subject itself (up to the low C♯ on the third beat of m. 2), the program will assign all the notes to a single melodic line. However, if the following measures are added, the program will reinterpret the opening two measures as constituting two voices, as indicated by the dotted lines in figure 8.19. Since two melodic lines have to be created anyway, it is preferable by the pitch proximity rule to assign the lower notes of the first two measures to one line and the upper notes to another line, thereby avoiding the leaps required by the single-line analysis. While we would probably not consider the two-voice analysis of the subject to be "the" correct analysis (if only one analysis is allowed), it surely captures an important aspect of the passage, and one which—as the program indicates—emerges more strongly in retrospect: When the second line enters at the end of m. 2, it can quite easily be heard to connect with the low C♯ of the first line. The effect is that the two lines of m. 3 grow organically out of the subject.

8.7 Ambiguity in Meter

While synchronic ambiguity certainly occurs with some aspects of musical structure, there are other aspects in which the potential for ambiguity appears to be much more limited. The prime case in point is metrical structure. Consider a simple pattern such as figure 8.20; this pattern could be heard with the strong beats on either the G's or the E's (or perhaps other ways as well). Prior contexts could easily be constructed to reinforce one structure or the other; and even hearing the pattern in isolation (repeated indefinitely), it is not difficult to impose either structure at will, or to switch back and forth between them. What is almost impossible, however, is to hear both structures simultaneously. In general, it is difficult to entertain two metrical structures at once, even in cases where either one can readily be entertained on its own. This does not mean that synchronic ambiguity is never present in meter; I pointed to the hemiola as a case where it occurs. One difference between these two cases (figure 8.11 and figure 8.20) is that in the hemiola case, only one level of meter—the measure level—is different between the two interpretations, whereas in a case such as figure 8.20, the tactus as well as all higher levels are different.

Figure 8.20

In fact, it appears to be generally true that the cases where synchronic ambiguity arises in meter involve *higher-level* metrical ambiguity, where the two competing structures share all but the highest levels in common. Indeed, ambiguity in hypermeter is a widely occurring phenomenon that has been the subject of considerable study in music theory (Lester 1986; Schachter 1987; Kramer 1988; Rothstein 1989; Kamien 1993; Temperley in press-a). Often, this involves two hearings of a passage in which either even-numbered or odd-numbered measures can be heard as metrically strong. An important special case of this is "melody-accompaniment conflict," where the accompaniment conveys one meter and the melody another; this occurs with some frequency in the music of the Classical masters (Kamien 1993) and Mendelssohn (Rothstein 1989). I have argued elsewhere that hypermetrical ambiguity is particularly common in closing themes of sonata form movements—the theme just before the end of the exposition. Figure 8.21 gives one example, and also illustrates the phenomenon of "melody-accompaniment conflict." The prior context clearly establishes a pattern of even-numbered measures strong; the change in the accompaniment at m. 72 also reinforces this (this is due to the "first-instance-strong" clause of the parallelism rule—the fact that a new accompaniment pattern begins in m. 72, and is repeated in a varied form in m. 73, favors m. 72 as strong). On the other hand, a melodic phrase clearly begins at m. 73, favoring a strong beat there by the grouping rule and hence an "odd-strong" interpretation. I feel that both of these hearings are present in my experience of the passage.[8]

Despite the potential for ambiguity at higher levels of meter, the resistance of lower-level meter to ambiguity remains an important and puzzling fact. In principle, the absence of synchronic ambiguity in a certain kind of structure could be modeled fairly easily. We could simply say that the analysis present to awareness was the highest-scoring one, and (barring an exact tie) there can only be one highest-scoring analysis at any moment. Jackendoff (1991) proposes a similar solution, suggesting that there is a "selection function" which is continually monitoring all the available analyses and choosing one or another analysis as the preferred one. However, as we have seen, most kinds of musical structure *do* appear to allow synchronic ambiguity, in that multiple analyses may be simultaneously present to awareness. Lower-level meter is, in this respect, the exception to the rule. Why some aspects of perception should be more susceptible to ambiguity than others is an interesting question, deserving further study.

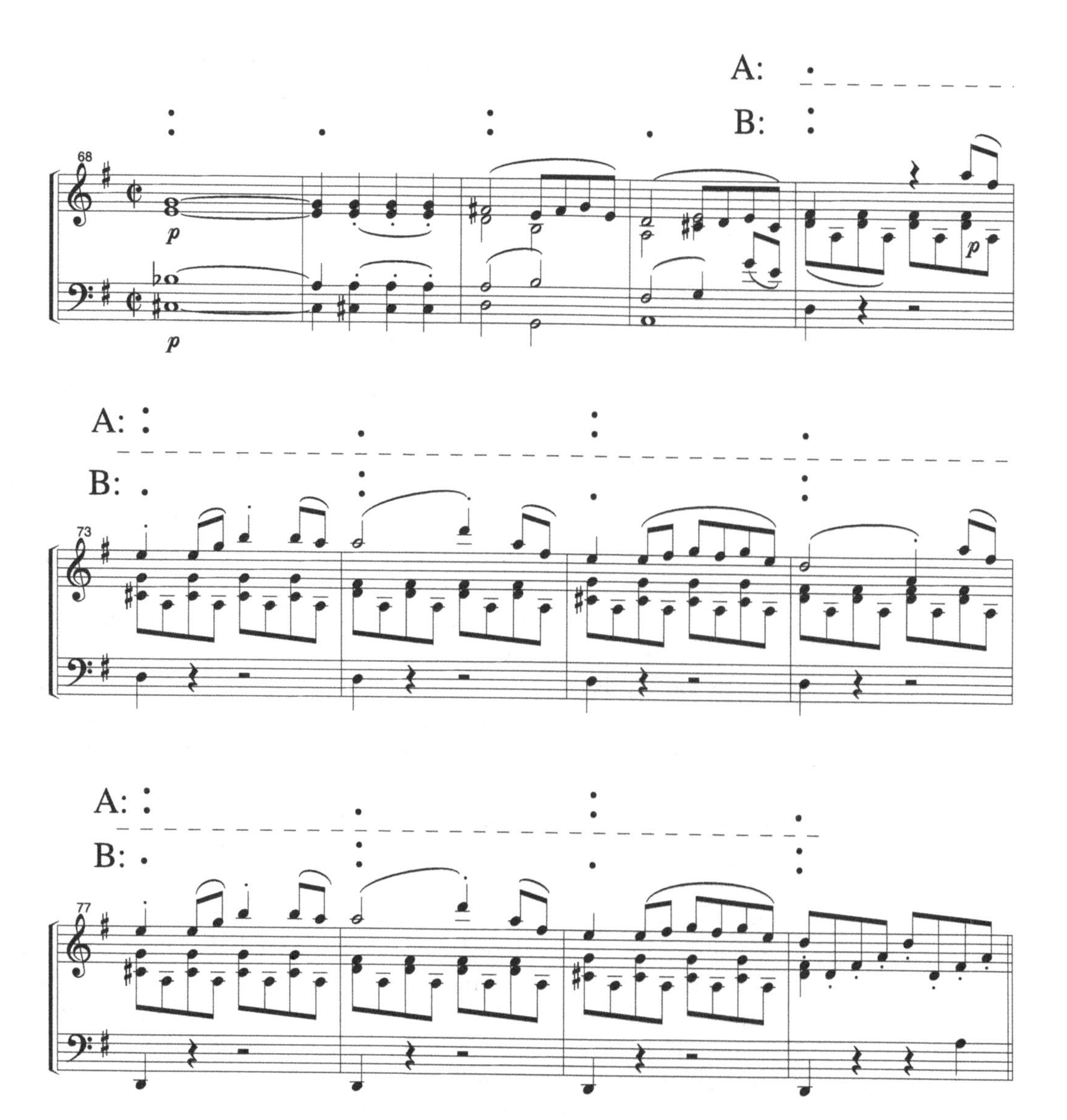

Figure 8.21
Haydn, String Quartet Op. 76 No. 1, I, mm. 68–80.

8.8 Expectation

In listening to a piece, we are not only building up our analysis of what we have heard so far (which may or may not be revised at a later time); we are also forming expectations about what will occur in the future. It is these expectations that allow the experience of surprise, which is such a vital part of music—an unexpected rhythmic jolt in a Beethoven sonata, or an unforeseen harmonic turn in a Chopin mazurka. Some theorists have argued that expectation, and the way expectations are denied or fulfilled, is an essential part of musical experience and enjoyment (Meyer 1956, 1973; Narmour 1990). While I will not explore this issue in depth here, preference rule systems do offer an interesting and promising way of accounting for expectation. This relates to the fact, already mentioned, that preference rule systems produce a score for any analysis that they generate; some analyses satisfy the rules well and are therefore high-scoring, while other analyses violate one or more of the rules and are low-scoring. Since most music within the style allows at least one analysis that is relatively high-scoring (a point I will develop further in chapter 11), we expect that future events in a piece will allow a high-scoring analysis as well. In the case of meter, for example, we would expect events that are reasonably well-aligned with whatever metrical structure has previously been established. In the case of key we would expect events that are reasonably compatible with the key-profile of whatever key is currently in force, thus permitting a high key-profile score without the penalty of a modulation. In the case of contrapuntal structure we would expect events relatively close in register to current events, allowing the pitch proximity rule to be satisfied without the initiation of new voices. Perhaps musical expectation is governed, in part, by the search for a continuation that satisfies these various constraints. By this view (or perhaps by any view), musical expectation is a kind of composition, since it involves the actual generation of music. In chapter 11 we will explore further how preference rules might be viewed as constraints on the production of music, rather than merely on its perception. (Actually, even the analysis of already-heard music can be seen to involve a kind of expectation. When we impose an analysis on a passage, this means in effect that we expect that analysis to be the one ultimately chosen. It is for this reason that having to revise that analysis later can involve a kind of surprise—as it surely does; surprise occurs because our expectations have been denied.)

Among the most important work in the area of expectation has been that of Schmuckler (1989). Schmuckler performed experiments in which subjects were played part of a composition (a Schumann song), followed

Table 8.1
Piston's table of usual root progressions

Context chord	Often follows	Sometimes follows	Seldom follows	Chords not considered
I	**IV, V**	VI	II, III	VII
II	**V**	**VI**	I, III, IV	VII
III	**VI**	IV	II, V	**VII**, I
IV	**V**	**I**, II	III, VI	**VII**
V	**I**	IV, VI	**II**, III	VII
VI	**II**, V	**III**, IV	I	VII
VII	**III**	—	—	I, II, IV, V, VI

Source: Piston 1987, as encapsulated in Schmuckler 1989.
Note: For each diatonic chord, the other chords are listed which often follow it, sometimes follow it, or seldom follow it. Chords which represent motion by fifths are shown in bold.

by a series of possible continuations; they were asked to rate how well the continuations matched their expectations, and their responses were analyzed. The melody and the piano accompaniment were used separately in the first two experiments, to test expectations of melody and harmony independently; these were then combined in the third experiment. In terms of the melody, it was found that the expectedness of continuations tended to conform with Krumhansl's key-profiles (discussed earlier in chapter 7), in that pitch-classes with high values in the key-profile for the current key of the piece were highly expected. This, then, is an example of the general reasoning advocated above: in terms of the current approach, the more expected events were those permitting a more "high-scoring" analysis. In the second experiment, the piano accompaniment was used, with different continuations based on different harmonic progressions. Schmuckler compared subjects' expectations with the table of common harmonic progressions found in Piston's harmony textbook (1987). (Piston's table—as represented by Schmuckler—is shown in table 8.1.) A strong correlation was found between the expectedness of a harmony and its appropriateness according to Piston's table (given the preceding context). While the model of harmony presented in chapter 6 has little to say about progression, one point is worth noting here. It can be seen from Piston's table that root motions by fifths—that is, motions of one step on the line of fifths, such as I-V or I-IV—tend to be judged as common. Out of the twelve possible "fifth" motions between two diatonic roots, seven are in the "often" category and three in the "sometimes" category.[9] Recall that motion by fifths is preferred by the harmonic model due to the harmonic variance rule (HPR 3), since this maximizes the proximity of roots on the line of fifths. In this way, Schmuckler's data suggest that listeners' expected continuations tend to be "high-scoring" under the current model of harmony. Both in terms of key and harmony, then, Schmuckler's experiments offer tentative support for the idea that listeners expect continuations which are high-scoring in preference rule terms.

Another approach to harmonic expectation has been pursued by Bharucha and Stoeckig (1987), using a well-established psychological paradigm known as "priming." This paradigm is based on the idea that if something is heard under conditions where it is highly expected, it should be more quickly and accurately processed. In Bharucha and Stoeckig's study, subjects were played a C major triad (the "prime"), followed by another triad (the "target") which might or might not be mistuned (by a fraction of a semitone); subjects had to discriminate the tuned targets from the mistuned ones. Subjects' speed and accuracy on this task depended strongly on the relationship of the target to the prime; in particular, the targets were more quickly and accurately judged if they were close to the prime on the circle of fifths. For present purposes, the circle of fifths can be reinterpreted as a range of points on the line of fifths centering around the prime, as shown in figure 8.22.[10] Again, these findings accord well with the current model: by the harmonic variance rule, chords which are closer to previous chords on the line of fifths will be more "high-scoring," and thus more strongly expected.

Other work on expectation has focused on the factors governing melodic expectation apart from harmony and tonality. As noted above, the current model of counterpoint predicts that, at a given point in a melody, a note will be expected which is close to the current note in pitch. This has in fact been shown experimentally: in judgments of the expectedness of melodic continuations, pitch proximity proves to be a significant factor. However, other factors may also play a role. Narmour (1990) has proposed a complex theory of melodic expectation, which has been quantified and greatly simplified by Schellenberg (1997). Schellenberg proposes that experimental data about melodic expectation can be modeled well by two factors: pitch proximity and "reversal," which is a change of direction after a large interval. However, von Hippel and Huron (2000) have argued persuasively that "reversal" is simply a consequence of the fact that melodies tend to be confined to a certain range;

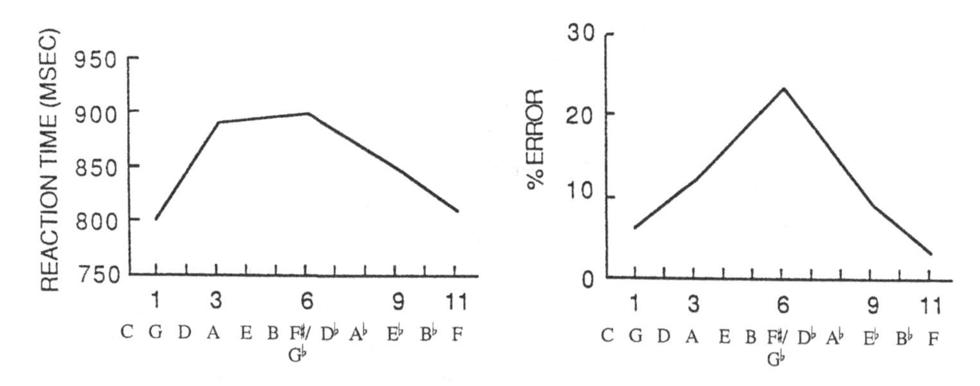

Figure 8.22
Data from Bharucha 1987. Subjects were asked to identify a "target" chord as tuned or mistuned, following a "prime" chord. The horizontal axis indicates the circle-of-fifths position of the target, with the prime at zero; thus both position 1 and position 11 are one step away from the prime. I have added TPC labels to represent the corresponding line-of-fifths position for each circle-of-fifths position (assuming a prime of C). (For each circle-of-fifths position, the line-of-fifths position was chosen that was closest to the prime.) Reprinted by permission of the Regents of the University of California.

a large leap is likely to land near the extreme of the range, and thus will most likely be followed by a return to the middle. Possibly this idea could be incorporated into the contrapuntal analysis model proposed in chapter 4: in grouping notes together into lines, we may be seeking not only to minimize note-to-note intervals, but also to keep each line within its established range.

One serious problem should be addressed regarding expectation and revision. It is, again, a widely held view that the way music plays with expectations—sometimes fulfilling them, sometimes denying them—is an important part of its impact and appeal. The problem is that the denial of expectations would only seem to be possible with an unfamiliar piece. With a piece we know well, we know what analysis will ultimately prove correct (even if it is not the most preferable one at the moment it is heard), and we know what notes are going to follow, so no surprise is possible. And yet it certainly seems to be possible to go on enjoying pieces after we know them well. Jackendoff (1991) has proposed a way out of this dilemma, building on the modularity theory of Fodor (1983). Jackendoff suggests that music perception may occur within what Fodor calls a "module," an encapsulated cognitive system that does not have access to knowledge stored elsewhere—for example, any representation of a piece we might possess in long-term memory. Thus the module is always processing the piece as if it was being heard for the first time—generating expectations that we know will be denied, wandering naively down garden paths, and so on.[11] This is an intriguing proposal, for it suggests a way of reconciling the idea that expectation is central to musical fulfillment with the fact that that fulfillment can endure over many hearings of a piece. On the other hand, it is surely true that our experience and enjoyment of a piece is affected by familiarity to *some* extent: with enough hearings, the appeal of a piece will begin to fade. This suggests that our processing of a piece may be affected by our long-term knowledge of it to some degree. In any case, the modularity hypothesis deserves further consideration and study.

9
Meter, Harmony, and Tonality in Rock

9.1 Beyond Common-Practice Music

Once a theory has been tested with reasonable success in the domain for which it was originally intended, it is natural to wonder whether it is applicable elsewhere. In the following two chapters, I move beyond "common-practice music"—European art music of the eighteenth and nineteenth centuries—to investigate the applicability of the preference rule approach to other musical idioms. In this chapter I consider rock, focusing in particular on three aspects of the infrastructure: meter, harmony, and key. In the following chapter I examine the validity of the metrical and grouping models with respect to traditional African music. Rather than using computer implementations and quantitative tests (as in previous chapters), I will proceed in an informal fashion, consulting my own intuitions as to the analyses of rock songs and considering informally what principles might give rise to them. (My approach to African music will be similar, though there I will be relying mainly on the intuitions of ethnomusicologists rather than my own.)

To anticipate my conclusions, I will argue that the aspects of rock and African music in question can largely be accommodated within the models I have proposed, though some important modifications are needed. I should emphasize, however, that where cognitive principles are found to operate across styles, no claim whatever is implied that these principles are universal or innate. To draw any conclusions about this would require study of many more than three musical idioms. This is especially true in the case of rock, since part of its roots are in common-practice music (or at least are shared with common-practice music), so whatever similarities may exist between these styles may be due to historical rather than innate factors. Rather, to the extent that commonalities across styles are found, I will remain neutral as to their origins. It is interesting to examine how widely a musical theory can be applied, quite apart from issues of innateness or universality.

The study of styles outside the Western art music tradition presents challenges that do not arise with common-practice music. In common-practice music, a piece is written by a composer in the form of a notational score, and is then played by a performer; the score is generally taken to represent the piece, and is thus the main object of attention and analysis. In rock, by contrast, there generally is no score; rather, we are more likely to identify the piece (or rather the "song") as a particular performance, captured in a recording. Because scores are visual, and because they consist of discrete symbols, they submit easily to analysis and discussion. Moreover, as noted earlier, scores often provide explicit information about infrastructural representations (meter, spelling, and to some extent key, contrapuntal structure and phrase structure). However, what first appears to be a problem may well be a blessing in disguise. Though the structures conveyed by a score may often coincide with our hearing, in some cases they may not, and indeed may sometimes exert an unwanted influence on our hearing. To take one example, with a common-practice piece we need never ask what the metrical structure is; we simply look at the notation. With rock, there is no notation, so we must ask ourselves what metrical structure we actually *hear*; we must let our ears decide. Of course, such judgments may be somewhat subjective, so that we should be cautious about treating them as hard facts (though this problem arises with common-practice music as well). I will hope that my intuitions about the meter of rock songs (and other things) are shared by most other readers, although it would not surprise me if there were occasional disagreements.

The input representation assumed in this chapter and the following one is the same as in previous chapters: a "piano-roll," showing pitches represented in time. Such a representation may seem problematic for musics outside of the Western art music tradition. In the first place, a piano-roll input assumes quantization of pitch in terms of the chromatic scale. For some kinds of music, the chromatic scale may not be the appropriate scale to use for quantization; in other cases, such as rock, though the fundamental scalar basis is the same as in common-practice music, there are inflections of pitch (e.g. "blue notes") which cannot be represented in a quantized chromatic framework. This is, indeed, a limitation of the current approach which I will not attempt to address here.

Another problem with the piano-roll representation, as was discussed in chapter 1, is that it ignores timbre. While timbre is surely an important aspect of all kinds of music, some authors have argued that it is of particularly central importance for rock (Tagg 1982, 41–2; Middleton 1990, 104–5; Moore 1993, 32–3). While this may be true, I will maintain that the kinds of structure at issue here—meter, harmony, and key—depend mainly on pitch and rhythm, and thus can largely be inferred without timbral information.

9.2 Syncopation in Rock

With regard to meter, I will argue that the metrical principles of rock are, for the most part, highly similar to those of common-practice music. However, there is one important difference, concerning *syncopation*. Syncopation refers to some kind of conflict between accents and meter. (I use the word "accent" here in the sense of phenomenal accent, as described in chapter 2: a phenomenal accent is any event that favors a strong beat at that location, such as a note, long note, loud note, or stressed syllable.) Syncopation occurs frequently in common-practice music; figure 9.1 shows two examples. In each case, several weak beats

Figure 9.1
(A) Beethoven, Sonata Op. 28, I, mm. 135–43. (B) Beethoven, Sonata Op. 31 No. 1, I, mm. 66–69.

A

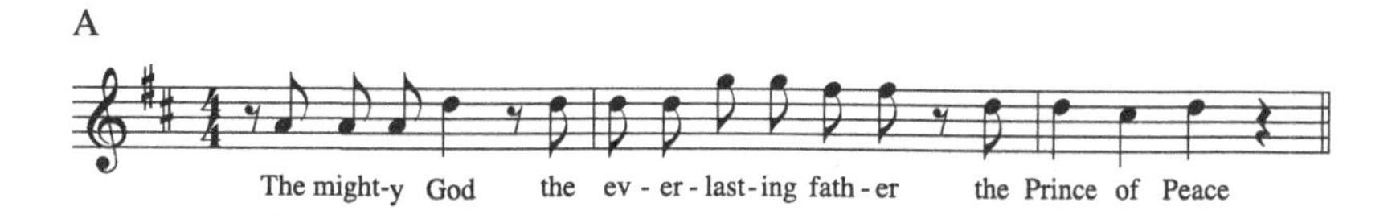
The might-y God the ev - er - last-ing fath - er the Prince of Peace

B

Lit-tle dar-ling It's been a long cold lone - ly win - ter

C

When I find my-self in times of troub-le Moth-er Mar - y comes to me

Figure 9.2
(A) Handel, "For Unto Us a Child is Born," from *The Messiah*. (B) The Beatles, "Here Comes the Sun." (C) The Beatles, "Let It Be."

are accented by long notes in the right hand, creating conflicts with the prevailing meter. However, syncopation in rock is of a rather different nature, and requires special treatment under the current model.

Consider the melodies in figure 9.2. The first is from Handel's *Messiah*; the second and third are from Beatles songs, "Here Comes the Sun" and "Let It Be." In all three cases, the metrical structure corresponding to the notated meter is clearly implied by the accompaniment (which is not shown here). This seems to be accounted for well under the metrical preference rule system proposed in chapter 2: the notated meter is the preferred one because it aligns strong beats with events rather than rests (MPR 1), particularly long events (MPR 2) (e.g., the bass notes in "Let It Be"), as well as with changes in harmony (MPR 6). The problem is with the melodies themselves. (The melodies are shown in figure 9.3 with their associated metrical structures; the stress pattern of the lyrics is also shown in an informal way.) An important preference rule in vocal music, mentioned briefly in chapter 2, is that there is a preference to align strong beats with stressed syllables of text (MPR 8). We could express this a little more precisely by saying that, in general, the metrical strength of a syllable should correspond with its degree of stress; for example, a stressed syllable should be metrically stronger than an adjacent unstressed one. In the Handel, the melody appears to adhere well to this principle; given the

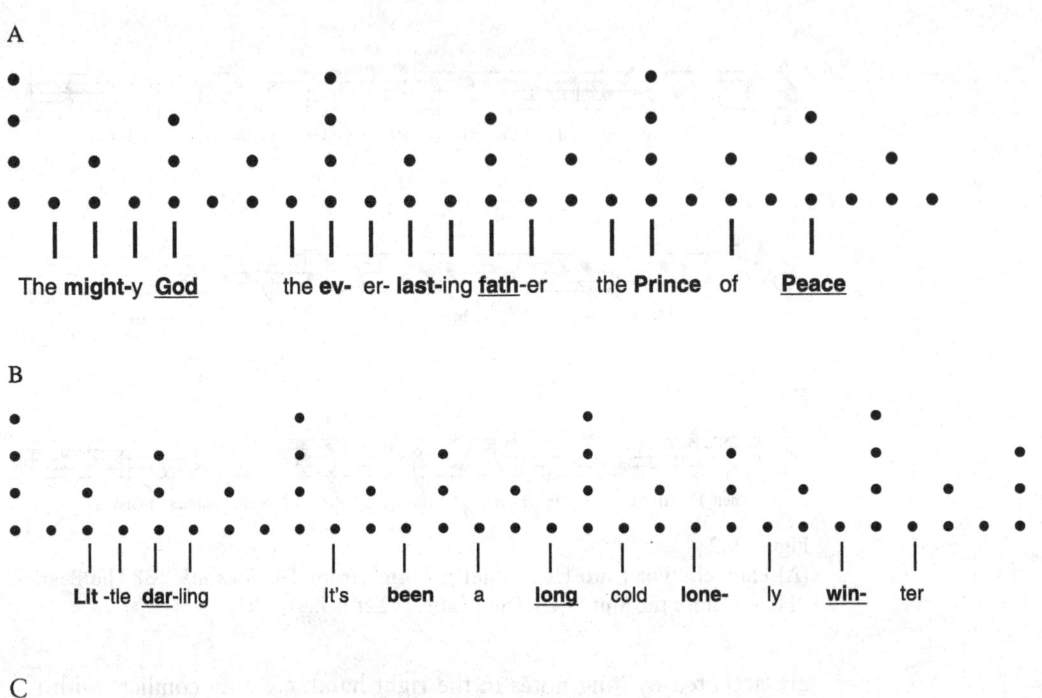
A
The might-y God the ev- er- last-ing fath-er the Prince of Peace
B
Lit -tle dar-ling It's been a long cold lone- ly win- ter

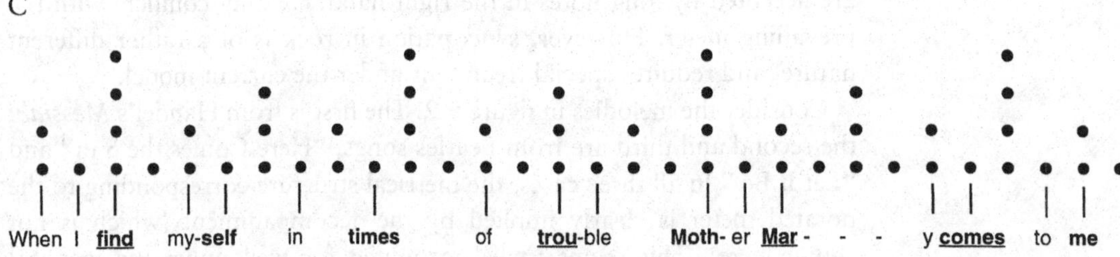
C
When I find my-self in times of trou-ble Moth- er Mar - - - y comes to me

Figure 9.3
The three melodies shown in figure 9.2, along with their metrical structures. The stress pattern of the text is also shown; stressed syllables are in bold, and highly stressed syllables are in bold and underlined. (A) "For Unto Us a Child is Born." (B) "Here Comes the Sun." (C) "Let it Be."

notated meter, stressed syllables and strong beats coincide almost perfectly. In the Beatles examples, however, the melody and the metrical structure appear to be in severe conflict. In some cases both stressed and unstressed syllables fall on very weak beats (as in "Here Comes the Sun"); in other cases, unstressed syllables are actually metrically stronger than adjacent stressed ones (for example, the word "myself" in "Let It Be"). (Since other syllables in each melody reinforce the notated meter—such as the words "When I find" in figure 9.3c—there is conflict not only between the melody and the accompaniment, but also between different sections of the melody.) There are conflicts with regard to other preference rules as well. For example, the event rule (MPR 1) would prefer a structure where every event-onset falls on a strong beat; yet in the second phrase of figure 9.3b, "It's been a long cold lonely winter," exactly the opposite is the case. The length rule (MPR 2) prefers structures where longer syllables fall on strong beats; in figure 9.3c, however, "times" and "Mar-" are relatively long but fall on weak beats. In the Handel melody, by contrast, both of these rules are followed fairly strictly, as they are in most common-practice music.

The phenomenon of rock syncopation would seem to present a problem, or at the very least a challenge, for the current metrical theory as it stands. In rock, it appears, the perceived metrical structures often involve severe violations of the preference rules. Of course, the mere fact that our perception of a piece involves some violation of the preference rules is not necessarily a reason to question the model. Preference rule violations occur often in common-practice music as well, and are an important source of musical interest and tension, as I will discuss further in chapter 11; indeed, syncopation itself is a common phenomenon in common-practice music, as shown in figure 9.1. In the case of rock, however, this view of syncopation seems untenable. It would suggest that rock syncopations—apart from adding a small degree of tension or instability—do not really affect our judgments of a song's meter; they are heard as conflicting with the meter, but are disregarded or overruled. But the syncopations in figure 9.2b and c do not seem incompatible with the notated meter; there is very little sense of ambiguity or instability in these melodies. Indeed, it seems quite likely that even if one of these melodies was heard unaccompanied, the meter that was inferred would be the correct one. Moreover, other possible settings of these melodies clearly *do* seem incompatible with the notated meter. Consider the alternative setting of the "Here Comes the Sun" lyric in figure 9.4 (assume, as before, that the notated meter is clearly implied by the accompaniment).

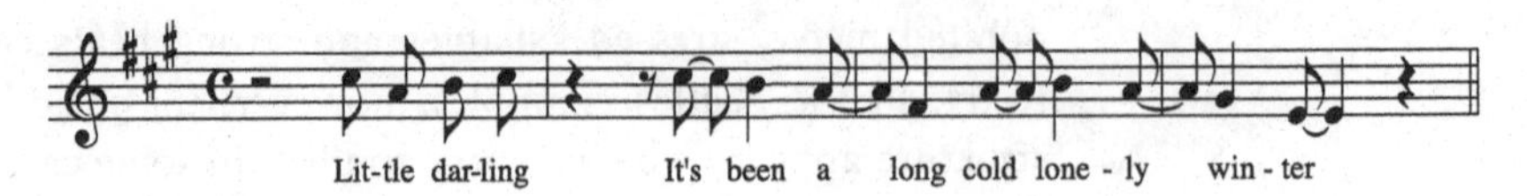
Lit-tle dar-ling
It's been a long cold lone - ly win - ter

Figure 9.4
An ungrammatical setting of the "Here Comes the Sun" lyric.

This setting clearly seems incompatible with the notated meter; in the context of an accompaniment implying the notated meter, it would, I think, sound quite wrong. This suggests that we do not simply disregard, or override, melodic syncopations in our judgments of meter; indeed, syncopated rhythms often seem to reinforce the meter of a song rather than conflicting with it.

An alternative approach would be to view syncopation as some sort of deviation from an underlying representation. Inspection of the melodies in figure 9.3b and c reveals a consistent pattern: each of the syncopated events occurs on a weak beat just before a strong beat, and generally the more stressed events occur before the stronger beats. For example, in figure 9.3b, the unstressed syllable "it's" occurs just before a quarter-note beat, while the stressed syllable "been" occurs just before a stronger half-note beat. If the syncopated events are shifted forward (to the right), the stress patterns and metrical structures can be made to align almost perfectly. Perhaps, then, a shifting process of this kind is what allows us to hear the melodies as supporting the meter, despite their apparent conflict with it.

One way of modeling this more formally is by positing a underlying "deep" representation of events which may differ from the "surface" representation. Both the deep and surface representations are pitch-time representations of the kind assumed in earlier chapters. The surface representation represents durations exactly as they are heard; in the deep representation, however, syncopated events are "de-syncopated"—that is, shifted over to the strong beats on which they belong. It is then this deep representation which serves as the input to further processing. We therefore posit the following rule:

Syncopation Shift Rule. In inferring the deep representation of a melody from the surface representation, any event may be shifted forward by one beat at a low metrical level.

How do we determined whether a particular event is to be shifted forward, and which level it is to be shifted at? The answer is simple: the solution is chosen which most satisfies the metrical preference rules. (All the rules are to be taken into account here, not just the one pertaining to text.) We therefore either leave an event where it is, or shift it forward by a beat at some level, depending on which resulting deep representation better satisfies the preference rules.

Two points of clarification are needed here. By an "event," I mean a single note; thus it is possible for one note to be shifted, while other simultaneous notes (such as accompaniment chords) are not. Secondly, when we shift an event forward, do we shift the attack-point forward (thereby making the event shorter), or do we shift the entire event forward? This is a rather difficult question. Shifting only the attack-point forward might reduce the duration of the event to zero (or it might even make the attack-point occur later than the end of the event); on the other hand, shifting the entire event forward might cause it to overlap with the following event in the melody. This problem might be addressed in a "preference rule" fashion: we prefer to shift the end of the event forward unless it causes an overlap with the following note. However, I will not address this problem further for now; we will simply adjust the durations of events in an *ad hoc* manner.[1]

Let us apply this model to the phrase from "Here Comes The Sun," as shown in figure 9.5. The placement of the syllables in the metrical diagram represents their position in the surface representation; the lines show how they are shifted (in some cases) to different positions in the deep representation. In the first half of the phrase ("Little darling"), the surface representation of the melody satisfies the linguistic stress rule well—the stress level of events corresponds nicely to their metrical strength—so no shifting of events is necessary. Now consider the syllables "It's-been." According to the linguistic stress rule, stressed syllables should fall on stronger beats than unstressed syllables; thus "been" should fall on a stronger beat than "it's." If we leave the syllables exactly as they are, they have the same metrical strength: both are on beats at level 1 (i.e., the lowest level). But if we shift "been" forward by one beat, it now falls on a beat at level 3; thus "been" is stronger than "it's." In this case it makes no difference whether we shift "it's" forward to the following beat (level 2) or leave it where it is; either way, "been" is still stronger. By this technique, the syllables can be shifted forward such a way that every pair of adjacent syllables satisfies the linguistic stress rule, as shown in figure 9.5. As mentioned, if we consider only the linguistic stress rule here, it is not necessary to shift unstressed syllables forward. However, the event rule (preferring events to coincide with strong beats)

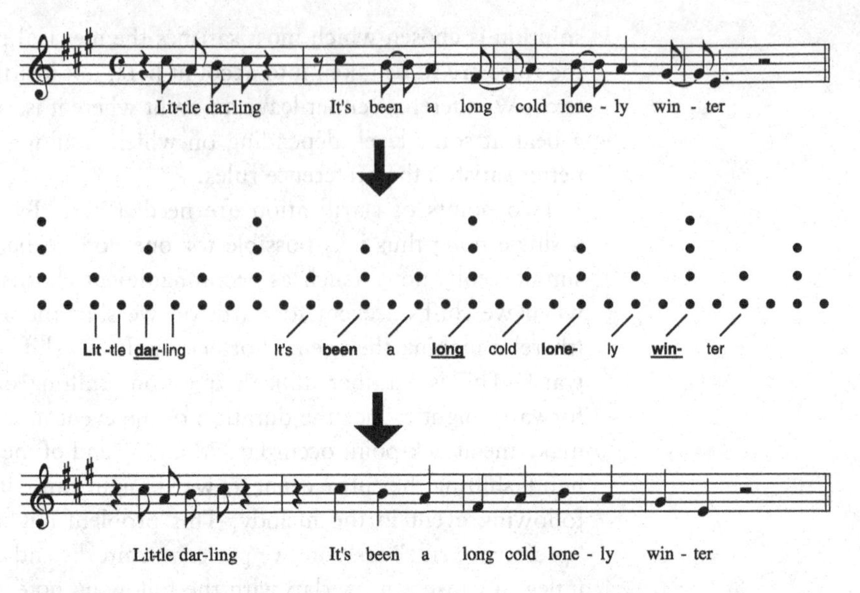
Lit-tle dar-ling It's been a long cold lone - ly win - ter
Lit -tle dar-ling It's been a long cold lone- ly win- ter
Little dar-ling It's been a long cold lone - ly win - ter

Figure 9.5
"Here Comes the Sun," showing syncopation shifts.

favors not only associating "been" with the following beat (reinforcing the linguistic stress rule), but also "it's," and every other syllable in the phrase. This leads to the deep representation shown in figure 9.5.

"Let It Be" presents a slightly more complicated example (figure 9.6). Here, several events are shifted forward by a beat at the sixteenth-note level, such as "self" and "comes"; however, two events, "Mar-" and "me," are syncopated at the eighth-note level. This presents no difficulty for the current model. The syncopation shift rule says only that an event may be shifted forward by one beat at a low metrical level; there is nothing to prevent shifts at different metrical levels within the same piece or even the same phrase. As in "Here Comes the Sun," there are several events whose placement in the surface representation fully satisfies the linguistic stress rule, but which are shifted forward so as to satisfy the length rule and event rule. It may be noted, however, that this was not done in every possible case. The second syllables of "trouble" and of "mother" would satisfy the rules more if they were shifted forward (placing the syllables on stronger beats), but I have left them unchanged; why? My intuition here is that these cases do not represent syncopations, perhaps because the second syllables of "trouble" and "mother" are naturally short and therefore fit well with the original rhythm. There is

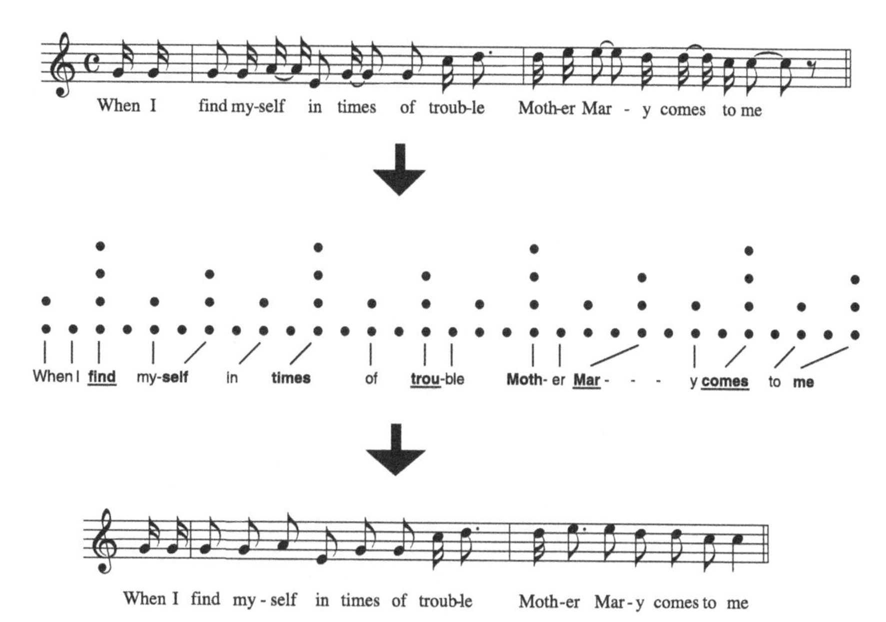
When I find my-self in times of troub-le Moth-er Mar - y comes to me
When I find my-self in times of trou-ble Moth- er Mar - - - y comes to me
When I find my - self in times of troub-le Moth-er Mar-y comes to me

Figure 9.6
"Let it Be," showing syncopation shifts.

clearly some indeterminacy here; it is not always obvious what the deep representation for a melody should be. While I will not attempt to solve this problem here, I do not believe it is intractable. Essentially, we normalize the surface (where possible, given the rule stated above) so as to resolve the most severe violations of the preference rules; other minor conflicts are perhaps left as they are.

Now let us examine how the syncopation rule might work with the ungrammatical setting of the "Here Comes the Sun" lyric, from figure 9.4 (assuming, again, that the notated meter is communicated by the accompaniment). Consider just the second half of the line, shown in figure 9.7. Again, we will examine the syllable pair "It's-been." The stress pattern dictates that "been" should be stronger. But notice that there is no legal way of shifting these syllables forward such that the preference rules are satisfied. As they stand, both syllables fall on weak beats; the event rule would prefer shifting both of them forward to stronger beats. But if they are shifted forward, the metrical strength of "it's" will exceed that of "been," thus violating the linguistic stress rule. If "been" is shifted forward but not "it's," the linguistic stress rule will be satisfied but not the event rule. In other words, by the current model, there is no way of

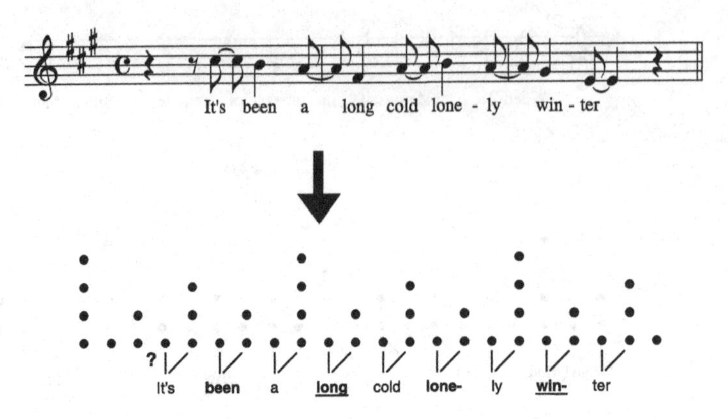
It's been a long cold lone - ly win - ter
?
It's been a long cold lone- ly win- ter

Figure 9.7
Ungrammatical setting of the "Here Comes the Sun" lyric, showing possible syncopation shifts.

shifting events in this example so that both of these rules will be satisfied. Thus the model predicts that this setting of the text would be heard as conflicting with the notated meter, as I believe it is. In this case, then, the model is able to distinguish a good setting of the line from an bad one.

9.3 Applications and Extensions of the Syncopation Model

This simple model proves to be quite useful in exploring the uses of syncopation in rock. In this section I will examine some general consequences of the model and some constraints on the way syncopations are used. I will then briefly consider the aesthetic functions of syncopation.

Syncopation could be viewed as a conflict between stress and meter. We "resolve" the syncopation, where possible, by shifting an event forward to the following beat. The cases we have considered so far all involve shifting an event forward to a beat which is stronger than the beat it falls on in the surface representation. Could a syncopation also be resolved by shifting an unstressed syllable to a weaker beat? Such situations do arise, but under somewhat limited circumstances. The only time they occur is in cases where an unstressed syllable is immediately followed or preceded by a syncopated stressed one, as in the phrase from "Hey Jude" shown in figure 9.8. In the last word of this phrase, the first syllable "bet-" is clearly stressed relative to the second syllable "-ter," yet the beat of the second syllable is stronger than the beat of the first. In order to resolve this, we might shift "bet-" forward to the strong beat that follows it. But this does not solve the problem; "bet-" and "-ter" are

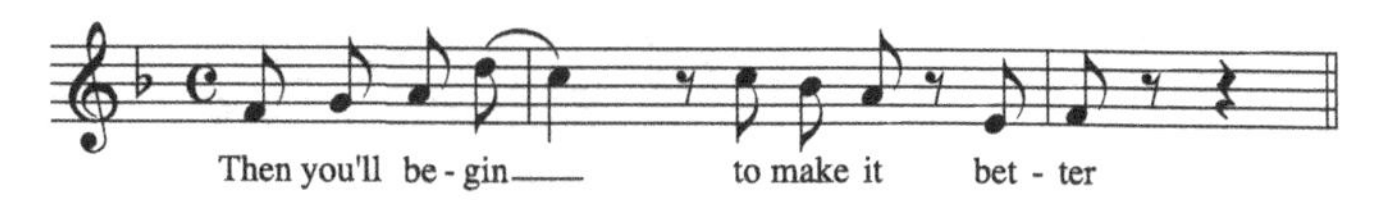
Then you'll be - gin——
to make it
bet - ter

Figure 9.8
The Beatles, "Hey Jude."

I bet - cha wond - er how I knew

Figure 9.9
Marvin Gaye, "I Heard It through the Grapevine."

a.
I - mag - ine there's no heav - en
b.
I - mag - ine there's no heav - en

Figure 9.10
John Lennon, "Imagine," showing metrical shifts.

now equally stressed. Moreover, both syllables now occur on the same beat: a conceptually troubling situation (we return to this point below). Clearly the way to resolve this is by associating "-ter" with the following *weak* beat. This kind of shifting is fairly common in rock, but it rarely occurs a number of times in succession; that is, we rarely encounter a series of syllables in which each syllable is shifted forward to the beat of the following one. The phrase from Marvin Gaye's "I Heard It through The Grapevine" shown in figure 9.9 is an extreme example.

A phrase from "Imagine," by John Lennon, presents another interesting case (figure 9.10a). Here, "heav-" is stressed relative to "no," but falls on a weaker beat; we can resolve this conflict by shifting "heav-" forward to the following beat at the eighth-note level. As the rules stand, there is no reason to shift the second syllable of "heaven." However, shifting the first syllable but not the second would cause the first syllable of "heaven" to fall on a beat *after* the second syllable! The more natural

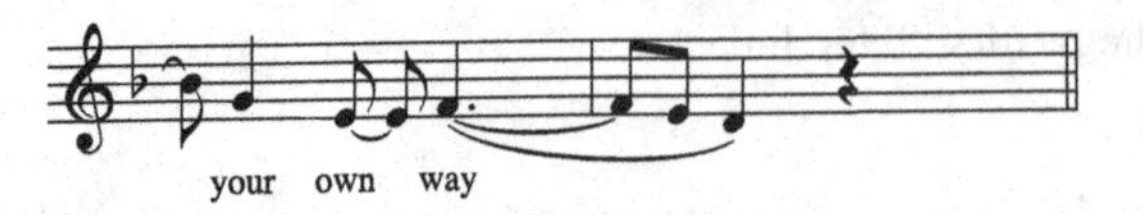

Figure 9.11
Fleetwood Mac, "Go Your Own Way."

solution would be to shift the second syllable of "heaven" forward by an eighth-note as well, as shown in figure 9.10b. In general, it seems counterintuitive to have situations where two events in a melody fall on the same beat in the deep structure, or where the order of events in deep structure differs from their order in surface structure. We can enforce this with a simple added rule:

Deep Representation Ordering Rule. The order of events in a line in the deep representation must be the same as their order in the surface representation.

The shifting of events from strong beats to weak beats (or from one beat to another of equal strength, as in "Imagine"), while not infrequent, is not nearly as common as shifting from weak beats to strong beats. In general, it only seems to occur in situations where the ordering rule would otherwise be violated (as is the case in figures 9.8, 9.9, and 9.10). The relative rarity of "strong-to-weak" shifting is not dictated by any rule, but is simply an emergent feature of the model; from the point of view of the preference rules, there is rarely any motivation for shifting events from strong beats to weak ones.

I mentioned that it is quite possible for events within a piece or phrase to be syncopated at different levels. This raises a further question: do we ever find a single event that is syncopated at more than one level; for example, at both the eighth- and sixteenth-note level? This would mean that the event was shifted forward to a beat three sixteenths after the beat it occurred on. This phenomenon does in fact occur, but only very occasionally. One example is found in the Fleetwood Mac song, "Go Your Own Way," shown in figure 9.11. In our understanding of this passage, I would suggest, each note is first shifted to the following eighth-note beat; then the final note of each phrase is further shifted to the following

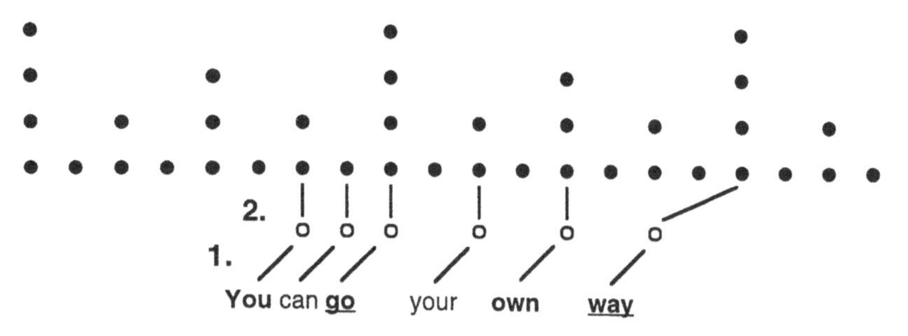

Figure 9.12
"Go Your Own Way," showing two-stage metrical shifts.

A

B

Figure 9.13
(A) The Beatles, "Day Tripper." (B) The Beatles, "Hey Bulldog."

quarter-note beat. (This two-stage shifting process is shown in figure 9.12.) (Another example is the Nirvana song "Come as you Are," on the phrase "And I DON'T have a GUN"; here again, "gun" seems to be syncopated at both the eighth- and quarter-note levels.) Such cases are rare, however, and one feels that they are "stretching the rules" of the style.

While all the examples discussed so far have used vocal music, syncopations may also occur in instrumental lines. Two examples of syncopated instrumental lines are shown in figure 9.13, from the Beatles songs "Day Tripper" and "Hey Bulldog." The motivation for considering these lines as syncopated is essentially the same as in the case of vocal melodies. As they stand, the lines appear to create metrical conflicts between certain events of the melody and the notated meter (implied, as usual, by the accompaniment and by other parts of the melodies themselves). In "Day Tripper," for example, the D at the end of the first measure and the following F♯ are both long notes on relatively weak beats; the event rule

And in my hour of dark - ness she is stand - ing right in front of me

Figure 9.14
The Beatles, "Let it Be," first verse, third line.

would thus favor treating these events as metrically strong. But again, these melodies do not seem incompatible with the notated meter; the logical explanation is that the syncopated events are understood as occurring on the following beat.

Earlier I argued against the view that the function of syncopations in rock is to add metrical tension or conflict. One might ask, then, what functions do they serve? This is a very difficult question, but I wish to offer a few thoughts. One basic attraction of syncopation is that it allows for a great variety of surface rhythms. I mentioned earlier that in common-practice music, certain metrical preference rules tend to be followed to a high degree, particularly the event rule, which prefers events (rather than rests or continuations of events) to occur on strong beats. (When I say that a rule is followed in a style, I mean that the music is usually written so that an analysis can be formed for it which satisfies the rule to a high degree. This idea will be fleshed out more fully in chapter 11.) Of course, the rule is not followed absolutely (so that events are *always* on strong beats and rests *never* are), but it is followed to a much greater degree than in rock. As a result of this, many rhythmic patterns found in rock simply never occur in common-practice music. For example, I would challenge anyone to find a melody with anything like the rhythm of the first line of "Let It Be" in common-practice music. In rock, the preference rules may well apply as strongly as they do in common-practice music, but they apply only to deep representations; the surface rhythm is allowed considerable freedom, due to the syncopation shift rule. The variety of rock rhythm is apparent also in cases where a single melodic line is repeated with different lyrics. In such cases, the syncopations of the melody often vary slightly from one repetition of the line to another (although the deep representation normally stays the same). "Let It Be" is an example of this; compare the line shown in figure 9.2c—the first occurrence of this melodic phrase—with the second occurrence, shown in figure 9.14. In such cases, however, I would argue that the goal is not so much variety, but rather, fitting the melody to the rhythm of the lyrics in the optimal way. Even if two lines have the same stress pattern,

a.
Such a mean old man
Such a mean old man
b.
Such a dir-ty old man
Dir-ty old man

Figure 9.15
The Beatles, "Mean Mr. Mustard." (a) Last line of first verse; (b) last line of second verse.

the most natural rhythms for them may not be the same. In the case of "Let It Be," the word "mother" is naturally sung with the first syllable very short; however, the word "standing"—at the analogous place in the second line—is naturally sung with the first syllable longer. Through syncopation, such variations can be accommodated—in this case, by shifting the syllable "stand" one sixteenth-beat earlier—while preserving the same underlying rhythm for each line.

The basic model of syncopation proposed above is quite a simple one, and many of its uses are quite straightforward. However, syncopation also has the potential for more ambitious and complex uses, and these are sometimes found in rock as well. One such use occurs in cases of metrical shift from one structure to another. If we consider the surface rhythm of a melody such as "Let It Be," we find that in a number of cases stressed events occur three beats apart: for example, "FIND my-SELF." We might think of this as implying a latent triple meter in the song, which may or may not be exploited. In some cases, this latent triple meter *is* exploited and there is an actual move to triple meter; and in cases where this happens, the fact that the triple meter has been anticipated by the syncopated rhythms of the duple meter section makes the transition smoother than it might otherwise be. The Beatles' song "Mean Mr. Mustard" provides an example (figure 9.15). The first verse (the last line of which is shown in figure 9.15) is in duple meter; as usual, the syncopations feature stressed syllables three eighth-notes apart ("MEAN old MAN"). In the second verse, however, the latent triple meter of the vocal line "takes over," as it were, and becomes the underlying meter of the song. The transition to triple meter here seems rather natural and seamless; in part, I think, this is because it has been subtly anticipated by the syncopated rhythms of the melody. (Other elegant examples of duple-triple shift are found in the Beatles' "Two of Us," Led Zeppelin's "The Ocean" and "Over the Hills and Far Away," Pink Floyd's "Have a Cigar," Rush's "Entre Nous," and Soundgarden's "Spoon Man.")

To account for syncopation in rock, I have argued, it is not necessary to posit a much greater degree of metrical conflict in rock, nor is it necessary to introduce fundamentally different preference rules. Rather, rock syncopations can be explained through the introduction of one simple (though rather fundamental) extension of the model proposed earlier: a "deep representation" in which syncopations have been resolved. (The model I propose might also be applicable to jazz and other popular idioms, but I will not explore that here.)

It might seem extravagant to posit two complete input representations of a piece, just for the sake of handling syncopation. Alternatively, one might assume only a single representation, which initially reflects the rhythms of the surface, but is modified in accordance with the syncopation shift rule. However, there is reason to think that the surface representation is more than an ephemeral stage of processing which is lost once syncopation-shifting has taken place. As noted earlier, surface rhythms perform important aesthetic functions, providing rhythmic variety and also playing a role in cases of metrical shift. On the other hand, the "deep representation" is important too; it not only serves to resolve metrical conflicts, but also appears to provide the input to harmonic analysis, as I discuss below. Thus there is reason to suppose that both surface and deep representations are "kept around" and are able to play a role in higher levels of musical processing.[2]

9.4 Harmony in Rock

How well does the harmonic model proposed in chapter 6 apply to rock? It is important to note, first of all, that rock is clearly a harmonic style, one in which harmonic structure is present. A rock piece is composed of a series of harmonies—entities implied by groups of pitches; harmonies in rock are characterized first and foremost by roots, although other information is also important, such as major/minor and triad/seventh.

Many rock songs are handled quite effectively by the rules proposed earlier for common-practice harmony. Consider figure 9.16, the Who's "The Kids are Alright" (first verse), and figure 9.17, Pink Floyd's "Breathe" (first verse). The harmonic structure in both cases seems clear, and is as indicated by the chord symbols shown above the staff. (Rather than just including roots—as the program discussed in chapter 6 would

Figure 9.16
The Who, "The Kids are Alright," first verse.

Figure 9.17
Pink Floyd, "Breathe," first verse.

do—I also include other harmonic information such as mode and extension. Major chords are indicated with capital letters, minor chords with capital letters plus "m.") The harmony in both cases is conveyed most clearly by the bass and rhythm guitar (not shown), which present each chord in a straightforward fashion. If we consider the melodies of these songs, it can be seen that they are compatible with the harmonic structure under the current model, in that all tones are either chord-tones of the current root, or are closely followed stepwise (see HPR 1 and HPR 4). For example, in "The Kids are Alright," the A's and D's in mm. 1–2 are chord-tones of the root D ($\hat{5}$ and $\hat{1}$, respectively); the B is not a chord-tone, but is followed stepwise. In mm. 3–4, the D on "guys," the C♯ on "my" and the A in m. 4 are chord-tones; the other notes are ornamental, and are followed stepwise, as they should be. In "Breathe," too, it can be seen that all notes *not* followed stepwise are chord-tones of the current harmony.

The harmonic structure of both of these songs clearly adheres to the strong beat rule (HPR 2), since harmonic segments are generally heard to begin on strong beats. A problem arises, however. Consider the A at the end of m. 10 in "The Kids are Alright," part of an E chord-span. This note is not a normal chord tone of E. Since it is eventually followed stepwise, it could be considered ornamental; however, it seems more intuitively right to regard it as part of the A harmony in m. 11. Beginning the A chord-span on the last eighth-note of m. 10 would be undesirable due to the strong beat rule. The solution to this problem, as discussed in the previous section, is to regard the A of m. 10 as a syncopated note, which is understood as really "belonging" on the following strong beat, following the syncopation shift rule. Then it is possible to begin the A major segment on the downbeat of m. 11, while still including the A. The same applies to the D in m. 11 and the A in m. 12 (as well as several other events in this example); each of these notes is understood as falling on the following strong beat. This suggests, then, that harmonic structure as well as metrical structure takes as input a "deep representation" in which syncopations have been removed (see section 9.2). The role of the deep representation in harmonic analysis is, of course, a further reason for proposing it. Without it, we would have to conclude that the harmonic rules for rock were fundamentally different from those of common-practice music; with it, the same rules appear to accommodate to both styles. Adjusting the syncopations in "The Kids are Alright" has another benefit as well: it can be seen that, after syncopation-shifting, the melodic events on strong beats are mostly chordal rather than ornamental tones, just as is the case in common-practice music. (Of course, metrically strong ornamental tones do sometimes occur in rock, as they do in the common-practice idiom; an example is the B at the end of m. 3, which is metrically strong after shifting.)

An important question to ask about any harmonic style is what scale degrees (relative to the root) are permitted as chord-tones. As noted in chapter 6, the primary chord-tones in common-practice music are $\hat{1}$, $\hat{5}$, $\hat{3}$, $\flat\hat{3}$, and $\flat\hat{7}$ (HPR 1). However, this is not true of all harmonic styles. In jazz, there is a wider variety of possible chord-tones; $\hat{2}$ (usually known as $\hat{9}$) and $\hat{6}$ (or $\hat{13}$) are widely used, and others as well. In rock, the common chord-tone degrees appear to be more or less the same as in common-practice music. That is, most notes have a relationship of $\hat{1}$, $\hat{5}$, $\hat{3}$, $\flat\hat{3}$ or $\flat\hat{7}$ with the current root, unless they are ornamental (closely followed stepwise). It can be seen that all of the "must-be-chordal" notes (notes not followed stepwise) in "The Kids are Alright" and "Breathe" are of these types. There are cases of added sixths, ninths, and other tones used as chord-tones in rock, but such cases occur occasionally in common-practice music as well.[3]

While a problematic note can always be labeled as an added sixth, fourth, or whatever, there are some cases in rock where this solution is not very satisfactory.[4] Consider the opening of "A Hard Day's Night," shown in figure 9.18a. According to common-practice rules, this melody is somewhat in conflict with the underlying harmony (presented, as usual, by the bass and rhythm guitars). The second D of the melody (on "day's") conflicts with the C major harmony; the D on "work" clashes with F major. Rather than trying to explain these events as chord-tones

Figure 9.18
(A) The Beatles, "A Hard Day's Night." (B) U2, "Sunday Bloody Sunday."

of the current harmony, it might make more sense to characterize the melody as an elaboration of a G7 chord, which proceeds somewhat independently of the changing harmonies in the accompaniment. Another example is seen in figure 9.18b, from U2's "Sunday Bloody Sunday." Here too, rather than explaining the F♯ at the end of the second phrase as a major seventh of the underlying G chord, it is perhaps better to regard the melody as elaborating a B minor harmony over a changing accompaniment.

Other problematic cases arise from rock's use of pentatonicism. Many rock melodies are fundamentally pentatonic, using either the major pentatonic scale shown in figure 9.19a, or the minor one shown in figure 9.19b. Two examples are shown in figure 9.20. These melodies pose difficulties for the current model; while the implied harmony in both cases seems clear, the melodies contain events that conflict with these roots, since they are not chord-tones of the root and are also not followed stepwise. For example, the B's in figure 9.20a are incompatible with a root of D, as are the G's in figure 9.20b. There are two possible ways this situation could be handled. One would be to regard the pen-

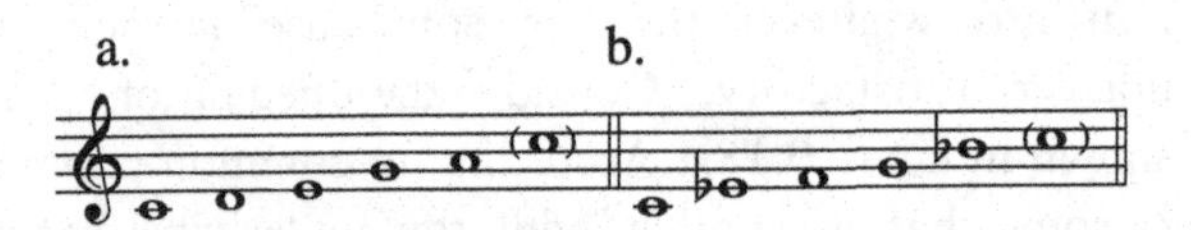

Figure 9.19
(a) The major pentatonic scale; (b) The minor pentatonic scale.

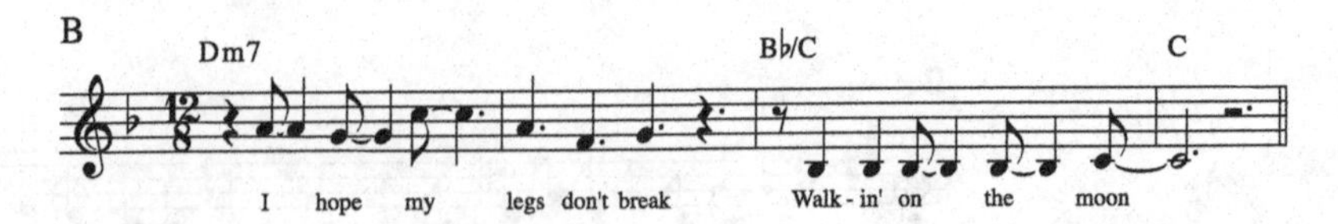

Figure 9.20
(A) Creedence Clearwater Revival, "Proud Mary." (B) The Police, "Walking on the Moon."

tatonic scale as analogous to the diatonic scale in common-practice music, so that stepwise motion on the pentatonic scale is sufficient to qualify an event as a potential ornamental tone; for example, the motion from B to D in figure 9.20a is stepwise on the pentatonic scale, so the B can be regarded as ornamental. However, this does not solve the problem of the second G in figure 9.20b, which is not closely followed stepwise even in pentatonic terms. A more radical solution would be to treat the pentatonic scale as a kind of chord, in which all scale degrees are stable chord tones not requiring resolution; thus the G in figure 9.20b could be treated as a chord-tone of D.

The phenomena of melody-harmony conflict and pentatonic harmony represent problems for the current model, but they arise relatively rarely. On the whole, the basic principles of harmonic structure—that is, the principles governing the harmonic implications of pitch-events—appear largely the same in rock and common-practice music. We have said nothing about the whole issue of progression in rock: the logic whereby one harmony succeeds another. As Moore (1992, 1995) has shown, the rules for progression in rock are clearly quite different from those of common-practice music: in particular, the V-I cadence which serves such a central role in common-practice music is virtually absent in rock. However, the issue of progression is really beyond our purview, just as it was in the case of common-practice music (see section 6.7).

9.5 Modality and Tonicization in Rock

A number of authors have commented on the modal character of much rock music. By saying that rock is modal, we mean that it uses the diatonic scale, but with the tonal center at different positions in the scale than in the customary "major mode" of common-practice music. For example, given the C major scale as the pitch collection, a common-practice piece would typically adopt C as the tonal center.[5] A rock song using the C diatonic scale might adopt C as the tonic as well (thus using the Ionian mode), but it might also adopt G (the Mixolydian mode), D (the Dorian mode), or A (the Aeolian mode). Moore (1993, 49) has noted that the most commonly used modes in rock are precisely these four: the Ionian, Dorian, Mixolydian, and Aeolian. Table 9.1 shows a number of well-known songs which are mainly or entirely in each of these modes. It will be useful to think of modes as represented on the "line of fifths," as shown in figure 9.21.

While many songs remain within a single mode, it is also common—perhaps even normative—for songs to shift freely between the four

Table 9.1
Some well-known rock songs entirely (or almost entirely) in a single mode

Ionian
"I Wanna Hold Your Hand," Beatles
"The Kids Are Alright," The Who
"Turn, Turn, Turn," Byrds
"Bad Moon on the Rise," Creedence Clearwater Revival
"Go Your Own Way," Fleetwood Mac
"My Best Friend's Girl," The Cars
"Start Me Up," Rolling Stones
"Jump," Van Halen
"Hero of the Day," Metallica
"Time of Your Life," Green Day

Mixolydian
"The Last Time," Rolling Stones
"Paperback Writer," Beatles
"So You Wanna Be A Rock 'n' Roll Star," Byrds
"Rebel Rebel," David Bowie
"Lights," Journey
"Hollywood Nights," Bob Seger
"That's Entertainment," The Jam
"Pride (In the Name of Love)," U2
"Elderly Woman Behind the Counter in a Small Town," Pearl Jam
"No Rain," Blind Melon

Dorian
"Somebody to Love," Jefferson Airplane
"Born to be Wild," Steppenwolf
"Battle of Evermore," Led Zeppelin
"Another Brick in the Wall Part II," Pink Floyd
"Girls on Film," Duran Duran
"White Wedding," Billy Idol
"Big Time," Peter Gabriel
"Spoon Man," Soundgarden
"Walking on the Sun," Smash Mouth
"Blue on Black," Kenny Wayne Shepherd

Aeolian
"Welcome to the Machine," Pink Floyd
"One of These Nights," Eagles
"Rhiannon," Fleetwood Mac
"Don't Fear the Reaper," Blue Oyster Cult
"Rain on the Scarecrow," John Mellencamp
"Walking on the Moon," Police
"Sunday Bloody Sunday," U2
"People Are People," Depeche Mode
"Losing My Religion," R.E.M.
"Smells Like Teen Spirit," Nirvana

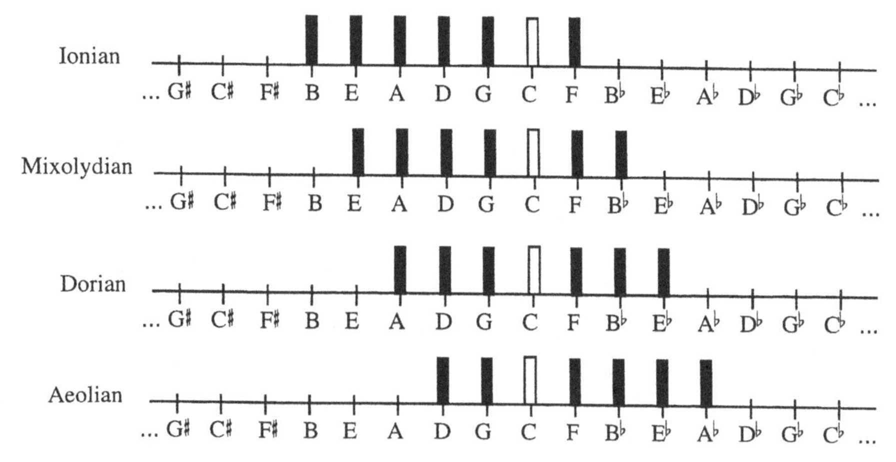

Figure 9.21
The common modes of rock music (assuming C as a tonal center in all cases).

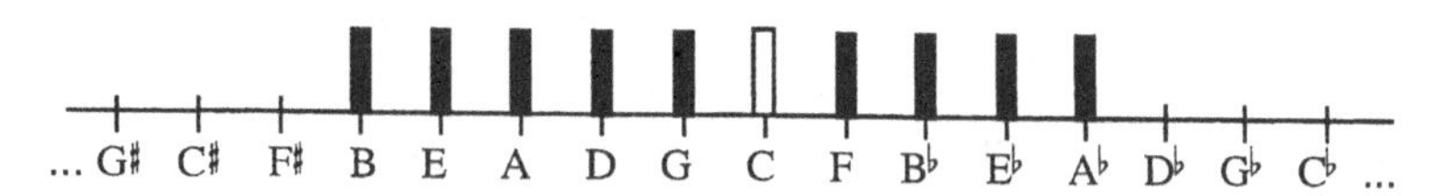

Figure 9.22
A "supermode," formed from the union of the four common rock modes.

common modes. In some cases, a shift in mode will occur between sections: the verse might be in one mode, the chorus in another. For example, in the Beatles' "A Hard Day's Night," the verse is primarily Mixolydian (though with touches of $\flat\hat{3}$ and $\hat{7}$); the bridge ("Everything seems to be right") is entirely Ionian. In Supertramp's "Take the Long Way Home," a Dorian verse is followed by a Ionian bridge. Other songs feature even more fluid motion between harmonies of different modes, so that the presence of any single mode—as a governing pitch collection—becomes questionable. The Rolling Stones' "Jumpin' Jack Flash" features the progression D-A-E-B; Heart's "Barracuda" features an alternation between E and C. It can be seen that no single mode contains all the pitches of these progressions. However, rock songs rather rarely venture beyond the pitches provided by the four common rock modes. That is, the $\flat\hat{6}$ scale degree is common (provided by the Aeolian mode), but the $\flat\hat{2}$ and $\sharp\hat{4}$ are not.[6] We could imagine a kind of "supermode," as shown in figure 9.22, comprising the union of the four common rock modes; such a collection provides the pitch materials for most rock songs.

The approach to key-finding proposed in chapter 7 was based on the key-profile model of Krumhansl and Schmuckler. Under that model, the distribution of pitches in a piece or passage is tallied and compared to an ideal distribution for each key. Let us examine this idea with respect to rock. The aim is simply to develop a satisfactory method for determining the tonal center of a piece.[7] We might first consider a "flat" key-profile, along the lines of that proposed by Longuet-Higgins and Steedman, in which each of the members of the diatonic scale has an equal value (and chromatic pitches have lower values). This clearly will not work for rock, where the pitches compatible with (for example) a tonic of C include not only the C major scale (the Ionian mode), but also those of the F, B♭, and E♭ major scales (the Mixolydian, Dorian, and Aeolian modes centered on C).[8] Another possibility would be to assign high values to all scale degrees in any of the four common rock modes—in effect, treating the "supermode" shown in figure 9.22 as the scale collection. Then, pitches in any of the common rock modes would be recognized as compatible with the corresponding tonic. (Note that each tonal center has a unique supermode: the supermode for C contains the line-of-fifths positions A♭ through B, the supermode for G contains the positions E♭ through F♯, and so on.) This would work for songs which use all the pitches of the supermode. However, many do not; many only use a single diatonic mode, and would therefore be compatible with several different supermodes. Clearly, there is more to tonic-finding in rock than simply monitoring the scale collection in use.

A more promising approach would be to use something like the Krumhansl-Schmuckler profiles (or my modified version, presented in chapter 7), in which the values for degrees of the tonic triad are higher than those of the other diatonic degrees. This captures the intuition that we expect tones of the tonic triad to occur more prominently than other diatonic degrees. This could then be combined with the "supermode" for each tonal center, yielding the "tonic-profile" shown in figure 9.23. The profile has moderately high values for all the pitches within the supermode, reflecting the fact that all of these pitches are compatible with the corresponding tonic. However, the profile has especially high values for members of the major and minor tonic triads ($\hat{1}$, $\hat{5}$, $\hat{3}$, and ♭$\hat{3}$—or, in the current case, C, E, E♭ and G). Consider a song in D Dorian; the pitches of this mode (i.e., the C major diatonic scale) would be compatible with several tonic-profiles (C, G, D, and A), but the members of the D minor triad would hopefully occur more often than others in the scale, and this would favor D over the other tonics.

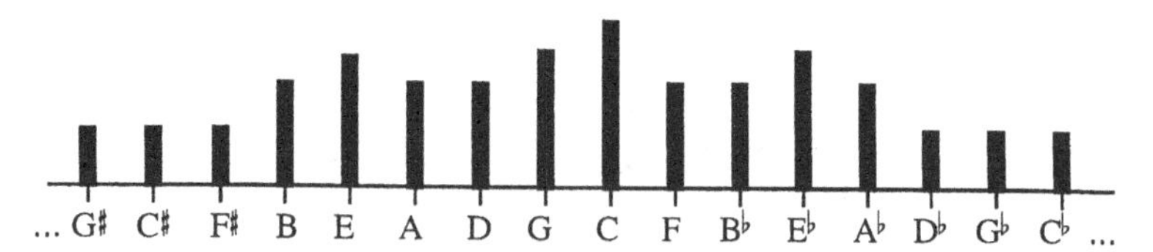

Figure 9.23
A possible tonic-profile for rock (assuming C as a tonal center). The tonic has the highest value; then other degrees of the major and minor triads; then other members of the "supermode"; then all other TPCs.

Figure 9.24
Jefferson Airplane, "Somebody to Love."

Preliminary consideration suggests that this approach to tonic-finding in rock might work well. In many of the modal songs listed in table 9.1, the tonic triad is very prominent both in the harmony and the melody. Indeed, it is presumably this that allows us to identify the tonic out of the several possibilities allowed by the scale collection. Jefferson Airplane's "Somebody to Love" is an example; as indicated in table 9.1, this song is in Dorian mode, using an E diatonic scale but with a tonal center of F♯. While the harmony of the verse involves F♯m-B-E chords, it spends much more time on F♯ minor than on the other two harmonies; the melody, meanwhile, is centered on the first and third degrees of the F♯ minor triad (figure 9.24). (The melody also uses occasional C's, thus moving outside the E diatonic scale.) As another example, the Beatles' "Paperback Writer" remains almost exclusively within the C diatonic scale, but its heavy emphasis of the G major triad in the harmony and melody establishes G as the tonal center (implying Mixolydian mode).

While the tonic-profile approach appears to work in many situations, there are also cases where it does not. Consider the two progressions in figure 9.25. Both are common progressions in rock (allowing for trans-

a.
b.
E A D A
A D A E

Figure 9.25
Two common rock progressions.

F G/F FM7 G/F F G/F FM7 G/F
And though I know all a-bout those men
Still I don't re-mem-ber

Figure 9.26
R.E.O. Speedwagon, "Keep on Loving You."

position, of course). (An example of the first is The Romantics' "That's What I Like About You"; an example of the second is Bruce Springsteen's "Rosalita.") Both involve the same chords, and each chord is present for the same amount of time in both cases. Even the order of the chords is essentially the same, if we imagine the progressions repeating many times. Yet the first progression seems to imply E as the tonic more strongly, the second one A. (Of course the melody often affects the tonal implications as well, but the point here is that even the progressions taken by themselves carry different implications.) The important difference seems to be that, in the first case, E is in the position of greatest metrical strength (assuming a two-measure level of hypermeter, with odd-numbered measures strong), whereas in the second case A carries greater metrical emphasis. In cases such as this, then, metrical strength appears to be an important criterion in tonicization. We could perhaps accommodate this into the tonic-profile model by weighting each pitch event in the input profile according to its metrical strength.

While the importance of the tonic triad in tonicization seems clear, the role of the supermode might seem more questionable. Since each supermode contains 10 pitch-classes, the only effect of the supermode in the tonic-profile is to treat the $\flat\hat{2}$ (or $\sharp\hat{1}$) and $\sharp\hat{4}$ (or $\flat\hat{5}$) scale-degrees, relative to a particular tonic, as incompatible with that tonic—that is, as favoring other tonics instead.[9] For example, the presence of D♭ or F♯ is treated as incompatible with a tonal center of C. This does not mean that a tonal center of C is impossible in a song containing F♯'s, but it does imply that F♯'s are heard as destabilizing to a tonic of C. Is this really true? I believe it is, at least in many cases. Consider figure 9.26, R.E.O. Speedwagon's "Keep on Lovin' You." In some respects, the tonal center most favored here would appear to be F. An F pedal is maintained in the bass; an F harmony occurs on the downbeat of each measure; and the melody is largely centered on F major, beginning and ending on degrees of this triad. Yet to my mind, the sense of F as tonal center is quite tenuous. The reason, I submit, is the strong presence of B, both in the harmony and the melody.[10] As B is $\sharp\hat{4}$ relative to F, hence outside the F supermode, it undercuts F as a tonal center. Given the harmonic strength of F major, however, it is difficult to entertain any other tonal center. The result is that the passage is quite tonally ambiguous.

The progression F-G is not uncommon in rock; other examples include Fleetwood Mac's "Dreams," Walter Egan's "Magnet and Steel," Tom Petty's "Here Comes My Girl," and the Human League's "Don't You Want Me," all of which repeat this pattern extensively. (Again I am allowing for transposition here: in essence, the progression consists of two alternating major triads a whole step apart. For the current discussion I will assume that all of these songs are transposed so that the progression is in fact F-G.) In all of these cases, the presence of B weakens F as a tonal center, despite the harmonic emphasis on F major. It is interesting to note that, in all but one of the songs just mentioned, the F-G progression of the verse is followed in the chorus by the strong establishment of another tonal center rather than F (this tonal center is C in "Keep on Lovin' You," "Here Comes my Girl," and "Magnet and Steel"; it is A in "Don't You Want Me"). This suggests that perhaps the writer of the song did not regard F as the tonal center in the verse. The exception is "Dreams," which maintains its F-G progression—and hence its ambiguous tonality—throughout the song.

10
Meter and Grouping in African Music

10.1 African Rhythm

It is a common belief that there is something special about African rhythm: that it is unusually complex and sophisticated, and plays an unusually central and important role in African music. This view is reflected, also, in scholarly work on African music. In the large body of research in this area, considerable attention has been given to rhythm, including a number of studies which are devoted exclusively to the subject. In this chapter, I will examine African rhythm from the preference rule perspective, with the following questions in mind: How well can African rhythm be reconciled with the model of rhythm proposed earlier (by which I mean, primarily, the models of meter and grouping proposed in chapters 2 and 3)? What similarities and differences emerge between African rhythm (as it is viewed by ethnomusicologists) and Western rhythm? I will argue that the preference rule approach applies with considerable success to African music, bringing out important commonalities between African and Western rhythm and providing a valuable new perspective on African rhythm itself. There are also important differences between African and Western rhythm, but I will argue that these differences, too, are greatly illuminated by the preference rule approach.

As in the previous chapter, my use of preference rule models in what follows will be informal, rather than quantitative. It should be borne in mind, also, that the metrical preference rule system assumed here—at least in its qualitative outlines—is largely drawn from Lerdahl and Jackendoff's *Generative Theory of Tonal Music* (1983) (the grouping theory is also somewhat indebted to *GTTM*, though less so). In light of this, the current chapter can largely be seen as an attempt to apply the rhythmic theory of *GTTM* (with the modifications suggested in previous chapters) to traditional African music. Since I myself am not an expert in African music (though I have considerable familiarity with it), the evidence I will consider is drawn almost entirely from the work of ethnomusicologists: observation, transcription, and analysis of African music and music-making, based in many cases on years of first-hand involvement and participation.

A word is needed about the musical domain under consideration in this chapter. Among the ethnomusicological studies discussed below, very few can truly be considered general studies of African music (Nketia's *The Music of Africa* is a notable exception). Most concern themselves primarily with the music of a single cultural group; the Ewe of Ghana have received particular attention. However, a number of authors extend their conclusions, to some degree, to traditional sub-Saharan African music in general (Jones, Koetting, Pantaleoni, Chernoff, Kauffman, Arom, Agawu); Jones (1959 I, 203–29), in particular, argues at length for the "homogeneity of African music." Other authors limit their claims to a single group (Blacking, Locke, Pressing); but none deny that there are significant musical commonalities across cultural groups in sub-Saharan Africa. This chapter will concern itself with the general features of African rhythm, and will necessarily neglect the many interesting differences between groups and regions in Africa brought out by these studies. The research discussed here also explores a number of different genres. But for the most part, it is concerned with two: songs—often children's songs or work songs—generally sung without accompaniment, or only with clapping; and drum ensemble pieces, performed with an orchestra of percussion instruments as well as singers and dancers.

The models put forth in previous chapters have assumed that music perception begins with a "piano-roll representation," showing the timepoints and pitches of events. In much African music, however, percussion plays a central role. Percussion sounds generally do not have a determinate pitch (though sometimes they do), and are distinguished mainly by timbre—an aspect of sound which is ignored by the pitch-time representation. Given the informal approach of this chapter, it is not necessary for us to develop a rigorous solution to this problem, but it is worth considering how the problem might be solved. One possibility is simply to imagine that the pitch dimension of the input representation—or at least one region of it—is not quantitative (representing pitches), but rather qualitative, representing distinct timbral sounds, as shown in figure 10.1. We could imagine the axis divided into timbral "parts," repre-

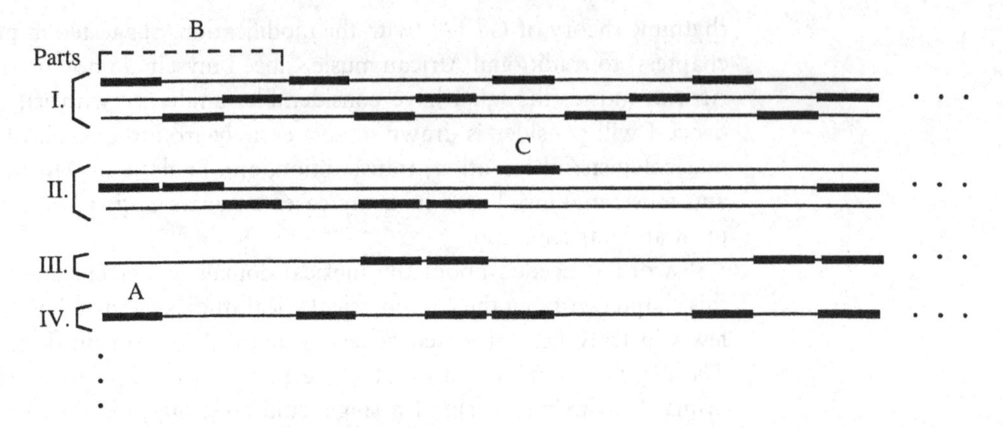

Figure 10.1
A hypothetical example of a "timbral input representation"—a format suitable for representing music in which sounds are distinguished mainly by timbre rather than pitch.

senting different instruments, with some parts perhaps including several different sounds (again, distinguished timbrally). With this modification, the metrical and grouping preference rules (which are our primary concern here) can mostly apply in a straightforward fashion. Strong beats are preferred where there are event-onsets, just as before (MPR 1)—the more event-onsets there are at a point, the better a beat location it is; onsets of louder events are preferred as beat locations over quieter ones (MPR 7). The length rule (MPR 2) can also be applied in a modified form, stating that strong beats are preferred on onsets of events that are not closely followed by another event *within the same part*—for example, events A and C in figure 10.1. (The idea of "registral inter-onset-interval" does not apply, since it assumes the existence of pitch.) The parallelism rule (MPR 9) applies straightforwardly in terms of rhythmic parallelism —that is, there is a preference for the metrical structure to be aligned with repeated rhythmic patterns; pitch parallelism does not apply, but one could imagine a kind of *timbral* parallelism, involving a repeated pattern of sounds within the same instrument (or among different instruments), such as segment B in figure 10.1. In terms of grouping, the phrase structure preference rules—relating to IOIs and OOIs (PSPR 1), the number of events within groups (PSPR 2), and metrical parallelism (PSPR 3)—can be transferred just as they are, applying to events within a timbral "part."

10.2 Meter in African Music

Perhaps the most basic question to ask about African rhythm is this: does African music have meter, as we have understood the term? There is almost unanimous agreement among scholars that it does.[1] An important figure in this regard is Waterman (1952), who suggested that African music involves a "metronome sense," an underlying pulse which is felt but not constantly expressed; this idea has been affirmed by a number of other scholars. Chernoff (1979, 49–50, 96–8) and Locke (1982, 245) cite Waterman's view with approval. Jones (1959 I, 19, 32, 38, 40) speaks on a number of occasions of an underlying regular beat, which is often not explicit but is present in the mind of the performer and can easily be supplied if requested. Agawu (1995, 110) remarks on the "secure metronomic framework" underlying the complex rhythms of the surface (see also pp. 189, 193); Nketia (1963, 65, 86–7) expresses a similar view, using the term "regulative beat." Pantaleoni and Koetting also seem to assume some kind of structure of beats in African music; however, they differ with the other authors cited above as to the nature of this structure, as we will see.

According to the current model, Western music has several levels of beats, each one selecting every second or third beat from the level below; to what extent are these multiple levels of meter present in African music? The comments of these authors about meter, metronome sense, and the like usually relate only to a single level of beats. Locke (1982, 221–2) claims that it is the dotted-quarter beats in Ewe drum music that form the "primary metric accents." Jones (1959 I, 19, 32, 38, 40), Blacking (1967, 157–8), Chernoff (1979, 48), and Pressing (1983b, 5) also seem to have mainly the quarter or dotted-quarter beat in mind. Tempo indications in these authors' transcriptions suggest that the quarter or dotted quarter corresponds to a pulse of 80–170 beats per minute (Jones 1959 II, passim; Blacking 1967, passim; Locke 1982, 221). (Although decisions about how to express durations in terms of note values are, of course, somewhat arbitrary, there seems to be general agreement as to the appropriate durational range for each note value.) Not all authors accept the dotted-quarter pulse. The "regulative beat" proposed by Nketia (1963, 78–8, 85–7, 91) appears to be the half-note or dotted-half-note level; the Ewe bell pattern (figure 10.2), for example, has two regulative beats per cycle. Koetting (1970, 122–3) and Pantaleoni (1972a) maintain, particularly with respect to Ewe dance music, that the eighth-note level constitutes the only regular pulse.[2] In Koetting's words (1970, 120): "The fastest pulse is structurally fundamental, there being no standard substructure internal to it or between it and any pattern as a

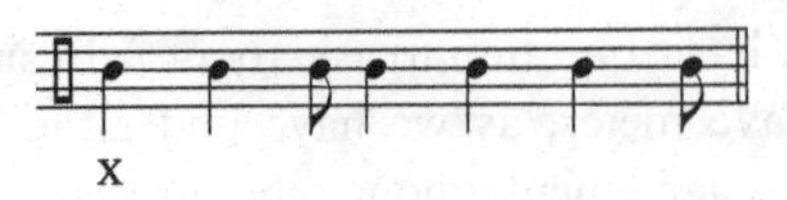

Figure 10.2
The "standard pattern" of Ewe drum ensemble music.

whole ... The fact that the repetitions of the fastest pulse often group themselves into 'gross' pulses or beats is ... incidental." However, this is clearly a minority view: the strongest support is given to the dotted-quarter level. The range of 80–170 beats per minute commonly cited for this level is close (though somewhat faster) to the range for the tactus proposed in chapter 2 for common-practice music: roughly 40 to 150 beats per minute.[3]

There is some consensus, then, that the most salient level of meter in African music is in the same range as the most salient level in Western music. What about levels of meter below the tactus; what evidence is there for these? As noted, Koetting and Pantaleoni argue for the eighth-note level as the *only* level of meter in drum ensemble music. Other authors acknowledge the eighth-note level as forming a level of basic time-unit—sometimes called a "density referent"—from which rhythmic patterns are composed (Jones 1959 I, 24; Nketia 1974, 127, 168–9; Kauffman 1980, 396–7). The reality of lower metrical levels is often reflected in more subtle ways as well. For example, the standard bell pattern of Ewe dance music (sometimes simply called the "standard pattern"), shown in figure 10.2, is often expressed as a series of integer values: 2-2-1-2-2-2-1 (Pressing 1983a, Rahn 1987). As noted in section 2.1, this kind of "quantization" appears to imply the existence of a lower-level pulse.

Regarding higher levels of meter, the picture is less clear, but some evidence can be found. Nketia's comments about the importance of the half-note pulse have already been noted. One clue to ethnomusicologists' views on higher metrical levels—although it should be used with caution—lies in the way their transcriptions are notated. Among the scholars who notate Ewe drum ensemble music—much of which is built on the bell pattern shown in figure 10.2—a 12/8 time signature is generally used; in Western terms, this implies a four-level metrical structure, with beats at the eighth-note, dotted-quarter, dotted-half, and dotted-whole-note levels (figure 10.3). Moreover, there is universal agreement in these transcriptions that the position marked X in figure 10.2 represents the

Figure 10.3
The "standard pattern," showing the metrical structure implied by a 12/8 time signature.

"downbeat"; it is also usually described as the "beginning" of the pattern.[4] The assumption of a "measure level" of meter is also reflected in other references to the "downbeat" or the "main beat" (Chernoff 1979, 56; Agawu 1995, 64). Song transcriptions are usually notated in 6/8, 2/4, or 4/4, again implying at least one level above the tactus. However, the evidence for higher levels is somewhat less conclusive. (We should note that, in general, the evidence for meter in African music points entirely to duple or triple relationships between levels, with every second or third beat being a beat at the next level up; other metric relationships—such as quintuple—seem virtually nonexistent.)

To summarize, there seems to be broad support for a tactus-like level of meter in much African music; there is strong evidence for lower levels, and some evidence for higher levels as well.[5] Despite general support for the idea of meter in African music, some authors show ambivalence on this issue, notably Jones. In his extensive transcriptions of African songs and drum ensemble pieces, Jones often uses different barlines in different lines. In the transcription excerpt shown in figure 10.4, for example, the Gankogui (bell) and Axatse (rattle) pattern are notated with one barring pattern (which is completely regular throughout the piece); this apparently indicates the "underlying beat" Jones describes. However, the parts for the supporting drums (Sogo, Kidi, and Kagan) have a different barring, out of phase with the bell-rattle barring; the clapping pattern has another; and the Atsimevu (solo drum) and vocal line are given their own highly irregular barring. These barrings are mysterious. One might suppose they represent the meter that would be perceived if the individual lines were heard in isolation. This is of interest, since the metrical implications of individual lines presumably contribute to the metrical implications of the texture as a whole—although if Jones and others are correct, the secondary meters of these lines are not usually strong enough to override the underlying meter. But if this is what the barrings represent, it is odd that they are often highly irregular—for example, the bar-

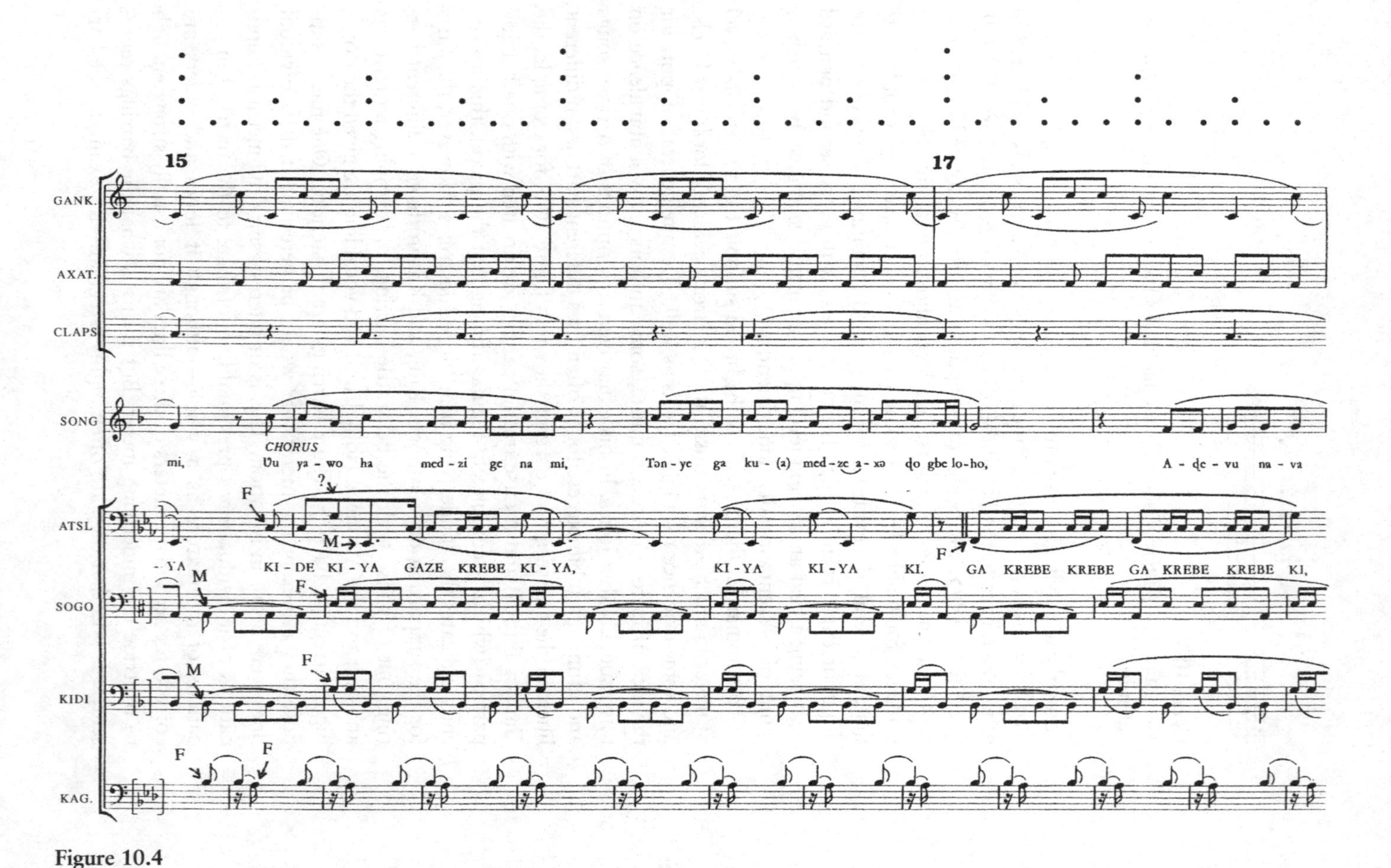

Figure 10.4

Excerpt from the Nyayito dance, transcribed by Jones (1959). The staff lines of the drum parts, as well as showing approximate pitches, indicate different kinds of strokes (see Jones 1959 I, 67). Notes indicating "free" (louder) strokes are marked F; notes indicating "mute" (quieter) strokes are marked M. The underlying 12/8 metrical structure is shown above the staff (added by me). Jones indicates a tempo here of quarter = 113.

ring of the vocal line here, and in many other cases. Another possibility is that what the barlines represent is not meter, but grouping; but this is problematic too, as I will discuss.[6]

10.3 How Is Meter Inferred?

How is meter inferred in African music? While I can find no extended discussion of this issue, a number of comments are made about the way meter is affected by musical cues. Most often, these comments relate to ways in which surface events conflict with or undermine the underlying meter. Agawu on numerous occasions notes things in the music that give rise to "contradictions" or "tension" with the prevailing meter (see Agawu 1995, 64, 68, 110, 192; 1986, 71, 79). Chernoff (1979) notes that a player can vary his part to a certain extent, but not so much that it destroys the beat (see pp. 53–60, especially 58; 98; 121); similarly, Locke and Pressing discuss at length how instruments of the drum ensemble can create conflicts with the main beat. Regarding the specific factors influencing the perception of meter, however, only a few passing remarks are found. Agawu (1995, 64, 68) notes the effect of duration—a long note on a weak beat creates tension. Parallelism is also mentioned: when a repeated pattern occurs, there is a tendency to hear a metrical structure that is aligned with it (Locke 1982, 233). (Locke here is referring to repeated timbral patterns in a percussion ensemble—something which might well be considered a kind of parallelism, as discussed earlier.) Several authors also discuss the relationship between word stress and meter. The consensus here seems to be that, while there is usually some correspondence between stressed syllables and strong beats, there are frequent conflicts between the two as well (Blacking 1967, 165; Nketia 1974, 182–3; Agawu 1995, 192).

While these authors give little discussion to the factors involved in African meter, we can examine this ourselves, using the transcriptions provided (which, we must assume, accurately show the pitches and durations of the music), along with their time signatures and barlines (indicating the perceived meter).[7] Figure 10.5 shows a children's song transcribed by Jones. (As discussed earlier, Jones sometimes uses different barrings in different lines, as is the case here. Let us assume that the time signature and barring of the clapping line represents the underlying meter of the melody. The clapping line here is not used in performance, but was provided on request by an informant.) Four possible metrical structures are shown above the melody. Assuming structure A is the correct one, and is preferred over the others, can this be explained using the prefer-

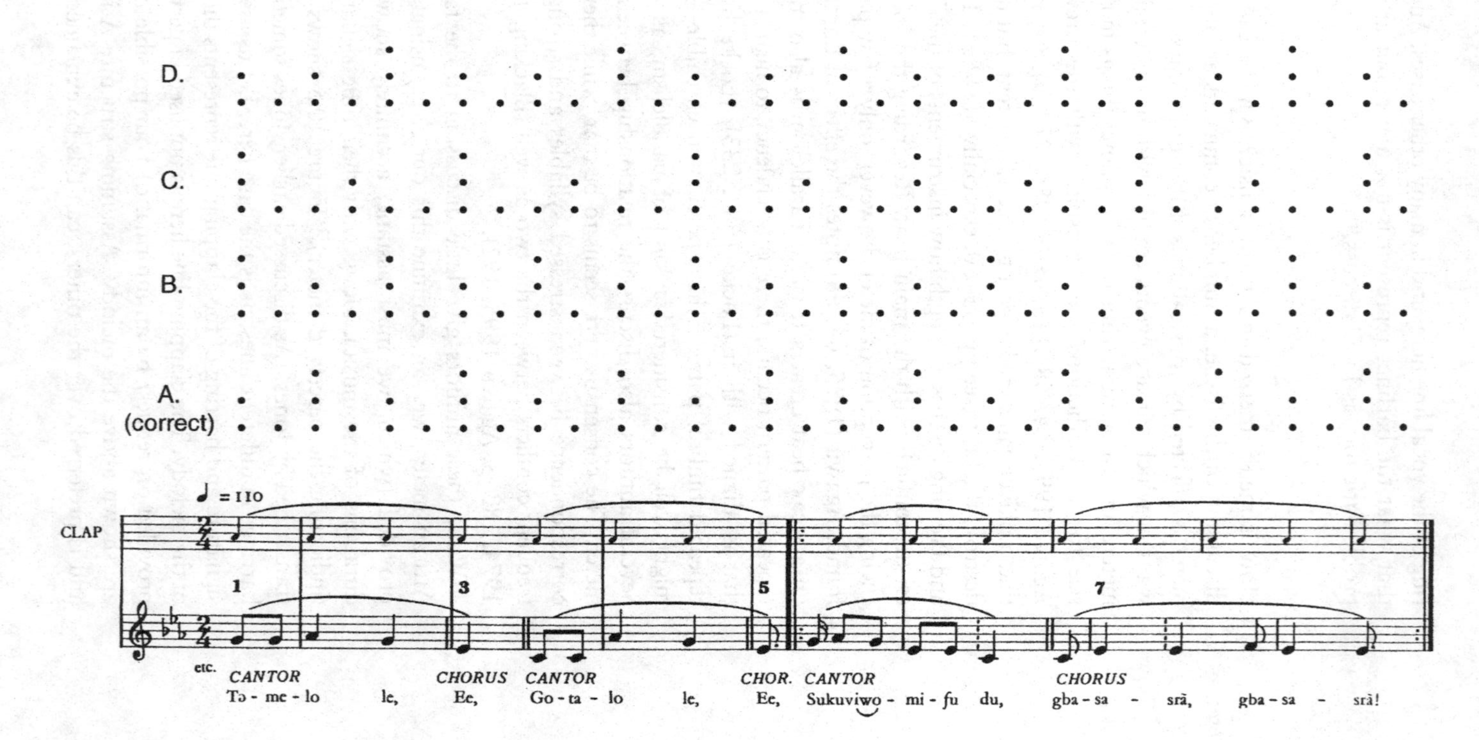

Figure 10.5
Children's Song No. 2 from Jones 1959, showing four alternative metrical structures.

ence rules proposed in chapter 2? Consider just mm. 1–5. By the event rule—preferring structures where strong beats coincide with events—structures A, B, and D are favored over structure C (which would imply a 6/8 hearing), since in structure C, two level 2 beats do not coincide with events (level 1 being the lowest level). The event rule eliminates many other possible structures not shown here. The length rule (preferring structures where longer notes start on strong beats) favors structure A over structure B, since by structure A all the level 3 beats coincide with quarter notes, whereas in structure B some do not. However, structure D is equal to structure A by the length rule; in structure D, too, all the level 3 beats coincide with quarter-notes. The preference for structure A over structure D seems to be due to parallelism; by structure A, the two occurrences of the two-measure pattern at the beginning of the melody are aligned with the meter in the same way. If we proceed to the second half of the melody, we find that the situation is less clear-cut. Measures 7–8 are somewhat in conflict with structure A, since two level 2 beats do not coincide with events (and two quarter-note events fall on very weak beats). But we can assume that the previously established metrical structure is strong enough to persist here, with the conflicting events of mm. 7–8 simply adding a degree of tension and interest. In this case, then, the event rule, the length rule, and parallelism appear to be sufficient to infer the correct metrical structure. (Linguistic stress may also be a factor, but I will not consider this here.)

In drum ensemble textures, the situation is of course more complex. Consider figure 10.4—a fairly typical excerpt from the Ewe Nyayito dance, as transcribed by Jones. (The notation of the percussion parts generally indicates an approximate pitch; in some cases, as indicated, it also carries information about the type of stroke.) The time signature at the beginning, as well as numerous comments by Jones and others, indicates that 12/8 is the underlying meter here; this is shown above the staff. How well does the passage accord with this metrical structure? The clapping pattern clearly implies the dotted-quarter pulse; moreover, the longest clap occurs on the strongest beat of the measure. (As discussed earlier, the "length" of an event is assumed to be the time interval between the event's onset and the onset of the following event in the same part.) The Gankogui (bell) pattern here establishes the 12/8 meter quite strongly (compared with the more common pattern of figure 10.2), since all the dotted-quarter beats are marked by bell notes (and two of them by long bell notes); the Gankogui and Axatse (rattle) both favor a 12/8 periodicity by parallelism. The Sogo and Kidi (supporting drum) patterns clearly favor a 6/8 periodicity, but they accent the third eighth-note of each dotted-quarter rather than the first; the louder "free beats" carry phenomenal accents relative to the "muted beats." (Recall from chapter 2 that a "phenomenal accent" is anything which gives emphasis—and therefore metrical strength—to a point in the music: this could include an event-onset or the onset of a loud or long event.) The Kagan (another supporting drum), similarly, favors a 3/8 periodicity, but avoids the first eighth of each dotted-quarter. The Atsimevu (master drum) does not particularly favor the notated meter; while the repeated "high-low" pattern supports it somewhat by parallelism and by the length rule, the (unusual) 5/8 parallelism towards the end strongly undercuts it. The vocal line seems rather ambiguous metrically; again, linguistic stress may be a factor here (Jones's barlines suggest that it is implying quite a different metrical structure from the notated one). The overall effect will of course depend on the relative prominence of the different instruments, the vocal line and master drum being the most prominent elements here. On the basis of the current model, the passage seems to support the 12/8 meter shown more than anything else, but there are definite elements of conflict.[8] Of course, there might be sections in a piece where the ensemble, on balance, was in conflict with the underlying meter (due perhaps to cross-rhythms in the master drum); but one assumes the underlying meter is supported most of the time, in order for it to be conveyed and maintained.

Parallelism is a factor of particular importance in this passage. Note that each of the levels of the underlying meter is reinforced by parallelisms in one line or another: the dotted-quarter level by the Kagan, the dotted-half-note level by the Sogo and Kidi, and the dotted-whole level by the Gankogui, Axatse, and clapping. (This relates to the earlier discussion of higher levels of meter. If we assume that parallelism is generally a factor in African meter, the pervasive use of parallelisms at the 6/8 and 12/8 level is further evidence that these higher metrical levels are present.) The role of parallelism is complex, however. It was argued in chapter 2 that parallelism relates primarily to *period*, rather than phase; that is, it favors a particular time interval between beats, rather than a particular placing of those beats relative to the music.[9] In terms of phenomenal accents, the supporting drum parts in the Nyayito passage are sharply in conflict with the notated meter, in that they align few accented events with strong beats; but because (through parallelism) they suggest the same period as the notated meter, they give that meter at least partial support.

All in all, the view of African meter in these studies accords well with the current model. There is almost unanimous agreement that something like Western meter is present in African music; and African metrical structures appear to involve several levels in duple or triple relationships. Regarding the criteria for choosing metrical structures, support can be found, in these authors' comments as well as in my own analyses, for several of the factors discussed in chapter 2—in particular, the event rule (MPR 1), the length rule (MPR 2), the stress rule (MPR 7), the linguistic stress rule (MPR 8), and parallelism (MPR 9); and there is no evidence for important determinants of meter which are not covered by these rules. We should remember that the current model is primarily a model of mental representations in the mind of the listener; one might wonder how much basis we have for conclusions about this in the case of African listeners. To a large extent, my conclusions here are based simply on ethnomusicologists' intuitions about how African music should be heard. But it seems fair to assume that these authors—many of whom (unlike myself) have had years of immersion in African music—developed a hearing of the music which was similar to that of Africans themselves. In some cases, other kinds of evidence were sought as well; most notably, both Jones and Blacking asked native informants to indicate the meter of a number of pieces by clapping, sometimes establishing more than one metrical level by this means. (Some of Blacking's results are of particular interest, and will be discussed below.)

10.4 Western and African Meter: A Comparison

The reader may at this point be becoming uneasy. Much has been said about the commonalities between African and Western rhythm, but little has been said about the differences. This is intentional; for one problem with the research discussed above is its exaggeration of the differences between Western and African rhythm. Consider Waterman's comment (1952, 211–12), regarding the "metronome sense":

> The assumption by an African musician that his audience is supplying these fundamental beats permits him to elaborate his rhythms with these as a base, whereas the European tradition requires such close attention to their concrete expressions that rhythmic elaboration is limited for the most part to mere ornament. From the point of view of European music, African music introduces a new rhythmic dimension.

Chernoff goes even further, drawing a series of stark, qualitative contrasts between African and Western listeners (1979):

We begin to "understand" African music by being able to maintain, in our minds or our bodies, an *additional* rhythm to the ones we hear ... In African music, it is the listener or the dancer who has to supply the beat: the listener must be *actively engaged* in making sense of the music (pp. 49–50).

"[T]he Western and African orientations to rhythm," Chernoff concludes, "are almost opposite" (p. 54; see also pp. 40–2, 47–54, 94–7). Even where no explicit comparison is drawn with the West, discussions of "metronome sense," "regulative beat" and the like tend to carry the implication that this is something distinctively African. But as we have seen, the basic framework which these scholars propose for African music—a framework of regular beats, which the events of the music establish but may then conflict with and deviate from in pursuit of musical interest and tension—is, at least in its basic outlines, very similar to that proposed in chapter 2 for Western music. If "metronome sense" merely means the ability to infer and maintain a pulse that is not always directly reinforced by the music—or perhaps sometimes is even in conflict with the music—this is surely a commonplace ability among Western listeners.[10]

This point is worth emphasizing. Waterman's claim that the rhythms of Western music require "concrete expression"—because, he argues, the fundamental beats are not supplied by the listener—suggests that, in general, the metrical strength of each event has to be explicit in the event itself.[11] Yet examples can be found everywhere in common-practice music where the meter of a segment cannot be determined from the segment alone, but must be inferred from the larger context. Consider figure 4.1: there is nothing about mm. 5–6 that inherently suggests triple meter; one might just as easily hear these measures as being in duple meter, with every second quarter-note strong (perhaps preferably, due to the repeating two-quarter-note motive). In figure 8.5, likewise, there is surely no way of knowing from mm. 17–20 that the odd-numbered downbeats are strong relative to even-numbered ones (since there are only rests there); it is only the prior context that establishes this. As an even clearer example, the entire passage in figure 9.1a would surely not be interpreted according to the notated meter if heard in isolation (omitting the first chord); rather, the melody notes on the third beat of mm. 135 and 136 would no doubt be heard as strong. Again, the prior context is crucial in establishing the correct hearing. Each of these passages requires, for its correct interpretation, a structures of beats inferred from the larger context and imposed by the listener—essentially similar to what has been proposed for African music.

If the differences between Western and African rhythmic perception have been exaggerated, it is nevertheless true that there are differences. The studies discussed here give some examples of this. In some cases, the authors note examples where they suggest African and Western judgments of meter would be different; I also find that my own judgments sometimes differ from those of Africans (as indicated by the authors' time signatures and barlines). It is interesting to consider some situations where this occurs. A simple case is found in figure 10.5. According to Jones, this melody is heard with an unvarying quarter-note pulse; for me, however, there is an inclination to hear the final phrase as metrically irregular.[12] Similarly, there are a number of cases in drum ensemble music where I find myself being seduced by the cross-rhythms of the solo drum, losing the dotted-quarter pulse which is supposed to remain primary (this tendency among Western listeners to African music is noted by Locke [1982, 230] and Chernoff [1979, 46–7]). Another difference of a more general nature, although it is harder for us to appreciate experientially, concerns the setting of text. In common-practice Western music (*not* popular music!), the tendency for stressed syllables to be aligned with strong beats is quite strong; deviations from this rule would probably cause considerable metrical confusion. As mentioned earlier, several authors note that this rule is violated quite frequently in African music, so that stressed syllables are set on weak beats; although all seem to agree that there is *some* correspondence between linguistic stress and meter.

How can these differences in perception be explained? All of them have something in common: the Western perception involves shifting the metrical structure in order to better match the phenomenal accents, while the African perception favors maintaining a regular structure even if it means a high degree of syncopation. In preference rule terms, the African mode of perception gives relatively more weight to the regularity rule (MPR 3), and relatively less to the accent rules (those pertaining to some kind of phenomenal accent—specifically MPRs 1, 2, 7, and 8). One might argue that this is nothing more than a different way of expressing Waterman's "metronome sense" idea; and in a way, this is true. But the current formulation—as well as being more precise—views the difference as much less fundamental than it has generally been portrayed by Waterman and others. The principles involved in the two modes of perception are the same—Africans have no kind of "sense" that Western listeners do not have; all that differs is the relative weightings of the different rules.

Another possible difference between African and Western rhythm might be observed, though it is more conjectural; this relates to the strictness of tempo. Recall from chapter 2 that tempo can be viewed in terms of the intervals between beats at the tactus level. In some kinds of common-practice music (Romantic-period music in particular), considerable variation in tempo is employed for expressive purposes—sometimes called "rubato." My impression is that there is very little rubato in African music: in general, an extremely strict tempo is maintained.[13] This is confirmed by Jones, who states that African musicians perform with a much greater precision in timing than Western musicians (even Western musicians attempting to play quite strictly) (1959 I:38). It can be seen that the two differences mentioned here—the higher tolerance for syncopation in African music, and the lower tolerance for rubato—have a kind of complementary relationship; we will return to this in the next chapter.

10.5 Hemiolas and the "Standard Pattern"

A centrally important aspect of African rhythm, noted by a number of authors, is "hemiola": an implied shifting between two different meters, most often 3/4 and 6/8 (Jones 1959 I, 23; Nketia 1974, 127–8, 170; Agawu 1995, 80, 189–93). This is often reflected in a simple alternation between quarters and dotted-quarters, as in figure 10.6; this pattern is a common accompaniment pattern for songs (often expressed in clapping or work-related actions). In many other cases hemiola patterns are present in more elaborated forms. The preference rule view provides an interesting perspective on the rhythm in figure 10.6. Consider just the event rule (which seems to be the only rule expressing a strong preference in this case): what metrical structure does the preference rule model predict? At the tactus level, the pattern is ambiguous between a quarter-note or dotted-quarter-note pulse. At the next level up, however, the pattern clearly implies the dotted-half-note pulse shown as structure C in figure 10.6, since every beat in this pulse coincides with an event. It can be seen that a half-note pulse is not supported in this way (structure A), nor is any other phase of the dotted-half-note pulse (such as structure B). (By contrast, an undifferentiated string of quarters or dotted-quarters would be quite ambiguous in terms of the higher-level pulse it implied.) Similarly, at the lower level, the eighth-note pulse shown as structure F in figure 10.6 is clearly the favored pulse; other alternatives such as a dotted-eighth pulse

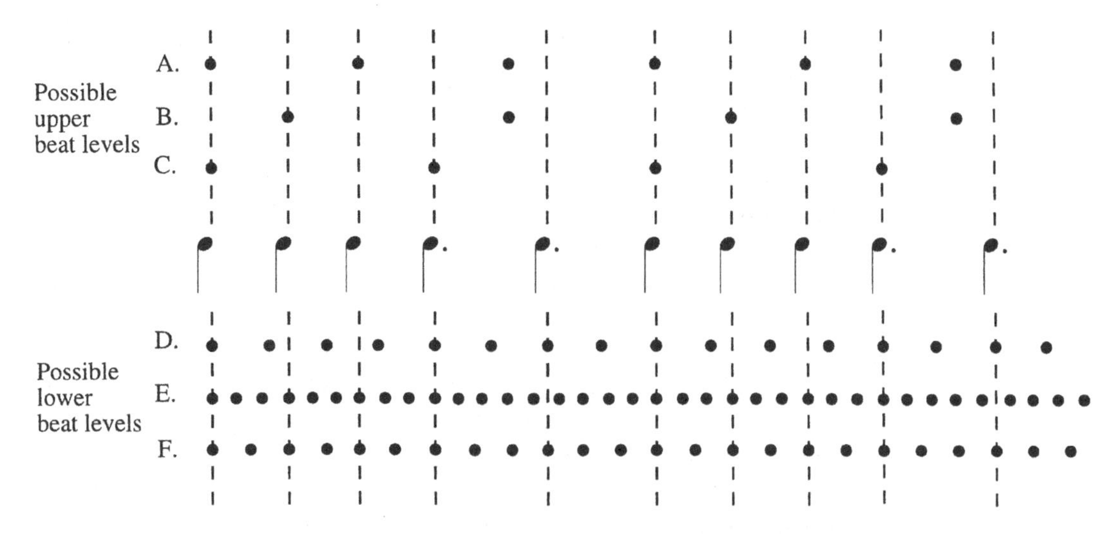

Figure 10.6
The "hemiola" pattern, showing possible beat levels.

(structure D) or a triplet-eighth pulse (structure E) coincide with fewer notes. These favored levels, structure C and structure F, correspond to the measure level and the eighth-note level in the way the pulse is normally notated. In short, this pattern conveys a three-level metrical structure, which is ambiguous in terms of its middle level but quite clear in terms of its upper and lower levels.[14]

Another pattern which deserves special attention is the "standard pattern" of Ewe drum ensemble music, shown in figure 10.2. The pattern is hardly a metrically stable one—relative to the 12/8 meter normally associated with it, or indeed relative to any other regular meter. In fact, as Pressing (1983a, 46–7) has shown, the standard pattern is almost maximally ambiguous, as it samples several different meters (12/8, 6/4, and 3/2)—and different phases of those meters—almost equally. One could argue, of course, that the standard pattern is satisfying precisely because of its metrical ambiguity and instability, but this is hardly an explanation for it. We might also consider whether the preference rule approach provides any insight into the fact that the position marked with an X is usually the downbeat. Of the twelve positions in the standard pattern, the X position is a relatively good candidate for a strong beat, since there is an event there, and a relatively long event (that is, a quarter-note rather than an eighth-note). But this still leaves five possibilities, and the preference rules seem to offer little basis for choosing between them.

It has been noted that the standard pattern is exactly analogous to the diatonic scale (Pressing 1983a; Rahn 1987, 1996). If we consider the pattern as a series of durational values, and we consider the diatonic scale as a series of intervals on the chromatic scale, we arrive at the same pattern in both cases: 2-2-1-2-2-2-1. Moreover, the *centricity* of the pattern is the same in both cases: the "strong beat" position in the Ewe rhythm corresponds to the "tonic" position in the diatonic scale. (I am assuming the "Ionian" mode of the scale here—the major mode in Western tonal music—although of course other modes of the scale are used as well; similarly, other "modes" of the Ewe bell pattern occur in some kinds of African music [Pressing 1983a, 57].) Recent work in pitch-class set theory has shown that the diatonic set has a number of interesting and highly unusual properties (Browne 1981; Clough and Douthett 1991; Cohn 1996). A detailed discussion of this work seems unwarranted here, since it has little connection to the current approach. The importance of this pattern—we might call it the "diatonic pattern" —in both pitch and rhythm suggests that the reasons for its success might, indeed, lie in quite abstract properties. However, it is one thing to show that a pattern has unusual properties, and another thing to show how these properties might explain the pattern's success; so far, the set-theoretical approach has had little to offer towards the latter question.

One property of the diatonic pattern should be mentioned, however, which is of some relevance to the current study: this is its asymmetry. In the diatonic scale, each position in the scale is unique with respect to its intervals with other positions; for example, the tonic is the only diatonic scale degree that lies a half-step above one scale degree and a perfect fourth below another (Browne 1981). Thus it is possible for one to orient oneself to the pattern simply from the intervals between the notes presented. In a whole-tone scale, by contrast, every step of the pattern is intervallically identical. This has possible relevance to rhythm, and particularly to meter-finding. Once it is conventionally established that a certain position in the standard pattern is the "downbeat," then one could orient oneself metrically to whatever is going on simply by locating that position in the rhythm and considering it the downbeat.[15] Whether the standard pattern serves this function for African listeners is unclear. If so, it suggests a factor in African meter perception which is quite unlike the other factors we have been considering here: a *conventional* cue to meter, which relies simply on the listener's knowledge that a certain position in the pattern is conventionally metrically strong. Again, however, there are many asymmetrical twelve-beat patterns, and it remains to be explained why the diatonic one achieved such unusual prominence. For now, the prevalence of the diatonic pattern remains something of a mystery.

10.6 "Syncopation Shift" in African Music

A comparison of Western and African rhythm brings us to the issue of syncopation. If syncopation simply means conflicts between the notes of a piece and the prevailing meter, then it is present—to some degree—in many kinds of music, including much Western music. However, a certain kind of syncopation is found in jazz, rock, and other kinds of recent popular music, in which highly accented events (e.g., long notes, stressed syllables) tend to occur on weak beats immediately before a much stronger beat. I argued in the previous chapter that, rather than being heard as metrical conflicts, these syncopated events are understood as "belonging" on the beat following their actual beat, and therefore can be seen as reinforcing the prevailing meter rather than conflicting with it.

Does this kind of syncopation occur in traditional African music? Waterman (1948) has suggested that it does; beyond this, the studies reviewed here do not address this question. However, some evidence can be found in their transcriptions. If "syncopation shift" were an important aspect of African music, we would expect to find many accented events on weak beats just before strong beats, but not so much just after strong beats. This does, indeed, appear to be a characteristic of some kinds of African music. Consider the melody in figure 10.7. Notice that several long (and therefore phenomenally-accented) events occur on the fourth sixteenth-note of a quarter-note: one sixteenth before the following quarter-note beat (these are marked with arrows). (Notice also the syllable "nee," which occurs on the fourth eighth-note of a half-note.) According to my syncopation shift model, which allows events to be shifted forward by one beat, these events would be understood as belonging on the following beat. Accented events occurring on the fourth sixteenth of a quarter-note are indeed quite common, suggesting that "syncopation shift" may be present in African music. One also occasionally finds accented events on the second sixteenth of a quarter—indeed, there are two examples in this melody (marked with X's), which we would not expect to find under the syncopation shift model; but these are much less common. Figure 10.8, from Jones, shows another kind of evidence for syncopation shift (again, we will assume the upper line represents the underlying meter). Consider the word "A-ba-ye"; according to Jones, the second syllable "ba" is stressed (Jones [1959 I, 32] notes

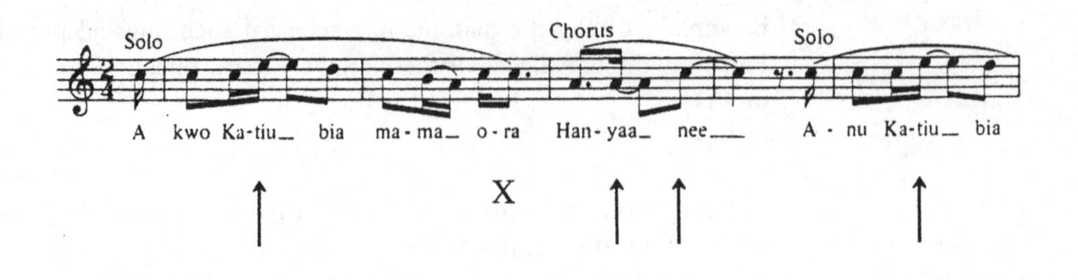

Figure 10.7
A Kasem melody, from Nketia 1974. Arrows indicate long notes on the fourth sixteenth of a quarter (or the fourth eighth of a half), supporting the syncopation shift rule. X's indicate long notes on the second sixteenth of a quarter. Reprinted by permission of W. W. Norton, Inc.

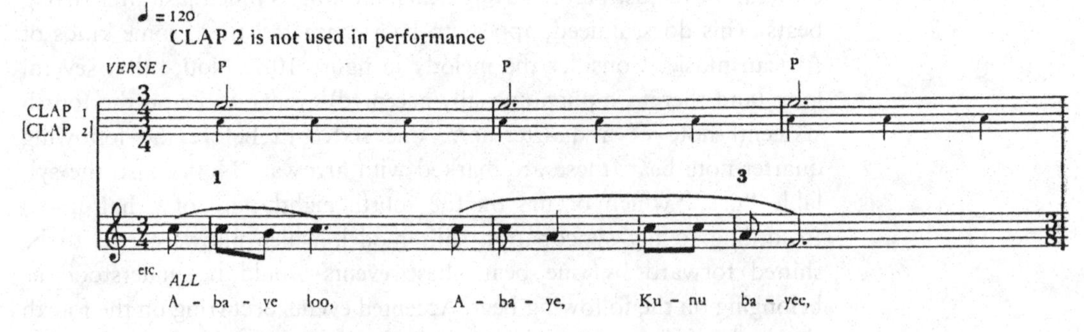

Figure 10.8
Children's song No. 6 (excerpt), from Jones 1959.

also that "ba-yee" is the same word with the first syllable elided). Thus the stressed syllable "ba" is metrically strong in its first two occurrences, metrically weak in the third. But in the third case, the strong syllable occurs just before a strong beat, making the following unstressed syllable metrically strong. (This is a common phenomenon in rock as well; consider figure 9.8 and the accompanying discussion.) Again, by the syncopation shift rule, this word is understood as being shifted over by one beat, and therefore aligned with the meter. Cases like this seem to suggest that some kind of "syncopation shift" is present in African music. If it is, then some of the events in African pieces which initially appear to conflict with the prevailing metrical structure may in fact be reinforcing it. This issue deserves further study. The crucial test might be in the setting of text. In rock, stressed syllables occur commonly on the fourth sixteenth of a quarter, but rarely on the second; one wonders if this is true in African music as well.

A final example presents something of a dilemma. In his studies of Venda children's songs, Blacking (1967, 157–60) notes that, in a number of cases where informants were asked to clap along with melodies (for which there was normally no clapping accompaniment), unexpected results were obtained. Figure 10.9 gives one case. The meter heard by Blacking was that indicated by his barlines; however, when asked to clap, subjects did so at the points marked by crosses, putting the strong beats one eighth-note earlier than expected. My own interpretation agrees with Blacking's; how can we account for the difference? Blacking's explanation is that the Venda's criteria for strong beats differ from ours: they place a higher priority than us on having each beat aligned with an event. (By Blacking's hearing, not all strong beats coincide with events; the fourth downbeat—not marked with a barline for some reason—is "empty.") In fact, the Venda interpretation is not implausible by the current model, since it locates the final long note of each phrase on a strong beat. But the idea that African listeners require events on strong beats more than Western listeners is exactly opposite to our earlier conclusions, and is manifestly untrue. Possibly the Venda simply differ from other Africans.[16] However, there is another possible explanation that we should consider. Perhaps the claps in this case do not coincide with strong beats, but rather occur on the weak beats just before the strong ones. This might be accounted for by the syncopation shift rule, since then each clap is understood as belonging on the following strong beat. In this case the claps reinforce the strong beats felt by Blacking rather than conflicting with them. Or—even if this is not a case of syncopation shift—it might

♩. = 100–112
1. Ná – ndí Mù – nzhē – dzī hà – eè – to!
2. To – 'Ḓá – ní ngè – nó rí tà – mbè – to!
6. To – Ná – ndí khwà – lí dzí dzù–ndè–nì – to!
3. To – Nṋé thí tâ mbí nà dī – thù – to,
4. To – Ḓì – thù ḽí ná má – bê – sú – to.
5. To – Ná – ndí Ṋé – tshí – vhú–ngū–lù– lù – to!

Figure 10.9
"Nandi Munzhedzi Haee" (excerpt), from Blacking 1967. Reprinted by permission of the University of Witwatersrand Press.

simply be that the claps do not coincide with strong beats. We should be cautious about admitting this possibility, since a major source of our evidence for meter rests on the assumption that informants' clapping—especially elicited clapping that is not usually done—is an indication of (some level of) the meter they perceive. Still, it is quite clear that clapping patterns do not *always* indicate strong beats. In some songs, clapping patterns are irregular (see Jones 1959 II, 2); in others, a single clapping pattern may be aligned with the song in different ways (Jones 1959 I, 17; Agawu 1995, 67–8); in some drum ensemble pieces, several simultaneous and conflicting clapping patterns may be used, despite the general assumption that only one primary meter is present (Jones 1959 I, 116). Allowing the possibility that elicited clapping patterns do not always indicate listeners' perceptions of meter adds a major complication to the empirical study of meter; but it is probably something we should consider.

10.7 Grouping Structure in African Music

It was argued in chapter 3 that grouping structure is best regarded as a segmentational structure, independent of meter. A model of melodic grouping was proposed, which derives a single level of groups for a monophonic input. This model involves three simple rules. First, there is a preference for grouping boundaries after long notes and rests (PSPR 1). Secondly, there is a preference for groups of a certain length, roughly eight notes (PSPR 2). Finally, there is a preference for beginning successive groups at parallel points in the metrical structure (PSPR 3). We should also recall a further rule relating to the interaction of meter and grouping; there is a preference to locate the strongest beat in a group near the beginning of the group (MPR 4).

To what extent does this model of grouping apply to African music? Here, we encounter some serious problems of terminology and notation. Several ethnomusicologists confuse grouping and meter, saying that barlines represent the grouping of notes. Blacking (1967, 35) says that "[b]arlines generally mark off the main phrases, and half-bars give some indication of the stresses and the grouping of the notes." Since Blacking clearly indicates elsewhere that his barlines represent meter (p. 162), he would seem to be equating grouping with meter. Agawu (1995), too, claims that his barlines represent grouping (pp. 71, 188, 200; see my note 6). The confusion between meter and grouping is also apparent in Jones, not so much in his comments but in the transcriptions themselves. As noted above, the fact that Jones uses differing barlines in different lines of the ensemble can sometimes be explained by taking the barlines as indicators of grouping, not meter. Note the barring of the clapping line in figure 10.4, for instance. In many cases, however, Jones's barlines do not appear to be aligned with plausible groups; for example, a barline is often placed after the first note of an apparent phrase. (It seems more likely that Jones's *slurs* indicate grouping, as they often do in Western music; and these are often not aligned with the barring.) I suspect that Jones's barlines indicate sometimes grouping and sometimes metrical implications of individual lines; this is unfortunate, since it is often not obvious how to interpret them.

These confusions between grouping and meter are especially strange, since a number of authors—including some of those just cited—observe that grouping is often *not* aligned with meter in African music. Agawu (1995, 64) comments that "in songs, as in drum ensemble music, phrases rarely begin on downbeats" (see also pp. 66, 110). Nketia (1963, 88), describing the phrasing of the vocal line in Ghanaian music, says: "It does not appear to follow any definite rule, though there is a marked

Figure 10.10
Unidentified Ghanaian song, from Nketia 1963. Reprinted by permission of Northwestern University Press.

preference for phrases which begin before and after the main beats of a gong phrase—that is off the regulative beat." Other remarks about the non-alignment of meter and grouping are cited below. In part, the problem may simply be one of terminology: what I have been calling meter, ethnomusicologists (sometimes) call grouping; what I have called groups, ethnomusicologists call phrases. There is somewhat more to it than that, however; the ethnomusicologists' terminology implies that meter involves a "grouping" of events, which I would argue is mistaken.

Aside from these general observations, what specific evidence do these authors provide of grouping structures in African music? The best source here is the phrasing slurs used in transcriptions, notably those of Jones, Nketia, and Pressing. It seems reasonable to take these as indications of grouping, as they generally are in Western music. Consider figure 10.10, an unidentified Ghanaian song from Nketia (1963, 90). The grouping here corresponds well with my own intuition about the grouping of this melody, and appears to be accounted for well by the current model. The main criterion involved appears to be a preference for long notes or rests at the ends of groups (PSPR 1). Inspection of the phrasing in Nketia and

Figure 10.11
From Pressing 1983b. A pattern used by the Kidi drum in the Ewe *Agbadza* dance. Slurs indicate the phrasing given by Pressing.

Jones's examples suggests that this is the predominant factor in African melody and ensemble music, as it is in Western music as well (at least at low levels). As for the phrase length rule, the average length of phrases in this melody is 5.9 notes, not far from the norm of 8 notes assumed in chapter 3 (PSPR 2). There is some evidence for parallelism as well. Pressing (1983b, 9) argues that the phrasing in drum ensemble music tends to be influenced by repeating patterns; in figure 10.11, for instance, the repeated five-note pattern encourages a grouping structure which is aligned with it. As noted in chapter 3, the current model can account for such effects indirectly: the parallelism encourages a metrical structure of the same period (MPR 9); this then encourages a grouping structure of the same period as well (PSPR 3).[17] Occasional evidence can be seen for other principles in grouping, such as similarity. In cases where a drum pattern involves alternating runs of two strokes (A-A-A-B-B-B-B-B, for example), there is a tendency to group them accordingly, placing a group boundary at the change of stroke (Jones 1959 II, 52, 73–4; Pressing 1983b, 9). (Timbral similarity as a factor in grouping is discussed in *GTTM*; in principle, it could be incorporated into the current model along with the "timbral" input representation proposed at the beginning of this chapter.)

Perhaps it should not surprise us that the grouping criteria for African and Western music are similar; as discussed in chapter 3, they reflect "Gestalt" principles of similarity and proximity which are known to apply to perception in general. However, there is evidence for one important difference between Western and African grouping. As noted, there is a tendency in Western music to prefer strong beats near the beginning of groups (MPR 4). In African music, however, several authors claim that the reverse is true: the norm is for strong beats to occur at the ends of groups. Chernoff (1979, 56) observes that, in African music, "the main beat comes at the end of a dynamic phrase and not at the beginning"; Jones (1959 I, 41, 84, 86, 124) notes this tendency as well. (Nketia's and Agawu's comments about the tendency of phrases not to *begin* on strong beats have already been cited.)[18] Inspection of the phrasing in Jones's and Nketia's transcriptions gives further support to these observations; although there are many exceptions, there is a common tendency for both vocal and instrumental phrases to end on strong beats. Figure 10.10 is representative: if we assume a two-measure level of meter with odd-numbered measures strong (this is supported by the clapping pattern), we find that many phrases end just after two-measure beats, although not all do. This suggests a further modification to the preference rules for African music: there is a preference for strong beats near the end of groups rather than near the beginning.[19]

10.8 Conclusions

I have argued here that—based on ethnomusicologists' analyses of African rhythm—African and Western rhythm are profoundly similar. The fundamental cognitive structures involved are the same, and the criteria involved in forming them are largely the same as well. I have also pointed to some differences: the differing weight given to the regularity rule versus the accent rules, and the difference in the preferred alignment between meter and grouping. These differences are certainly important, and they may often lead to quite different musical perceptions and experiences between African and Western listeners. But to my mind they are less striking than the underlying similarities between African and Western rhythm. Admittedly, there are other aspects to rhythm that we have not considered here, most notably motivic structure: the network of rhythmic patterns in a piece that are heard as similar or related. A comparison of Western and African music in this regard is beyond our scope (though I will treat motivic structure briefly in chapter 12).[20]

In closing, I wish to address a few criticisms that might be made of the preceding discussion. One might object, first of all, to my reliance on the conclusions of ethnomusicologists, rather than on my own experience of African music. This should not be seen as a denial of the importance of first-hand ethnomusicological research. On the contrary, it is an acknowledgment that such research provides a kind of musical insight and understanding that cannot be attained in any other way—from transcriptions or recordings, for example. I do assume that, by careful and informed reading of these experts' writings, one can to some extent share in the knowledge and understanding that they have gained; surely this is whole point of ethnomusicological writing. It is true that adding another stage in the interpretive process—from informant to ethnomusicologist to music theorist to reader—increases the probability of misunderstanding or distortion somewhere along the way. But if we are to have any real collaboration between the disciplines, we must allow this kind of division of labor to some extent.

A second possible criticism is that the current approach imposes a framework on African rhythm which has no direct support in what African listeners and performers *say* they are doing; Jones's and Chernoff's informants, for example, make no mention of anything like preference rules. That is to say, the framework proposed here is much more "-etic" than "-emic." While this criticism deserves consideration, there is an important point to be made in response. The current theory of meter purports to describe the hearing of the "experienced listener" to Western art music; this includes many people (both of the past and present) with no formal musical training or knowledge of music theory. Certainly few people of this kind would spontanenously describe their hearing of meter in terms of preference rules (without having studied the theory), any more than an African listener would. In this sense, the current theory is as "-etic" for Western listeners as it is for African listeners. Indeed, the whole premise of music cognition—and for that matter cognitive science in general—is that there are processes and structures involved in cognition to which we do not have direct introspective access, but whose reality can be established by other means. To what extent this assumption is valid is an open question; but I see nothing *ethnocentric* about taking this attitude with African listeners, since we take the same attitude with Western listeners as well.

A final criticism one might make of the current discussion is that it is extremely myopic. I have said a great deal about meter and grouping, but very little about their role in music, or the role of music in African society. It is certainly true that to obtain a full understanding of African music—or indeed any other music—one must consider its larger cultural context. In particular, an analysis of meter and grouping—in African music, Western music, or any other kind—tells us nothing, in itself, about why these things are important and how they contribute to the value of music. This is a perfectly valid criticism, one that applies as much to the rest of this book as to the current chapter. I strongly believe it is worthwhile to focus on the "purely musical" aspects of music (and the content of this book should give some idea of what I mean by this). However, the preference rule approach does in fact shed some interesting light on the higher-level "meanings" of music, in common-practice music and perhaps in African music as well. I will return to this issue in chapter 12.

11
Style, Composition, and Performance

11.1 The Study of Generative Processes in Music

A frequent complaint about research in music cognition is that it tends to be excessively focused on perception and listening, at the exclusion of other musical activities, notably performance, improvisation, and composition. Sloboda (1988, v) notes that, while a journal called *Music Perception* has been in existence for some time, a journal called *Music Performance* would hardly have enough material to sustain it. Sloboda cites several reasons for this bias, including the much greater number of music consumers as opposed to producers in Western society, and the problems of measurement and control that arise in experiments on generative musical processes. This imbalance has been rectified to some extent in the ten years since Sloboda's comments, particularly in the area of performance.[1] Still, it remains true that the bulk of work in music cognition is focused on listening. This study, so far, has only added to the preponderance of research concerned with perception at the expense of other activities. However, the preference rule approach can also be fruitfully applied to generative aspects of music; in this chapter I will explore how this might be done.

While music-psychological work may focus on the perception, performance, or creation of music, it is important to note that many aspects of music cognition may be common to all of these activities. An analogy with language is appropriate. Linguists assume that the basic structures of language—phonemes, syntactic constituents and trees, and semantic structures of whatever kind—are important to both producing and perceiving language. The point of linguistic communication, after all, is to convey some kind of information from the mind of the producer to the mind of the perceiver. Broadly speaking, this assumption seems warranted in the case of music as well. If we assume that certain mental representations—preference rule systems, in this case—are present in the minds of listeners to a certain style of music, it seems reasonable to suppose that they are also present in the minds of composers and performers. However, this reasoning leads us to a number of questions. If preference rule systems are indeed involved in generative processes, what insight do they provide into how these processes work? How do preference rule systems affect performance and composition, and how are they reflected in the products of these activities? Related to that, how might claims about the role of preference rule systems in generative processes be tested in a rigorous way? These are the questions to be addressed in this chapter, with regard to composition first and then performance.

11.2 Describing Musical Styles and Compositional Practice

The reader may have noted a certain sleight-of-hand in the argument of the book so far. I have generally characterized preference rule systems as models of people's mental representations of music as they listen to it—models, that is, of the *perception* of music. At the same time, however, I have freely characterized preference rule systems as applying—or sometimes as not applying—to actual bodies of music and musical styles. It was stated in the opening pages of the book, for example, that the preference rule systems proposed in part I are designed to "apply" to common-practice music, Western art music of the eighteenth and nineteenth centuries. In chapters 9 and 10, I investigated the applicability of some of these preference rule systems to rock and traditional African music, and proposed modified rule systems to accommodate these kinds of music. But again, preference rule systems are models of the *perception* of music, not of music itself. To use preference rule systems to describe actual *music* is to make a leap of reasoning which, at the very least, requires examination.

What does it mean, exactly, to say that a preference rule system applies to a piece or style of music? There is a certain intuitive rightness to saying that a preference rule system applies to some styles more than others. It seems reasonable to say, for example, that the harmonic preference rule system proposed in chapter 6 is not applicable to atonal music, or that the metrical rule system presented in chapter 2 does not apply to Gregorian chant and other music without meter. Perhaps the first impulse is to say, "A preference rule system applies to a style of music if music in the style can be analyzed with the preference rule system." But in principle, any piece of music can be analyzed with any preference rule system. We need some other way of defining what it means for a preference rule system to apply to a style.

One solution to this problem lies in the numerical scores outputted by preference rule systems. As noted in chapter 8, in analyzing a piece, a preference rule system not only produces a preferred analysis; it also produces a numerical score (and also numerical scores for individual segments of the analysis), indicating how well that preferred analysis satisfies the rules. I suggested earlier that these scores could be used to indicate the ambiguity of a passage. In cases where one analysis scores much higher than any other, the passage is fairly unambiguous; in cases where two or more analyses are roughly equally preferred, we can take this to represent an ambiguous judgment. We can also consider the actual magnitude of the score for the preferred analysis (or analyses). In some cases, one or more analyses may be found that satisfy all the rules relatively well; in other cases, none of the possible analyses may be very high-scoring. We can take the score of the preferred analysis as a measure of how well *the piece itself* scores on the preference rule system—or, in other words, as a measure of how well the preference rule system applies to the piece.

Consider a simple example. It was suggested in chapter 9 that the syncopations of rock music cause frequent violations of the metrical preference rules proposed for common-practice music. This is the kind of "sleight-of-hand" I was talking about earlier; if we take a preference rule system only as a model of perception, it is not obvious what it means to say that a piece violates the preference rules. Now, however, we are in a position to flesh out this statement. Consider figure 11.1 (a Beatles melody discussed in chapter 9), and suppose the common-practice metrical system of chapter 2 were applied to it (let us consider just the melody for now). The system searches for an analysis in which beats are roughly equally spaced (MPR 3), and strong beats are aligned with accented events—the primary sources of accent in this case being longer notes (MPR 2) and stressed syllables of text (MPR 8). In most common-practice melodies,

When I find my-self in times of troub-le Moth-er Mar - y comes to me

Figure 11.1
The Beatles, "Let it Be."

such an analysis is available. In figure 11.1, however, there simply is no analysis which satisfies these criteria. The analysis implied by the notation preserves regularity, but aligns several accented events with weak beats (see figure 9.3c). Alternatively, we could align *every* accented event with a strong beat, but this would sacrifice regularity. Thus the best analysis of the passage will receive a poor score from either the regularity rule or the accent rules, and thus a poor score overall (since the total score for the analysis sums the scores from all the rules). Now, however, suppose we apply the modified "rock" version of the preference rules, including the syncopation shift rule, which allows us to shift events to the right by one beat (as shown in figure 9.6). Now, a metrical structure is available which satisfies both the regularity rule and the accent rules, thus achieving a high score. In this way, the fact that the "rock" version of the rule system is more applicable to the melody than the "common-practice" version is reflected in the melody itself.

A similar line of reasoning applies to the case of meter in traditional African music. As I argued in chapter 10, one main difference between the metrical rules for common-practice Western and traditional African music seems to be that, in traditional African music, greater weight is attached to the regularity rule and less to the accent rules. Is this reflected in the two kinds of music in any way? If we consider an excerpt from an African piece—figure 10.5, for example—we find, again, a high degree of syncopation (though it is rather different from the kind of syncopation found in rock). For an African listener, the preferred analysis maintains a high degree of regularity; there is considerable violation of the accent rules (since not all accented events fall on strong beats), but since these rules carry less weight, the violations do not greatly reduce the total score. For a Western listener, however, the greater weight of the accent rules would cause large penalties for this analysis, leading to a low overall score. (A Western listener might prefer an analysis that satisfied the accent rules better, perhaps inferring a highly irregular tactus level. Such an analysis would of course receive penalties under the regularity rule; moreover, such an incorrect analysis would no doubt be severely penalized in other ways—for example, it would not permit "good" analyses to be found for other levels which maintained a regular relationship with the tactus.)[2]

By this approach, it could be shown quantitatively (though I will not attempt this here) that a rock or African piece violates the metrical rules of common-practice Western music—by analogy with language, we might say that it is "ungrammatical" according to those rules. One might wonder, is a common-practice piece "ungrammatical" by the rules of rock or African music? It is not obviously true that this would be the case. It might be, rather, that *both* African music and common-practice music were considered rhythmically grammatical to the African listener. However, consider the observation—noted in chapter 10—that common-practice music involves considerably more variation in tempo than African music.[3] The kind of rubato routinely used in performances of Classical and Romantic music is rarely found in African music. Jones notes that even Western musicians attempting to play in strict tempo seem inconsistent and wavering to African listeners. Rock, too, is usually performed with an extremely strict tempo. It seems, then, that common-practice music allows a greater degree of violation of the regularity rule than rock or African music does. Given the greater weight assigned by African listeners to the regularity rule, music with substantial rubato would presumably give rise to low-scoring analyses for such listeners; it might also lead to mishearings, in which tempo fluctuations were heard as syncopations.

One more example should be mentioned, which has so far been discussed only briefly. In chapter 9 I suggested that jazz harmony has rather different rules from rock and common-practice harmony. In particular, jazz allows a greater variety of chord-tones than either rock or common-practice music does. On the other hand, there is much less use of chordal inversion in jazz than in common-practice music; jazz chords are overwhelmingly in root position.[4] Presumably, these stylistic differences are reflected in the harmonic preference rules used by jazz and common-practice listeners: the compatibility rule for jazz allows more chord tones as compatible with a given root, but jazz also has a rule preferring the bass note as root, whereas common-practice music does not. Consider the chord shown in figure 11.2a. A common-practice listener would most

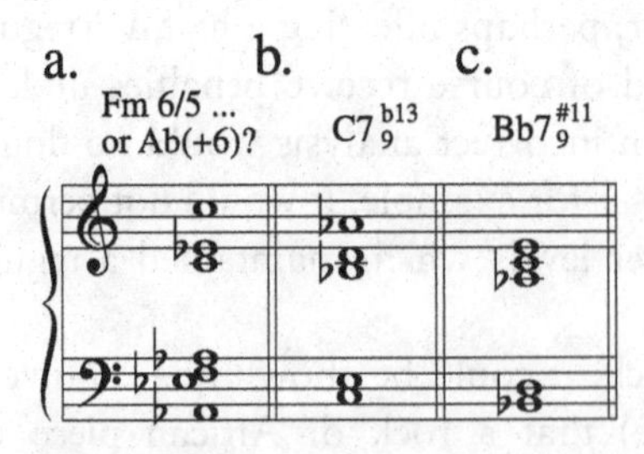

Figure 11.2

likely interpret this as an Fm 6/5 chord (probably leading to a V-I cadence in E♭ major); to a jazz listener, however, it would more likely be heard as A♭ with an added sixth, since added sixths are commonplace and first-inversion chords are rare. As with rhythm, this difference between jazz and common-practice harmony is reflected not only in perception, but also in jazz and common-practice *music.* A common-practice piece analyzed by jazz rules would score poorly, due to the frequent use of chordal inversion, while a jazz piece would score poorly by common-practice rules due to the use of "illegal" chord-tones.

In this case too, then, it can be seen that the harmonic differences between the jazz and common-practice rule systems are complementary. The wider palate of possible chord-tones in jazz is counterbalanced by a greater tolerance for inversion in common-practice music. One wonders why these complementary effects occur in different styles. It seems as if, for a given aspect of structure, there is some range of "total rule violation" which is considered acceptable in a given style (a difficult concept to quantify, and I will not attempt to do so here), and that this range tends to be rather consistent across styles. Relative freedom in one rule (for example, the allowance of syncopation in African music) tends to be balanced by relative strictness in another (the low tolerance for rubato). One possible reason for this is that if rule violation is too great, the correct analysis cannot be reliably inferred. For example, consider the chord C-E-B♭-D-A♭ (figure 11.2b). Under a jazz interpretation, root position is assumed and the root of the chord is C; D and A♭ are acceptable chord-tones ($\hat{9}$ and $\flat\hat{13}$, respectively) under the expanded vocabulary of jazz. However, if inversion were tolerated in addition, so that the lowest note could not be assumed to be the root, this chord could be mistaken for an inverted B♭9♯11 (see figure 11.2c). That is to say, there would be a danger of rampant harmonic ambiguity. A similar point could be made about rhythm; intuitively, it can be seen that if one allows great syncopation as well as great fluctuation in tempo (imagine figure 11.1 performed as if it were a Schumann lied), it could become quite impossible to find the beat.[5]

While we have focused so far on differences between styles, we should note that in each of the cases we have considered—harmony in common-practice music and jazz, meter in common-practice music, rock, and African music—there are profound commonalities as well. These commonalities across styles are reflected in their preference rule systems, just as the differences are. For example, the fact that non-chord-tones in both rock and common-practice music are usually followed stepwise is an important fact about the styles themselves, not just about their perception. On the other hand, there are also cases where a certain kind of rule system seems to be totally inapplicable to a certain kind of music. The key-finding system proposed earlier, for example, is clearly inapplicable to serial music (meaning that a serial piece would not generally yield a key analysis with relatively few modulations in which the key of each section was compatible with the pitches); moreover, it does not appear that *any* kind of key system is really relevant to such music (at least, none along the lines of the one proposed earlier).[6] Similarly, a non-metrical kind of music such as recitative would score poorly on the common-practice metrical system, because no analysis could be found in which regularly spaced beats were well aligned with accented events; nor does this genre seem to submit to any other kind of metrical system. The same might be said of other nonmetrical idioms such as Gregorian chant and the alap (slow introduction) of an Indian raga. In such cases the kind of musical structure at issue is simply not relevant to the style.

In this way, then, we can view preference rule systems as being reflected in actual musical styles and pieces. This suggests a possible way of quantifying and testing claims about musical styles. For example, it was suggested that the differences between jazz and common-practice harmony are reflected in their harmonic preference rule systems. If the current argument is correct, we should find that music from each style would generally score well under its own system but poorly under the other system. This is an empirical claim, which in principle could be tested. However, to test such a claim would be no easy matter. We would have to have some objective way of defining the style being described; for example, a style could be defined in historical terms, as music written in a certain geographical region during a certain period. And of course, the pieces to be analyzed would have to be represented in the appropriate input format—a formidable problem with many styles of music. I will not undertake any tests of this kind here, but leave them as interesting possibilities for the future.[7]

It is probably unrealistic to hope that a preference rule system could define the *sufficient* conditions for inclusion within a style. For example, there is much more to the harmonic language of common-practice music than what is captured by the harmonic preference rule system described in chapter 6. What a preference rule system gives us is some rather weak necessary conditions for acceptability within a certain style—here again, a rough analogy could be drawn with syntactic grammaticality in language.

The claims made here about styles of music could also be extended to *composition*. That is, beyond simply saying that a certain preference rule system characterizes Mozart's music, we could claim that these preference rules were actually acting as constraints on Mozart's compositional process. For example, it seems reasonable to suggest that metrical preference rules acted as fundamental, though flexible, constraints on Mozart's choice of rhythmic patterns, leading him to write the patterns he did and to avoid patterns such as that in figure 11.1; whatever other considerations went into his construction of rhythms (and undoubtedly there were many), the metrical preference rules had to be obeyed to a certain extent. To go from a claim about the objective properties of a piece to claims about the process of composing it is, of course, a significant and problematic step that should be taken cautiously.[8] However—as noted earlier—if we accept that the perception of meter is and was in Mozart's time guided by certain principles, it seems natural to suppose that Mozart's compositional process was informed and shaped by these principles and reflects them in certain ways. What I propose here is a suggestion of exactly *how* it was informed by these principles.

This way of using preference rule systems to characterize music is not without problems. If a grammatical piece (according to a particular rule system) is a high-scoring one, one might assume that an extremely high-scoring piece would be highly grammatical. However, consider what this implies. A maximally high-scoring piece according to the harmonic rule system proposed earlier would consist of a single prolonged harmony, with no harmonic change whatever (since this would be optimal under both the strong beat rule and the variance rule). It would also consist entirely of chord-tones, without any ornamental tones. Thus it might consist of a single repeated triad. Far from being "optimally grammatical," it would hardly be an acceptable piece of tonal music at all; it would also, of course, be extremely tedious and without any interest. (The reader might consider what kind of piece would be judged as maximally satisfactory by the other preference rule systems, and will find them to be similarly trivial and uninteresting.) One might argue that the unacceptability of such trivial pieces is accounted for by harmonic rules not covered by the current preference rule system. However, I believe a general case could be made that music which scores extremely high on a given rule system tends to be considered unacceptable (within the corresponding style). Consider a related example. The ethnomusicologist John Chernoff (1979) notes that, among musicians in traditional African drum ensembles, there is felt to be an optimal degree of rhythmic complexity, which it is the particular responsibility of the lead drummer to maintain; his drumming must not be so repetitive or stable as to be boring, but also must not be so wild and destabilizing as to overthrow the prevailing beat (pp. 53–60, especially 58; 98; 121). In terms of the current model, we could say that an acceptable African piece must avoid scoring either too low or too high from the metrical rule system.

This suggests that it may be oversimplified, in general, to judge grammaticality (even the fairly limited aspect of grammaticality that preference rule systems can capture) in terms of whether a piece exceeds a certain minimum score. Rather, it might be more accurate to say that pieces are judged as grammatical and acceptable when they fall within a certain range of scores. A score too low indicates excessive simplicity and lack of interest; a score too high indicates incomprehensibility and ungrammaticality.

As well as providing a basis for quantifiable claims about musical styles, then, preference rule scores may also account for judgments of acceptability within a style. It should be emphasized here that no claim is being made about any kind of absolute musical value. What is at issue here is the comprehensibility of a piece under a certain set of rules. It is probably uncontroversial to say that a listener who attempts to analyze serial music using tonal harmony will find it incomprehensible. But surely Schoenberg did not intend for his music to be heard this way, and it is hardly fair to judge it in these terms. Moreover, it is unclear how the kind of structural comprehensibility discussed here relates to aesthetic satisfaction. For some purposes—if the aim is to convey a sense of confusion, anxiety, chaos, or instability—it may be desirable to compose in such a way that, at least to some degree, deliberately resists easy comprehension within the perceptual framework the listener is assumed to have; and to perceive the music in this way may ultimately contribute to a rewarding experience for the listener. In short, the preference rule approach has, at best, only the most indirect bearing on the very complex issues of musical value and aesthetic reward.

11.3 Further Implications: Is Some Music "Nonmetrical"?

The argument so far is that different styles of music may reflect different preference rule systems and also that listeners to these styles may employ different rule systems in their perception of music. For example, the metrical perception of listeners of African traditional music may feature a different balance between the regularity and accent rules than that of Western listeners. This raises the question of where these rule systems are learned. The obvious answer is, they are learned from the music itself; and the current view offers a plausible explanation for how that might come about. Rock music itself encourages the listener to adopt the syncopation shift rule, because without it, no satisfactory analysis of rock pieces can be found. Jazz encourages listeners to entertain a broader range of chord-tones than common-practice music, because without doing so, the music cannot be analyzed in a way that satisfies the preference rules to a reasonable degree. African music requires a lower weighting of the accent rules relative to common-practice music, because if too much weight is given to these rules, no satisfactory analysis is possible.

If different rule systems can be learned from exposure to different styles, it is also possible that a listener familiar with more than one style might learn multiple rule systems—or "parameter settings"—and be able to apply them to different styles of music. How does such a listener know which rule system to apply? There might, of course, be external cues. In a jazz club, we might prefer the A♭ major (jazz) interpretation of figure 11.2a; in Carnegie Hall, the Fm 6/5 (common-practice) interpretation would seem more plausible. But it also seems likely that we determine the applicability of the rule system, in part, by actually trying to apply the rules. We might first apply the rules of common-practice harmony, but—if those did not lead to a satisfactory analysis—shift to the jazz rules instead. Again, the idea of numerical scores is important here: to say that a piece cannot be satisfactorily analyzed using a certain rule system means that it does not yield a high-scoring analysis according to that system.

The possibility that different groups of listeners hold different rule systems leads to still further questions. How much variation in rule systems can there be? What kinds of rules are we capable of learning? Are all the rules learned from exposure to music, or are some of them innate? As yet, there is little basis for drawing any conclusions about these very profound and controversial issues.

Suppose a piece does not yield a satisfactory analysis under any rule system available to us? This is a particularly interesting issue to consider in the case of meter. As noted earlier, it is easy to find pieces that do not yield a satisfactory metrical analysis, in that no analysis can be found in which accented events are consistently aligned with regularly spaced strong beats. Figures 11.3 and 11.4 provide two illustrations from twentieth-century music. While the eighth-note pulse in figure 11.3 is clear enough, any higher levels are, at best, highly irregular. In perceiving such music, do we simply suspend our metrical processing altogether,

Figure 11.3
Bartok, *Mikrokosmos*, Vol. 5, No. 133 ("Syncopation"), mm. 1–4.

Figure 11.4
Varèse, *Density 21.5*, mm. 1–5.

or do we do our best and infer some kind of very irregular or poorly supported (and thus very low-scoring) metrical analysis—something like the structure shown as "my analysis" in figure 11.3? The issue of whether all music has meter, and whether it is useful or cognitively valid to speak of highly irregular metrical structures, is a controversial one. Some authors have suggested that metrical structures must, by definition, have a high degree of regularity.[9] However, there are some reasons to suppose that, in cases like figure 11.3, we continue to infer some kind of irregular metrical structure, rather than none at all. First of all, it seems clear that we do not simply "turn off" our metrical processing in hearing such music; some kind of metrical analysis is always in operation. The evidence for this is that, even when a piece has established a norm of being completely nonmetrical, we will immediately notice any suggestion of a beat or regular pulse. This would be difficult to explain if we assumed that no metrical analysis was taking place. Other evidence comes from the consequences of meter for other aspects of structure. As I will suggest in chapter 12, metrical structure greatly affects motivic structure; in general, parallel segments must be parallel with respect to the metrical structure in order to be perceived. There is indeed a strong sense of motivic parallelisms in figure 11.3—one hears a repeated motive (though instances of the motive differ in length), suggesting a metrical structure with strong beats on each left-hand chord. (As another source of evidence, one might also consider the tendency for strong beats to serve as points of harmonic change, though that is less revealing in nontonal pieces such as these.) In figure 11.3, I would suggest that at least one level above the eighth-note level is present, as shown in "my analysis"; higher levels, are, I admit, doubtful.

As an experiment, the piece in figure 11.3 was given to the metrical program. The hope was that the program would produce something like my preferred analysis in figure 11.3, though assigning it a very low score. Disappointingly, the program did something quite different: it produced an entirely regular 4/4 structure, putting a downbeat on the third event and continuing in 4/4 from there. What accounts for the differences between the program's analysis and mine? The reason is parallelism: the tendency to assign similar metrical analyses to motivically similar segments. This is an important factor in metrical perception which (as discussed in chapter 2) the program does not recognize. As noted above, this passage contains a definite motivic parallelism, one which—despite the rhythmic irregularity—is in some ways extremely clear and strong. The motive in mm. 1–2 is not shifted in pitch-level as it repeats (as is more typical in common-practice music), but rather is reiterated at the same pitch level. Each instance of the motive begins with exactly the same left-hand chord; the right hand reinforces the pattern as well (despite the variation between one- and two-eighth-note groups), since each right-hand gesture ends with an F♯. The motivic pattern in mm. 3–4 is similarly obvious. Indeed, passages of this kind, with a fast pulse being grouped in irregular ways, very often feature strong motivic parallelisms (Stravinsky is very fond of this technique; many instances are found, for example, in the *Rite of Spring* and *Symphonies of Wind Instruments*).[10] If the intent is to convey an irregular meter, such parallelisms may be necessary, in order to counteract our strong preference for regularity. Without the motivic pattern, we might well favor a regular structure, just as the program does.

While one might argue that we always come up with some kind of metrical analysis of a piece that we hear—even an ostensibly nonmetrical piece such as figure 11.3—it is surely true that the phenomenological strength of this analysis can vary depending on its degree of acceptability.

Even if one accepts something like my preferred analysis of figure 11.3, it is not present to awareness as strongly as the preferred analysis of a straightforwardly metrical Mozart sonata. In the case of the Varèse excerpt (figure 11.4), perhaps, the acceptability of the preferred analysis is so low that it almost vanishes from awareness altogether. In such cases we might argue that the music really is—for all practical purposes—"non-metrical."

The issue of how "ungrammatical" music is perceived arises also in the case of harmony. As listeners, what do we do when confronted with a piece which simply does not submit to conventional harmonic analysis? Again, twentieth-century art music provides a wealth of examples for consideration. Some music from this period—such as much serial music—attempts to thwart any sense of harmonic or tonal structure, and often succeeds. However, much twentieth-century music employs the traditional structures of harmony and tonality to some extent, while also taking considerable liberties with them. In hearing such music—as with analogous cases of "quasi-metrical" music—I suggest that we generally do the best we can, applying the conventional rules and seeking the best analysis that can be found. Consider the chord progression in figure 11.5a, from Stravinsky's *Symphonies of Wind Instruments*. One is hard-pressed to find any analysis which accounts for all the notes of this passage as chord tones or conventional ornamental tones. The current model would predict that we seek an analysis which accounts for as many of the notes as possible; and I believe that is essentially what we do. My own preferred analysis is shown above the staff in figure 11.5a. Under this analysis, many of the events are chord tones; a few others are ornamental tones which resolve in a conventional stepwise manner (the lower F in chord 4 and the C in chord 6). However, a number of notes are not accounted for by this hearing, and simply remain as violations of the rules. It is interesting to consider what effect they have on our harmonic hearing of the passage, for they clearly do have some effect. The passage is reprinted in figure 11.5b, with all chord-tones (under my analysis) shown in small noteheads to draw attention to the non-chord-tones. The G's and E♭'s of chords 2 and 3 project an E♭ major triad—a kind of secondary harmony, in competition with the underlying F7 (though I think there are limits to how much such "bitonal" harmonic textures can be heard). The B in chord 1 is perhaps weakly heard as $\flat\hat{9}$ of B♭, spelled as C♭ (though the voicing is unusual in common-practice terms); likewise, the B in chord 3 (again as C♭) could be $\flat\hat{5}$ of F. Color is added to the harmonies by the mixture of major and minor thirds: the A

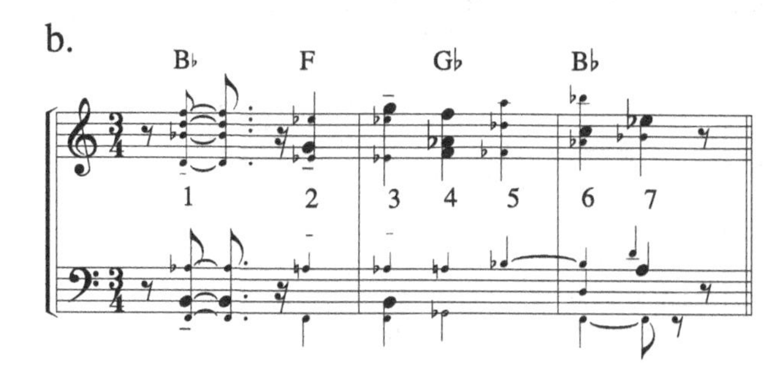

Figure 11.5
Stravinsky, *Symphonies of Wind Instruments*, Rehearsal 4+2 ff. (a) A reduction of the score, with chord symbols indicating my own harmonic analysis. (b) The same passage reprinted, with standard chord-tones ($\hat{1}$, $\hat{3}$, $\flat\hat{3}$, $\hat{5}$, $\flat\hat{7}$) shown in small note-heads.

and A♭ in chords 2 and 3 and the B♭ and A in chords 4 and 5. (I would argue that the spelling of the A in chord 4 is ambiguous; B♭♭ is favored since it allows a chord-tone relationship with the G♭, while A is favored by pitch variance, given the prevailing B♭ major context.) Another important feature of the passage is its tonal implications. Both chords 2 and 7 are contained within the B♭ major scale, giving the phrase a sense of tonal departure and return (despite the harmonic instability of the final E♭ in the melody). That the melody also remains within B♭ major helps to establish this as the prevailing tonality of the progression. The tonality of the inner chords is less clear, though the prominent use of G♭, B♭ and D♭ in chords 4 and 5 weakly projects G♭ as a tonic. Overall, the phrase reflects a trajectory of tonal stability, moving from clarity to ambiguity to clarity again. However, the B's (C♭'s) in chords 1 and 3 and the A♭'s in chords 1, 3 and 6 undercut the sense of B♭ major, giving the passage a subtle leaning towards the flat direction (perhaps to E♭ minor).

Figure 11.6
Boulez, *Structures 1a*, mm. 1–4.

11.4 Preference Rules as Compositional Constraints: Some Relevant Research

The idea of viewing preference rules as compositional constraints has interesting implications in the case of contrapuntal structure. Consider what it would mean for a piece to be grammatical—that is, to permit a high-scoring analysis—by the contrapuntal rules presented in chapter 4. It would have to yield an analysis in which all the notes were sorted into a small number of lines (CPR 2), with few white spaces (CPR 3) and few large leaps within lines (CPR 1). By this approach most common-practice music would be judged grammatical, as would most rock and jazz and much non-Western music. Some twentieth-century art music, by contrast, would be ungrammatical by this measure; consider, for example, the passage from Boulez's *Structures 1a*, shown in figure 11.6. It is hardly news that some music is inherently more contrapuntal than other music in this way; however, it is a nice feature of the current model that it naturally yields a measure of the "contrapuntality" of a piece. It should be emphasized again that whether a piece scores high or low says nothing about the value or success of a piece; for some composers' purposes—for example, if the intention is to create a pointillistic effect—it may be desirable to *prevent* a contrapuntal hearing.

This line of reasoning brings to mind the work of Huron (under review). Bringing together a large body of research by himself and others,

Huron shows that many rules of counterpoint in common-practice music relate closely to principles of auditory perception; other perceptual principles are reflected in aspects of compositional practice which are not dictated by explicit rules but have been verified statistically. These findings can be explained, Huron argues, by assuming that composers often seek to facilitate the process of auditory streaming. An example is pitch proximity. Huron cites a wealth of statistical evidence showing that composers tend to avoid large intervals within melodic lines; recent work has shown, furthermore, that this cannot be attributed solely to global constraints on the range of melodies (von Hippel 2000).[11] The perceptual avoidance of crossing streams is also reflected in composition: in Bach fugues, crossings among the voices are significantly less common than they would be if the same voices were aligned in a random fashion (again proving that avoidance of part-crossing does not merely result from the confinement of each voice to a distinct range). Thus both the preference for small intervals and the avoidance of crossing streams appear to be reflected in compositional practice.

While some of the contrapuntal principles Huron discusses relate directly to what I have called "contrapuntal analysis"—the grouping of notes into lines—in other cases the connection is less direct. An example is Huron's elegant explanation of the prohibition of parallel fifths and octaves. Simultaneous notes a fifth or an octave apart tend to fuse into a single note; moreover, two simultaneous voices tend to fuse into a single voice when they "co-modulate," that is, when they move by the same interval. The tendency for two voices to fuse should be particularly strong when these conditions are combined; and this is exactly the situation of parallel fifths and octaves. In other words, Huron suggests, the prohibition of parallel perfect intervals is designed to prevent two voices from fusing into one.[12] In this case, however, the perceptual principles involved relate to "simultaneous integration" rather than "sequential integration" (see section 4.2). Parallel perfect intervals are avoided not because they hinder the process of grouping notes into streams, but because they hinder (or confuse) the grouping of partials into notes: they create the risk of an inaccurate representation of pitch-events themselves (what I am calling the "input representation").[13] In any case, the general lesson is the same: that viewing perceptual principles as constraints on composition can provide valuable insight into compositional practice.

Another interesting application of preference rule models to the description of music is found in the work of Krumhansl. As noted in chapter 7, the Krumhansl-Schmuckler key-profile algorithm could be regarded as a simple preference rule system, since it involves considering different analyses of a piece (one for each key) and evaluating them numerically. The score for the most preferred key can then be regarded as a measure of the degree to which the piece fits any key—and, hence, the degree to which is the piece is "tonal."[14] In statistical analyses of a wide variety of tonal pieces, Krumhansl (1990, 66–74) found strong correlations between the input profile and the key profile for the "correct" key. That does not necessarily mean that the correct key would be the preferred one, but that is not our concern here; the point is that the pieces are judged to be tonal with respect to *some* key. Krumhansl also tested the model on one piece conventionally regarded as atonal: Schoenberg's "Farben." In this case, no key profile was found to have a statistically significant correlation with the input profile of the piece. Here, then, is tentative evidence that the Krumhansl-Schmuckler model accords well with conventional judgments as to what music is tonal and not tonal.

Mention should also be made here of studies by Krumhansl and her colleagues on Indian music. Castellano, Bharucha and Krumhansl (1984) performed "probe-tone" experiments using excerpts of North Indian classical music (for both Indian and Western listeners), similar to those described in section 7.3, and generated profiles for these. North Indian classical music involves a variety of different scales, or *thats*. The "*that*-profiles" generated from these experiments were found to correlate well with the pitch distributions of the context materials. One might suppose that these *that*-profiles are used as a means of tonal orientation in Indian music: in hearing a piece, Indian listeners apply each *that*-profile, oriented to each possible tonic, and choose the highest-scoring analysis as a way of determining the *that* and tonal center. It would be interesting to pursue a "*that*-finding" algorithm for Indian music along these lines; this might lead, in turn, to quantitative predictions about pitch organization in Indian music.[15]

11.5 Preference Rule Scores and Musical Tension

I have suggested that the score yielded by the preferred analysis of a piece under a particular preference rule system might tell us something about the grammaticality of the piece within a style. A piece yielding only a very low-scoring analysis may be incomprehensible; a piece yielding an extremely high-scoring analysis may be dull. This leaves a range of acceptable scores. One might wonder if differences in score within this range of acceptability tell us anything of musical interest. I believe they do, though here again my ideas are conjectural. I would suggest that relatively low scores can be a source of tension and instability, whereas high scores represent calmness and normality. A entire piece may be characterized by high or low scores in a particular aspect of structure; perhaps more interestingly, a piece may vary between high and low scores from one passage or segment to another, as a way of achieving a fluctuating trajectory of tension.

Consider the harmonic and TPC rules. (Recall that these are integrated into a single preference rule system.) What kind of musical devices would result in a low-scoring analysis? A passage featuring very rapid harmonic rhythm—that is, a passage in which the highest-scoring analysis that could be found featured a very rapid harmonic rhythm—would score poorly on the strong beat rule (HPR 2), since many changes of harmony would not coincide with strong beats. A passage with very syncopated harmony—with slower harmonic rhythm, but with changes of harmony generally on weak beats—would also receive a low score under this rule. A passage featuring very long ornamental dissonances—i.e. dissonances not quickly resolved by step—would score poorly on the ornamental dissonance rule (HPR 4); a piece featuring wide leaps on the line of fifths, that is moves to very remote harmonies, would score poorly on the harmonic variance rule (HPR 3). As for the pitch variance rule (TPR 1), a passage would incur high penalties by introducing new pitches which are remote from previous pitches on the line of fifths—for example, by moving to the scale of a remote key. All of these are well-known devices for creating musical tension and instability.

The first section of Schubert's *Moment Musical* No. 6 provides an illustration. Figure 11.7 shows the analysis of the TPC-harmonic program for this passage. The analysis seems largely correct. The program makes two errors in spelling (not shown in figure 11.7): it chooses F♭ (instead of E) on the third beat of m. 12 and E♭♭ (rather than D) in the German sixth chord of mm. 16–17. Questionable harmonic choices include the D♭ (rather than B♭) in m. 1, the F (rather than A♭) in m. 3, and the F♭ root for the German sixth chord in mm. 16–17. Table 11.1 presents the scores for the program's preferred analysis of this passage, showing scores for each rule on each measure. Consider the harmonic variance scores for the first 16 measures. (Recall that harmonic variance represents the distance of a segment's root from the "center of gravity" of previous roots on the line of fifths, weighted for recency.) It can be seen that harmonies whose roots are further from previous harmonies on the line of fifths receive larger harmonic variance scores (lower scores, in effect, since these scores act as penalties). The G-C segments in mm. 10–

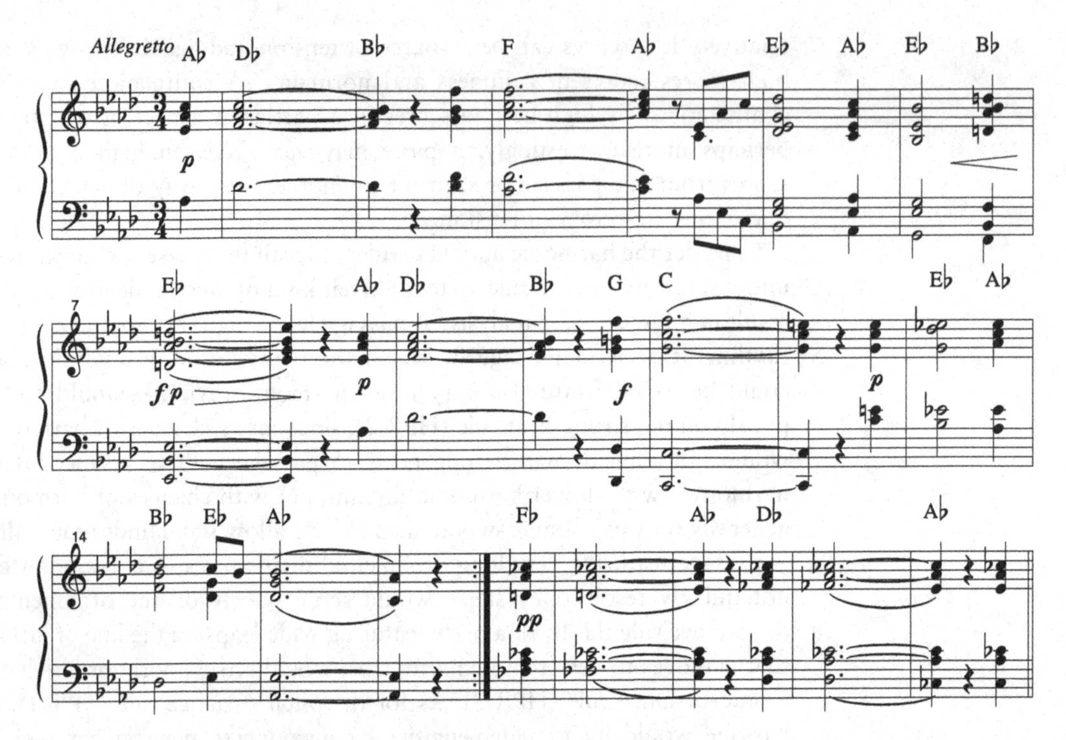

Figure 11.7
Schubert, Moment Musical No. 6, mm. 1–20.

12 incur a large penalty, reflecting the remoteness of these roots from previous roots; the F♭ chord in mm. 16–17 is also heavily penalized (although the choice of root here is dubious). In terms of pitch variance, the largest score occurs in the German sixth chord in mm. 16–17; this reflects the fact that the C♭ and F♭ introduced here are remote from the previous pitches, which mostly remain within the A♭ major collection. (M. 12 gets a high score as well; this is due to the E as well as to the erroneous F♭.) Thus both the harmonic and pitch variance scores reflect the very palpable sense of tension and surprise created by mm. 10–12 and mm. 16–17. Notice also the large ornamental dissonance penalty for mm. 7 and 15: this reflects another kind of tension, created by having several long ornamental dissonances on a strong beat.[16]

The possibilities for harmonic tension are exploited in a variety of ways in common-practice music. The opening of Beethoven's Quartet Op. 59 No. 1 is notable for the large number of metrically strong (and sometimes long) ornamental dissonances in the melody; these are indi-

Table 11.1
Scores of the harmonic-TPC program for preferred analysis of Schubert excerpt shown in figure 11.7

M. No.	Root(s)	Compatibility rule score	Ornamental dissonance penalty (−)	Harmonic variance penalty (−)	Pitch variance penalty (−)	Strong beat penalty (−)	Total
0	A♭	5.775	0.000	0.000	1.502	3.695	0.578
1	D♭	11.200	6.614	3.360	5.715	0.000	−4.489
2	B♭	6.720	2.869	7.052	3.981	0.286	−7.468
3	F	19.040	0.000	7.411	5.657	0.000	5.972
4	A♭	10.675	0.000	4.741	3.302	0.286	2.918
5	E♭, A♭	17.220	0.000	1.053	11.459	3.950	−0.758
6	E♭, B♭	21.945	0.000	1.593	8.937	3.954	7.461
7	E♭	17.920	19.843	0.180	12.148	0.000	−14.251
8	E♭, A♭	15.575	0.000	1.262	3.362	3.695	7.256
9	D♭	11.200	6.614	6.342	6.401	0.000	−8.157
10	B♭, G	−0.210	0.000	8.454	10.488	3.981	−23.133
11	C	20.160	6.614	9.120	6.794	0.000	−2.368
12	C	13.650	5.381	5.530	15.414	0.000	−12.675
13	E♭, A♭	16.170	0.000	6.789	8.191	3.695	−2.505
14	B♭, E♭	9.520	2.220	1.247	7.243	3.954	−5.144
15	A♭	8.960	19.843	4.243	7.634	0.000	−22.760
16	A♭, F♭	9.870	0.000	5.673	17.657	3.695	−17.155
17	F♭	10.745	6.614	9.746	27.396	0.000	−33.011
18	A♭, D♭	9.030	0.000	0.967	6.576	3.981	−2.494
19	D♭	6.720	0.000	0.274	10.366	0.000	−3.920
20	A♭	12.320	0.000	1.844	4.334	0.259	5.883

Note: For each measure, the roots of chord-spans within that measure are shown at left; to the right are shown the scores for the measure for each rule, followed by the total score for all the rules combined. Rules marked with "(−)" are penalties; thus scores for these rules are negative.

cated in figure 11.8. (Here the chord symbols above the staff show my own analysis, not the program's.) Non-chord tones such as the E in m. 1 and the B♭ in m. 3 are hardly unusual, but much more striking dissonances follow. The D in m. 5—which I hear as ornamental—resolves to an E which is itself ornamental; though one ornamental tone resolving to another is permitted by the ornamental dissonance rule, it incurs a large penalty since both notes are penalized. (The D-E-F-G-F figure of the cello is repeated in the violin in mm. 9–10, creating another dissonant clash, though this time it is the D and the F's that conflict with the C7 harmony below.) The cello in m. 7 outlines a C major triad, placing C, E and G on strong beats, while the A and C in the upper voices continue to imply F major. One could hear the cello's E and G as ornamental here, although the sense of a C major harmony is so strong that I am almost inclined to hear the A's in the viola as ornamental, resolving up to the following B♭'s, so that the C7 harmony in m. 8 really begins one measure earlier. In either case, the ornamental dissonance rule imposes high penalties here.[17] A similar conflict arises in m. 15, where the three degrees of the F major triad—the harmony to which the entire passage from m. 8 onwards seems to be leading—land on metrically strong beats (the tonic, F, being metrically strongest), though both F and A are ornamental over the C7 harmony. This playfully anticipates the joyful arrival on F major in m. 19. The longest dissonance of all, the D in m. 17—metrically strong at all levels, even the 2-measure hypermetrical level—provides a suitable climax to the passage. It is largely these accented dissonances that give the passage its color and character.

Figure 11.9 shows another kind of harmonic tension—the tension of harmonic rhythm. The move to A on a weak sixteenth-note beat in m. 79 is a startling gesture; Brahms heightens the tension further in the following two measures with several other harmonic changes on weak sixteenth notes. In several cases, it would be possible to analyze the harmonies here as ornamental—for example, the second chord in m. 80 could be treated as two lower neighbors, C♯ and E, instead of an A harmony—but in cases where two or three tones combine, a harmonic interpretation (if available) usually seems preferable over an ornamental one, as the current model predicts (see section 6.5). As a final example, figure 11.10 shows an extreme example of tension due to harmonic variance: the move from A minor on the downbeat of m. 181 to G♭ major on the (metrically weak) third eighth of the measure creates a leap of nine steps on the line of fifths. This is, again, accompanied by TPC tension due to the wide dispersion of events in line-of-fifths terms, from G♯ in m. 179 to

Figure 11.8
Beethoven, String Quartet Op. 59 No. 1, I, mm. 1–19. The asterisks mark notable ornamental dissonances. All dissonances are marked which are (a) quarter-notes (or longer notes) on strong quarter-note beats, or (b) eighth-notes on any quarter-note beats. The chord symbols indicate my own harmonic analysis.

Figure 11.9
Brahms, Violin Concerto, I, mm. 78–81.

Figure 11.10
Beethoven, String Quartet Op. 59 No. 3, II, mm. 178–85.

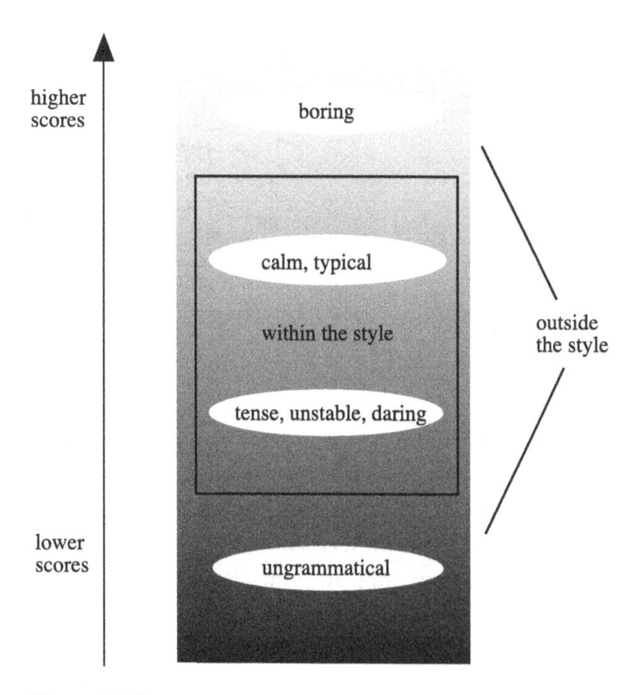

Figure 11.11
The numerical score of a piece yielded by the preference rule system for a certain style (for any aspect of structure) can tell us something about the status of the piece within the style. If the piece scores very high, it is likely to be judged as unacceptably boring within the style. If it scores somewhat less high it will be considered acceptable within the style; within this range, higher scores are considered calm and normal, while lower scores are tense or daring. Pieces with very low scores are likely to be considered ungrammatical or incomprehensible within the style.

G♭ in m. 181. (The beginning of m. 184 is a clear case of revision: first heard as a C7♭9 with D♭, it is then reinterpreted as A7♭9 with C♯.)

I suggest, then, that the scores produced by preference rule systems for segments of a piece may be a good indicator of musical tension. Within the range of acceptability, higher scores are associated with an effect of lower tension; lower scores mean higher tension. If we also include the range of unacceptably high or low scores, this produces a continuum from boredom to normality to tension to incomprehensibility (figure 11.11). Ways that other preference rule systems relate to musical tension can readily be imagined. A heavily syncopated passage creates metrical tension, since either the regularity rule or the accent rules will be violated;

Figure 11.12
Beethoven, Sonata Op. 31 No. 2, I, mm. 52–7.

a line with many leaps creates contrapuntal tension. A passage with a great deal of chromaticism stretches the key rules, since it probably will not permit a good fit with any single key-profile (at least not without frequent modulation). (As noted earlier, it is also likely to stretch the TPC rules, by preventing an analysis in which events are compactly represented on the line of fifths.)[18] A passage without clear phrase boundaries will convey tension by not permitting any satisfactory grouping analysis; this is a common feature of transitions and development sections, as noted in chapter 3.

When several of these kinds of tension are combined within a single passage, the effect can be dramatic. In figure 11.12, the move to the Neapolitan at mm. 55 combines harmonic tension (a leap of five steps in root motion, A to B♭), TPC tension (due to the close juxtaposition of B♭ and G♯) and extreme metrical tension (a long, dense sforzando chord on a weak beat, with nothing on the following strong beat). In figure 11.13, the tension is, first and foremost, contrapuntal, due to the frequent wide leaps in the melody. Hypermetrical disorder also contributes: the beginning of a group at m. 125 suggests a strong beat there at the two-measure level; the parallelism between mm. 127 and 129, however, suggests that m. 127 is strong at this level; the repeat of the descending two-note motive at mm. 130–1 (extended over two measures instead of one) then establishes m. 130 strong at the *one-measure* level, upsetting the previous "odd-strong" hearing; the "odd-strong" hearing is then restored at m. 135, forcing yet another reorientation.[19] One might also cite tension due to TPC variance here; however, this is more doubtful. The question is, do we really hear the leaps in mm. 127–32 as they are spelled—that is, as diminished sevenths? I would submit that, instead, we simply hear

Figure 11.13
Mozart, Symphony No. 40, IV, mm. 124–35.

them as major sixths (A♭-C♭, E♭-G♭, B♭-D♭, F-A♭—though in the last case the following context might encourage G♯); this is a preferable hearing by harmonic compatibility (allowing the sevenths to be interpreted as 1 and ♭3 of minor triads) and by TPC variance as well.

It is useful to consider the relationship between ambiguity and tension. In principle, they are quite distinct: ambiguity occurs when the two (or more) most preferred analyses are roughly equal in score (see section 8.5); tension occurs when the most preferred analysis (or analyses) is low-scoring. However, it can be seen that the two phenomena are often associated. In terms of key structure, for example, a chromatic passage is likely to be both ambiguous (since the pitch collection matches multiple keys to a roughly equal degree) and tense (since none of those keys provide the perfect match that would be yielded for a simple diatonic scale).[20] In figure 8.12b, the G's and E's not only serve to promote the sense of an E minor harmony; they also weaken the sense of B major at that point (since they are ornamental to that harmony). On the other hand, it is not clear that ambiguity and tension need always go together. The move to B♭ major in figure 11.12 causes harmonic tension, yet there is no ambiguity; no alternative analysis could reasonably be entertained. Conversely, a passage using a tonally ambiguous but non-chromatic pitch collection—such as C-D-E-F-G-A—might be ambiguous between two keys (C major and F major, in this case), yet fully compatible with both, and thus not particularly tense.

I certainly would not claim that the preference rule approach captures every aspect of musical tension. In particular, the current model hardly does justice to the complexities of harmonic and tonal tension. For example, the model says nothing about the greater tension of a second inversion triad compared to a root position one, or the unique ability of a perfect cadence to provide relaxation and closure. The model also tells us little about large-scale tonal tension. In an extended piece, a cadence in a nontonic key creates a high-level tension which can only be resolved by a return to the main key; the current model has no knowledge of this kind of large-scale departure and return. A more comprehensive model of tonal tension has been proposed by Lerdahl (1996). In Lerdahl's theory, events are assigned locations in a multi-level space, representing both chords and keys. Events are then assigned to a reductional tree structure, in such a way that the most stable events are superordinate in the tree, and events most proximate in the tonal space are directly connected. The tension of an event then corresponds roughly to its level of embedding in the tree.[21] It naturally emerges from this system that more harmonically remote events are more embedded, and thus more tense. Such a system is more adequate for the subtleties of tonal tension than the current approach. On the other hand—as Lerdahl's theory acknowledges—to generate such pitch-space and reductional representations in a rigorous way requires exactly the kind of infrastructural levels studied here: metrical structure, phrase structure, harmonic structure, and key structure. Achieving a model which could capture the full complexity and subtlety of tonal tension would no doubt require both lower-level infrastructural representations and hierarchical structures of the kind proposed by Lerdahl.

11.6 Performance

A certain view of musical communication has been assumed in the preceding discussion. It begins with a composer, who generates music under the constraints of various kinds of preference rules, and represents it in musical notation; this notation is then converted in performance into an aural "input representation" of the kind discussed earlier, consisting of pitches represented in time; this is then decoded by the listener, again with the aid of preference rule systems, who extracts the infrastructural representations (meter, harmony, and so on) imagined by the composer. However, there is something missing here. The conversion of musical notation into an aural input representation is not, of course, a deterministic, mechanical process, but is mediated by another human being, the performer (or performers). In this section I will consider what relevance preference rule systems have to the role of the performer in the communicative process.

First of all, we can assume that, in learning and perfoming a piece of music, a performer's own mental representation of the piece involves the same kinds of infrastructural levels that we have been assuming for listeners. If the performer first learned the piece (at least in part) by hearing it, then of course her understanding of it would entail structures of meter, harmony, and the like, as they would for any other listener. Even if her learning was reliant mainly on contact with the score, she would hear the music as she herself played the piece, and would thus be approaching it, to some extent, as a listener would. However, the score plays a role here that should not be overlooked. It was pointed out earlier that musical notation contains at least partial specification of several of the kinds of musical information discussed here. The metrical structure is almost completely specified (except for hypermetrical levels); the main key is specified, as is the spelling of the notes; the contrapuntal and grouping structures are often at least partly specified through notation as well. (Only harmony is normally completely absent in notation.) While these notated structures very often correspond to what listeners would hear, sometimes they do not. Moreover, as noted in chapter 8, there are many cases of ambiguity in all of these kinds of structure, and in such cases we can imagine that the composer's notation might tip the perfomer's understanding in one direction or another.

The importance of infrastructural levels in performers' mental representations of music has been confirmed in studies of music learning. Phrase structure plays a particularly central role in the learning process. In studies of sight-reading, Sloboda (1974) found that performers tend only to look ahead to the next phrase boundary. It has also been found that performers' errors reflect phrase structure; for example, when a note is played in the incorrect position, the correct location is more likely to be within the same phrase as the played location than outside it (even when controlling for the fact that within-phrase locations are closer) (Palmer and van de Sande 1995). This suggests that phrases serve as cognitive units for both learning and performance. Studies of errors in music learning also show the importance of harmonic structure. Sloboda (1976) presented pianists with unknown tonal pieces (by minor Classical composers) which contained a number of "misprints"—notes which

Figure 11.14
Part of an excerpt used in a study by Sloboda (1976). The excerpt (shown in Sloboda 1985) was a passage from a Sonata by Dussek, with certain notes altered by Sloboda. In this segment, for example, the F♯ marked with an asterisk was not in the original but represents one of Sloboda's alterations.

were altered to be a step away from the original, usually in such a way as to violate the rules of harmony by producing an unresolved dissonant sonority. For example, in figure 11.14, the score shown to the pianists contained an F♯ in the right hand on the downbeat of the second measure, as opposed to the correct note (presumably G). In many cases, performers unconsciously "corrected" the errors—though they were unfamiliar with the piece—showing that rules of harmony create powerful expectations for performers as they read music. The idea of "high-scoring" and "low-scoring" music could be brought to bear here; in figure 11.14, G is strongly preferable to F♯ for the downbeat of the second measure, since G permits a high-scoring harmonic analysis while F♯ does not.

Infrastructural levels, and the preference rule systems that give rise to them, also play an important role in performance expression. We should remember here that performance expression involves important parameters—particularly dynamics and (with some instruments at least) timbre—which are beyond the scope of the current model, so I will have little to say about them here. We will focus on an aspect of expression that the model *does* represent, namely timing. I suggested in chapter 1 that for any notated piece, once a tempo is chosen, the score can be taken to imply a quantized input representation; given the tempo, the onset-time and duration of every note follows automatically. As is well known, however, performers (in most cases anyway) do not even attempt to produce such a quantized representation. Indeed, the deviations from exact rhythmic regularity—large and small, conscious and unconscious—are a vital source of expressive nuance in performance.

A good deal of research has been done investigating the phenomena involved in expressive timing. It has been found, first of all, that metrical structure is an important factor in timing. Metrically strong events (as implied by notation) tend to be played longer than other events; they are also played slightly louder (Sloboda 1983; Palmer & Kelly 1992; Drake & Palmer 1993). Phrase structure is also reflected in expressive timing; tempo tends to decrease at the beginnings and endings of phrases, with more pronounced decreases at the boundaries of larger sections (Todd 1985, 1989). Again, a possible explanation for these phenomena lies in the numerical scores produced by the model for alternative analyses. It seems likely that the performance strategies discussed above serve to heighten the score of the correct analysis relative to others, thereby decreasing the ambiguity—increasing the clarity—of the piece. Metrically accented notes are played slightly longer than others, because this makes them more attractive as strong beat locations, according to the length rule (MPR 2). (For the same reason, making metrically accented notes slightly louder favors them as metrically strong by the stress rule [MPR 7].) The "phrase-final lengthening" phenomenon can also be explained in preference rule terms. Recall that a grouping boundary is favored at points of large inter-onset intervals (or offset-to-onset intervals). Slowing down at a phrase boundary increases the inter-onset-interval between the last note of one phrase and the first note of the next. Of course, it also increases the intervals between other nearby note pairs; why not increase just the single time interval right at the boundary? One possible reason is that tempo changes cannot be made too suddenly, otherwise the metrical structure will be disrupted. For example, given a sequence of eighth notes, making one of them (intended as a phrase boundary) longer than neighboring ones may cause its rhythmic value to be misunderstood. (This point is illustrated in figure 11.15.) Deforming the tempo gradually allows the crucial between-phrase IOI to be lengthened while maintaining a clear metrical structure.

Another interesting factor in performance is "melodic lead." In polyphonic textures, the notes of the melody tend to slightly anticipate nominally simultaneous notes in other voices. This phenomenon, which has been demonstrated both in solo (piano) and ensemble performance (Palmer 1996b), is not so easy to explain in preference rule terms. As Palmer has suggested, it probably relates more to the recovery of correct pitch information—that is, the formation of the input representation itself. Events which are performed perfectly synchronously can tend to fuse together; making them slightly asynchronous allows the individual pitches to be more easily identified.

Problem: The performer wishes to convey a phrase boundary at the fourth note interval (that is, the interval after the fourth note). How can this be done?

A.

If the notes are performed with even duration, then the fourth interval is not favored as a phrase boundary over any other location (disregarding meter, parallelism, etc.).

B.

?

If a gap is inserted after the fourth note, then this will convey a phrase boundary, by the gap rule (PSPR 1), since the IOI at the fourth interval is now lengthened. However, this may also distort the metrical structure, causing the listener to infer an extra eighth-note beat at this location. (Having no beat here would be penalized by the regularity rule [MPR 3].)

C.

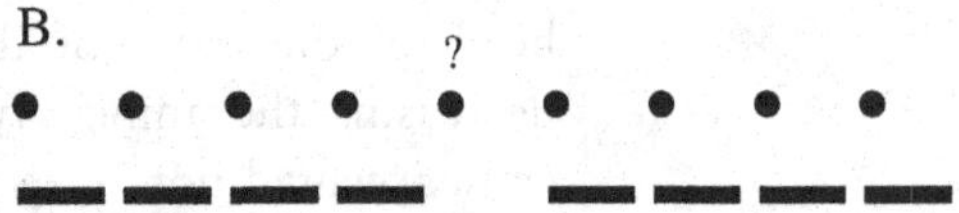

If the IOI's of the notes are altered gradually, slowing down as the fourth interval is approached and then speeding up again afterwards, the long IOI can be achieved without distorting the meter. (There is no reason for the listener to add an extra beat in this case.)

D.

Another way to achieve the desired phrase structure would be to shorten the OOI after the fourth note, while keeping the IOI unchanged. This leaves the meter unaffected (since meter generally depends only on IOI, not OOI), but favors the fourth interval as a beat location by PSPR 1.

Figure 11.15

The preference rule approach also makes predictions about expressive timing that have not, to my knowledge, been experimentally verified. It was argued in chapter 3 that an offset-to-onset interval (OOI) tends to imply a phrase boundary; that is, a given inter-onset interval (IOI) will be a better phrase boundary if it is filled mostly with "rest" rather than with "note." Consider, again, a sequence of eighth notes, and suppose the performer wished to convey one particular note as a phrase boundary; one way to do it—without disturbing the regularity of the IOI's—would be to increase the OOI at the phrase boundary, that is, by playing the last note of the first phrase more like a sixteenth-note followed by a sixteenth-rest (see figure 11.15). My own intuition is that performers do in fact do this quite routinely, but to my knowledge it has not been shown statistically (perhaps it is too obvious to require proof). The situation is complicated somewhat by the fact that slurs of the kind shown in figure 11.15 are ambiguous; they may also be taken as indications of articulation rather than phrase structure. A slur can mean that a group of notes should be played "legato"; and playing a phrase legato means, at least in part, playing with very short (perhaps even zero-length) OOIs within the phrase. One might argue that, if people do play legato passages with longer OOIs between slurs than within them, they are merely following the articulation directions rather than trying to convey the phrase structure. In any case, it seems clear that there is a fundamental connection between articulation and phrase structure, just as the gap rule predicts: playing notes with short or zero-length OOIs indicates that they are within a phrase, while longer OOIs suggest a phrase boundary. This fundamental convergence between articulation and phrase structure may explain why the slur developed its ambiguous meaning: playing a group of notes as a phrase, and playing them legato, often amount to the same thing.

A final issue to consider is the role of harmony and tonality in expressive timing. There is converging evidence that harmonically remote events tend to be played with decreasing tempo. By harmonically "remote," I mean far from the current region or center of harmonic activity. (This of course assumes some kind of spatial representation of harmonies; while theorists disagree on exactly what space is most appropriate—as discussed in section 5.2—there is considerable agreement on many points, for example, that harmonies a fifth apart are close together.) Palmer (1996a) found that in general, events with higher tension were played longer than other events—and the model of tension used by Palmer (that of Lerdahl) generally assigns higher tension to harmonically remote events. Indirect confirmation for this is found, also, in Sundberg's experiments (1988) with artificially synthesized expressive performances. In these experiments, the expressive timing of the performances was determined by rules; various rules were tried, and the musicality of the results was judged by expert listeners. One rule which generally produced satisfactory results was to decelerate on harmonies far from the tonic on the circle of fifths, and accelerate when closer to the tonic. Anecdotal evidence for this principle can also be found in composers' scores. Consider figure 8.13, from Brahms's Intermezzo Op. 116 No. 6; why does the composer indicate a sostenuto in mm. 7–8—especially since this passage, clearly located at the end of an eight-measure phrase, would be played with a slight ritard in any case? One possibility is that mm. 7–8 feature rather rapid harmonic rhythm (with chord changes on each quarter note), as well as significant motion on the line of fifths (both in terms of root and pitch collection), from D7 on the last beat of m. 6 to G♯ major on the downbeat of m. 8; the passage ends up a long way from where it was just a few moments earlier.

Why would performers tend to decelerate on remote harmonies? Consider the Brahms example. The harmonic variance rule (HPR 3) assigns penalties to each chord-span, based on its closeness to previous roots; this is based on the harmonic center of gravity, which calculates the mean position of roots on the line of fifths weighted for their recency in absolute time. At a faster tempo, the center of gravity at the G♯ segment at the beginning of m. 8 will be more affected by the previous D and B chords (in mm. 6–7) than it will be at a slower tempo (since they will be more recent); at a faster tempo, then, the penalty for this G♯ segment will be higher. The same logic applies to the pitch variance rule (TPR 1); at a moment of dramatic shift in pitch collection, pitch variance penalties will be higher if the tempo is faster. The strong beat rule, too, assigns penalties based on the absolute time interval between beats and the absolute length of chord segments. By all three of these rules, a passage such as mm. 7–8 of the Brahms would normally be quite heavily penalized (relative to the previous measures), giving a sense of heightened tension. Reducing the tempo alleviates this tension somewhat—intuitively speaking, it increases the time between m. 6 and m. 8, so that the harmonic changes do not seem so rapid and the motion on the line of fifths is not so abrupt. The same logic would apply to any case of motion to a remote chord or pitch collection. Here again, then, the idea is that performance nuance (whether initiated by the performer or indicated by the composer) serves to heighten the acceptability of the intended interpretation.

12
Functions of the Infrastructure

12.1 Beyond the Infrastructure

In chapter 1, I introduced the idea of a musical "infrastructure": a framework of structures underlying music cognition, analogous to the framework of water mains, power lines, and the like underlying the activities of a society. I justified the analogy, in part, by observing that the components of what I call the musical infrastructure, much like those of a societal infrastructure, generally do not provide direct satisfaction or enhance the quality of life in themselves, but are a means to some further end. Throughout this book, I have focused on the problem of generating infrastructural representations, saying very little about their role in the larger picture of musical experience. At this point, however, it seems appropriate to return to this issue. What is the *point* of meter, harmony, and the like? How do they contribute to what makes music valuable, enjoyable, and important to people? It is natural to ask, also, whether the particular hypothesis about the formation of infrastructural representations put forth in this book—the preference rule hypothesis—has any bearing on this issue. Do preference rule systems merely serve the task of generating infrastructural representations—a task which, for all that it matters, might just as well be performed some other way—or do they relate in some more direct way to the higher levels of musical experience and meaning?

These questions open the door to a subject of almost boundless scope and complexity. The meanings of music are many and diverse, and would take us into complex issues of cultural context. Nevertheless, I will venture some tentative answers as to how infrastructural levels, and in particular the preference rule approach to them taken here, relate to the "higher-level" aspects of music.

In speaking of "musical meaning," I am entering well-trodden and, perhaps, treacherous territory. Much has been written about the issue of meaning in music—whether music has meaning, and if so how it can be characterized.[1] For present purposes I will propose a definition of musical meaning which is straightforward and, I think, reasonably close to the way the term is generally understood. The meaning of something in a piece of music is simply its function, its role in the effect and impact of the entire piece. No further parallel with linguistic meaning is intended—though a comparison of meaning (as defined here) in language and music brings out some interesting similarities and differences, as I will explore in a later section.

12.2 Motivic Structure and Encoding

An important aspect of music not so far addressed is motivic structure: the network of segments in a piece heard as similar or related. Motivic analysis is essentially a process of *coindexing*—"labeling as the same"—various segments within the piece. Figures 12.1 and 12.2 give two examples. The first is straightforward, featuring a series of motives each heard twice. The second is more complex; the repetition of motive A forms a larger unit B which is itself repeated; the tail end of B then becomes a smaller motive, C, whose second half then becomes a still smaller motive. (In both cases, alternative analyses are possible, but the details do not concern us.) Motivic structure has been explored from a variety of perspectives, but the factors influencing motivic structure have not been widely studied.[2] In this section we will consider how motivic structure relates to the infrastructural levels discussed in previous chapters.

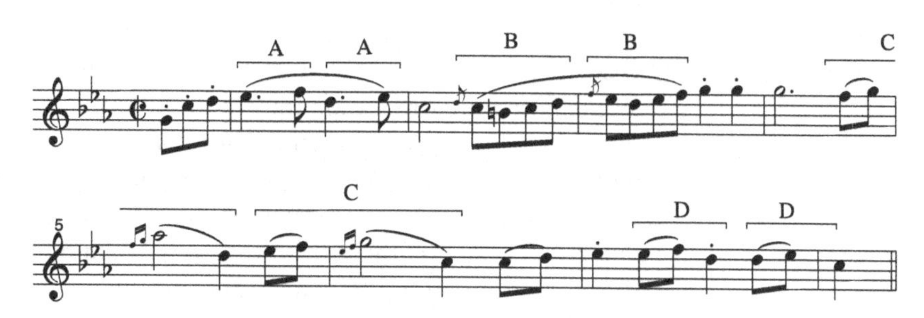

Figure 12.1
Beethoven, Sonata Op. 13 ("Pathetique"), III, mm. 1–8.

Figure 12.2
Beethoven, String Quartet Op. 59 No. 3, I, mm. 43–51.

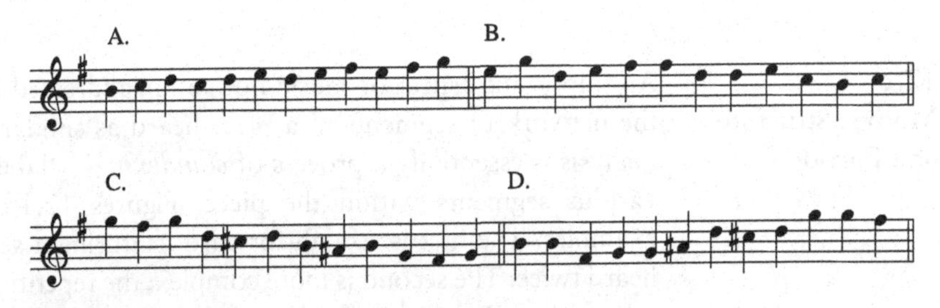

Figure 12.3
From Deutsch 1980. Sequences (A) and (B) have the same pitches, arranged in a different order; likewise sequences (C) and (D).

The most basic, and psychologically well-established, kind of motivic pattern is one of simple transposition, where an intervallic pattern is repeated at different pitch levels. The psychological reality of such relationships has been demonstrated in experiments by Deutsch (1980), in which trained listeners were played sequences of pitches, and asked to write them down. Some were structured sequences, such as figures 12.3a and c, based on a repeating intervallic pattern; other sequences featured the same notes as the structured sequences, but without any repeating pattern, as in figures 12.3b and d. Listeners reproduced the structured sequences much more accurately than the unstructured ones, indicating that the repeated patterns within figures 12.3a and c were being recognized. This points up an important fact about motivic structure: it is an aid to memory, allowing melodies to be encoded in a parsimonious way.

Deutsch and Feroe (1981) have proposed a system whereby motivic relationships might be represented in a hierarchical fashion. Motivic segments are encoded as patterns on various "alphabets," such as the chromatic scale, the diatonic scale, and major and minor chords; such patterns can then be nested hiererchically. For example, the melody in figure 12.3a can be encoded as follows (here I present a less formal version of Deutsch and Feroe's symbolic language):

$X = (^*, +1, +1)$G major (a pitch, followed by another pitch a step up, followed by another pitch a step up, all on the G major scale)

$Y = (^*, +1, +1, +1)$G major (a pitch followed by three ascending steps on the G major scale)

Z (the complete pattern) = Y, with each step of Y starting an instance of X, starting on the pitch B4.

In short, a three-note pattern on the G diatonic scale is repeated in a larger pattern also on the G diatonic scale. Figure 12.3c could be encoded in a similar way; here, a pattern on the chromatic scale is repeated in a larger pattern on a G major arpeggio.

Deutsch and Feroe's system recognizes the fact that tonal and harmonic structure play an important role in motivic analysis. More specifically, we could view tonal and harmonic analysis, in part, as a search for a parsimonious encoding of musical input. Each key or chord implies a certain alphabet (here we must assume that the root labels of chords are supplemented with major or minor as well). Applying a chord or key label to a segment means, in effect, choosing a certain alphabet as most appropriate for encoding the pitches of the segment. Even in melodies without internal repetition, there is evidence that harmonic and tonal alphabets, particularly diatonic scales, are used for melodic encoding. Studies by Cuddy and her colleagues show that diatonic melodies are more easily encoded than chromatic melodies (Cuddy, Cohen & Mewhort 1981); moreover, a melodic pattern is more easily recognized if it is presented in a compatible diatonic context rather than an incompatible one (Cuddy, Cohen & Miller 1979). Thus, while Deutsch's experiments suggest that a chromatic alphabet may sometimes be used in encoding as well, it appears that a diatonic alphabet can be used more readily than the chromatic one. Further evidence for the role of diatonic scales in encoding is that when a melody is heard in tonal transposition—that is, shifted along the diatonic scale—it is often judged to be identical with the original (Dowling 1978).

It is interesting to consider why encoding melodies using scalar or harmonic alphabets might be easier than encoding them chromatically. For one thing, we may simply be more fluent with scalar and harmonic alphabets than with the chromatic one, for whatever reasons of training and exposure. However, there are other reasons as well why scalar and harmonic alphabets might be preferred. In many cases, a scalar encoding is simply more efficient than a chromatic encoding. Consider figure 12.3a; in terms of the diatonic alphabet, the melody features a repeated pattern (* +1 +1), but in terms of the chromatic scale, each repetition of the pattern is different: the first is * +1 +2, the second is * +2 +2, and so on. (This, in fact, is the primary reason for thinking that listeners are using the diatonic alphabet to encode figure 12.3a; if the chromatic alphabet were being used, it is not clear why figure 12.3a would be any easier to learn than figure 12.3b.) There is a more subtle factor involved as well. Within a given pitch range, a harmonic or diatonic alphabet involves fewer steps than a chromatic one. Within the range of an octave, for example, a given harmonic alphabet offers only three possible pitches, whereas the chromatic alphabet offers twelve. In information-theoretic terms, a harmonic encoding of a melody contains less information than a chromatic encoding, and can thus be more efficiently stored. (Informally speaking, it is easier to remember a string of numbers when you know the only options are 0, 1 and 2, than when all ten digits are available.)

One could imagine the "alphabet" system used in much more complex ways than in Deutsch's original experiments. For example, a simple low-level pattern could be repeated in an arbitrary, unpatterned way. Consider motive A in figure 12.2, C-D-E-C: this motive first occurs in a descending quasi-arpeggio pattern, with instances starting on C, A, and D (mm. 43–4); this larger pattern is then repeated (mm. 45–6); then the motive is shifted up a step, to start on E (m. 47). Though the pattern of occurrence of the motive is somewhat complex, the motive itself still contributes to an efficient encoding. One could also imagine cases in which different alphabets (different arpeggios or scales) were used in different instances of the same pattern. Motive D in figure 12.2 provides an example: a three-note arpeggio pattern is imposed first on a D minor chord, then on G7 (or B diminished), C major, and F major. We should note also that Deutsch and Feroe's alphabet system is not the only aspect of pitch structure that contributes to motivic similarity. For example, similarities can be recognized between pitch sequences which are similar in

A

B

Figure 12.4
(A) Mozart, Symphony No, 40, I, mm. 1–5. (B) Mozart, Quintet in G Minor (K. 516), I, mm. 1–2.

contour, but not identical in any kind of intervallic pattern, or in cases where one sequence is an elaborated version of another—although it is less clear how these kinds of relationships might facilitate efficient encoding.

Deutsch and Feroe's model provides an elegant and powerful account of the role of pitch, harmony, and tonality in motivic structure. However, there is a rhythmic aspect to motivic structure as well, although this has been less widely studied. Motives generally have a rhythmic pattern as well as a pitch pattern. Indeed, some have *just* a rhythmic pattern; figure 12.4 shows two examples. In both of these cases, the similarity between the first and second phrases is primarily rhythmic (notwithstanding the repeated two-eighth-note pitch motive in figure 12.4a). It seems probable that, just as with pitch patterns, rhythmic patterns that involve repetition could be more easily learned than those that do not. It would be interesting to test this using purely rhythmic patterns (patterns using only a single pitch or non-pitched sound); this has not yet been done, to my knowledge.[3]

Since motivic patterns are clearly recognizable in both the pitch and rhythmic domains, it is natural to suppose that a repeating unit with congruent patterns of pitch and rhythm would be particularly salient. There is strong evidence that this is the case. When a pitch sequence with a repeating pattern such as figure 12.3c is played with pauses inserted, the repeating segment will be more easily recognized (that is, the melody will be more easily learned) when the pauses occur between occurrences of the pattern, as in figure 12.5a, rather than within occurrences, as in figure 12.5b; this has been demonstrated repeatedly (Handel 1973; Deutsch 1980; Boltz & Jones 1986). According to both Handel and Boltz

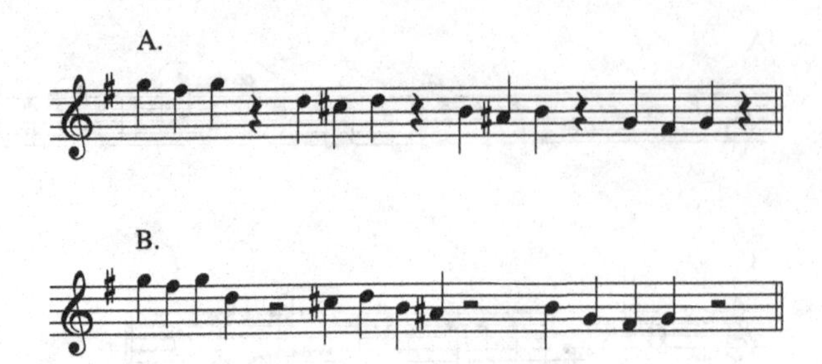

Figure 12.5
From Deutsch 1980.

and Jones, this finding suggests that the compatibility of pitch and rhythmic pattern is important: a melody will more easily be learned if the pitch pattern is repeated in synchrony with the rhythmic pattern. (Others would interpret these experiments differently, as I will explain.)

The role of rhythm in pattern recognition is generally acknowledged; what has not been so widely recognized is the crucial role that *meter* plays in motivic structure. In the first place, we should note that quantization is essential for rhythmic pattern recognition. We are perfectly capable of recognizing a repeated rhythmic pattern, even when the rhythms are performed somewhat imprecisely, as they almost always are. This is because we "quantize" the rhythms, representing them as integer multiples of a low-level beat. For example, motive C in figure 12.1 (disregarding the grace notes) consists of the durational pattern 1-1-4-2 (in terms of eighth notes). It was argued in section 2.1 that quantization is really an aspect of meter: we can think of quantization as imposing a low level of beats on a series of durations. What is important in motivic structure, then, is the repetition of quantized rhythmic patterns, rather than exact ones. (Another point worth noting is that what defines a rhythmic pattern is a sequence of inter-onset intervals, rather than a sequence of durations. Our recognition of the motives in figure 12.1 is not much affected by whether the notes are played staccato or legato—though perhaps it is affected slightly by this.)

Meter is also a factor in encoding at higher levels. As with Deutsch and Feroe's tonal alphabets, a metrical structure can be seen as a means of encoding a pattern hierarchically. A metrical structure can be thought of as a tree, as shown in figure 12.6, with terminals (endpoints of the branches) representing beats at the lowest level.[4] Each branch of the tree can be given a unique "address." A series of pitch-events can then be

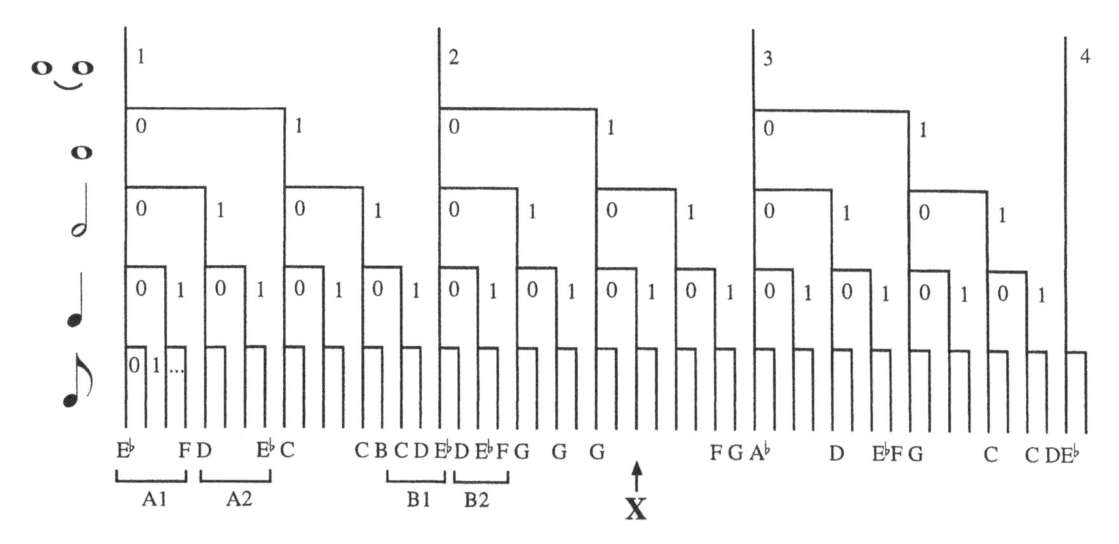

Figure 12.6
A metrical structure can be viewed as a binary (or ternary) tree. Every rhythmic location in the piece has a unique address, which can be read by listing the numbers of all the branches leading to that location. (Beats at the highest level —the two-measure level in this case—are simply numbered with integers.) For example, the address of the branch marked X is 21010. Segments are metrically parallel if the addresses they contain are the same, below a certain level (specifically, the level at which each segment contains only one or zero beats). For example, segments A1 and A2 are metrically parallel, since they both contain addresses ending in 00-01-10-11. (At the half-note level, they only contain one beat.) Segments B1 and B2 are not parallel; one contains 10-11-00, the other contains 01-10-11.

encoded by assigning them to branches of the tree, corresponding to their metrical locations (i.e., the locations of their onsets). I have argued elsewhere that a hierarchical representation of this kind, based on the metrical structure, plays an important role in the encoding of motivic patterns (Temperley 1995). At issue here is the kind of direct and obvious motivic relations that we seem to perceive almost automatically, such as those in figure 12.1. The model I propose is based on the concept of "metrical parallelism." We can define two segments as metrically parallel if they are similarly placed with respect to the metrical structure, so that strong beats fall in the same place in both segments. (In terms of the tree representation, two segments are parallel if the addresses they contain are the same below a certain metrical level. This is explained further in figure 12.6.) I then propose that, for two segments to be recognized as motivi-

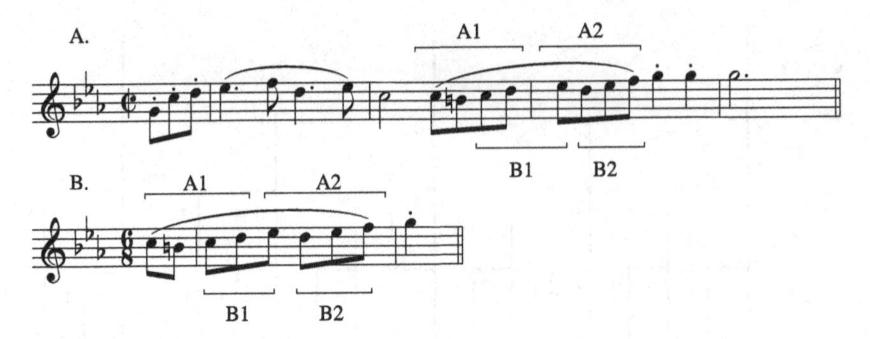

Figure 12.7

cally related, they must be (1) related in pitch pattern (in the way specified by Deutsch), (2) identical in rhythm (that is, they must have the same series of quantized inter-onset intervals), and (3) metrically parallel.[5] For example, consider the Beethoven *Pathetique* melody (reprinted in figure 12.7a). By the criteria just stated, segments A1 and A2 are recognized as motivically related, while segments B1 and B2 are not; while the latter pair of segments are identical in rhythm and related by transposition, they are not metrically parallel. Intuitively, this seems correct; the connection between A1 and A2 seems far more salient than that between B1 and B2. If one imagines the eighth-note sequence in figure 12.7a in a different metrical context, as shown in figure 12.7b, its effect is changed completely; now the connection between B1 and B2 is obvious, that between A1 and A2 virtually unnoticed.

We should recall that motivic parallelism is in itself a factor in metrical analysis; there is a preference for metrical structures in which parallel segments are similarly placed with respect to the meter (see section 2.7). Thus metrical analysis can be seen, in part, as a process of bringing out parallel segments which can be efficiently encoded. The role of meter in motivic structure deserves further study; so far it has received little attention, although it has been noted by several experimenters that changing the metrical context of a melody can greatly alter the way it is perceived (Sloboda 1985, 84; Povel and Essens 1985, 432).

By this view, harmonic and key analysis as well as metrical analysis emerge—in part—as a search for a parsimonious encoding of a piece. In a complex, well-written common-practice piece, the search for the optimal encoding is a dynamic process, requiring constant attention and providing a significant (though not insurmountable) perceptual challenge for the listener. The appropriate harmonic alphabet is constantly changing; less often, the diatonic alphabet and metrical framework may shift as well. The listener must also be constantly alert for patterns of repetition which may allow a more efficient encoding. Taken as a whole, this process—which of course may not always be completed after only a single hearing—can be seen as a search for order and pattern. This aspect of musical listening perhaps has something in common with processes such as reading a detective or mystery story, doing a jigsaw or crossword puzzle, or playing games such as charades or twenty questions; in each case, part of the appeal is finding the order or pattern in a highly complex body of information.

The temporal aspects of music perception discussed in chapter 8—revision and expectation—are relevant here as well. Our expectations involve, among other things, expectations of pattern: we might expect a particular motive to be repeated, for example. What actually happens may fulfill these expectations, or not—or it may represent *some* kind of continuation of a previous pattern, but not in the way we anticipated. A simple example is cases where the intervallic shape of a melody is what we expect, but the harmonic or scalar alphabet required to encode it is not. Consider figure 11.7: the melody in mm. 9–12 (C-C-B♭-F-F-E) repeats the scalar pattern of the melody in mm. 1–4; but due to the E (rather than E♭) in m. 12, the A♭ major scalar alphabet can no longer be used in encoding it—the F minor alphabet is called for instead.[6]

We should briefly consider the role of other aspects of the infrastructure in motivic analysis. Return to figure 12.5; it was mentioned that the greater ease of encoding the first example over the second was due—at least by one view—to the alignment of pitch and rhythm. In Deutsch's view (1980), however, this finding reflects the importance of grouping: segments will be more easily recognized as related if they are in clearly separated groups. In fact, Deutsch's study (as well as similar ones by Handel and Boltz & Jones) appear to confound rhythm and grouping; it cannot be determined whether the ease of learning figure 12.5a compared to figure 12.5b is due to the compatibility with the rhythm or to the grouping. It is undoubtedly true that rhythmic repetition aids encoding independently of grouping. Figure 12.8a is surely easier to learn than figure 12.8b; in both cases the grouping is clearly aligned with the pitch pattern, but figure 12.8a features exact rhythmic repetition while figure 12.8b does not. However, there is also evidence that grouping can affect pattern recognition independently of rhythm. In an experiment by Tan, Aiello and Bever (1981), already discussed several times, subjects asked

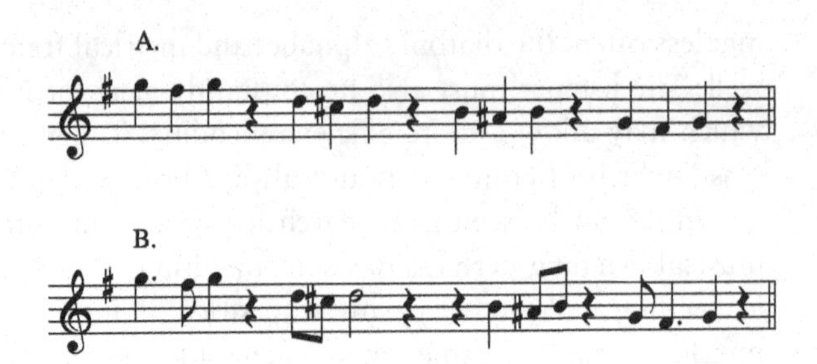

Figure 12.8

to identify a two-note probe in a melody did so more easily when the two notes occurred within a group, rather than across a group boundary. Since the notes of the pattern were all equal in duration, the fact that one two-note segment was more easily recognized than another cannot be attributed to rhythm. It appears, then, that—independently of rhythm—the interruption of a sequence of notes by a group boundary can prevent it from being identified as a motive.

A final factor in motivic structure is contrapuntal structure. It is readily apparent that a group of notes is much more likely to be heard as a motive if the events are within the same contrapuntal line. In a piece such as the Beethoven *Pathetique* third movement (whose melody is shown in figure 12.1), we do not perceive notes from the left-hand accompaniment as being part of the motives in the melody. This has also been demonstrated by studies, cited in chapter 4, showing that two interleaved melodies can be identified much more easily if they are widely separated in register (Dowling 1973a).

In short, several different levels of structure appear to act as direct and important inputs to motivic analysis: the pitch-time representation itself, as well as meter, grouping structure, contrapuntal structure, harmony, and key. Because motivic analysis is influenced by a variety of factors, it seems likely that it, too, would be amenable to a preference rule approach, though this will not be attempted here. It is natural to wonder whether motivic structure could "feed back" to influence the representations that contribute to it. It may be that an analysis that permits a "rich" motivic structure (one that contains many coindexed segments, and can thus be very efficiently encoded) is preferred over one that does not. The influence of parallelism on meter is one clear example of this; in other areas the influence of motivic structure is less obvious, though this would be an interesting area for study.[7]

12.3 Musical Schemata

A motive is generally confined to a single piece. Other kinds of musical patterns occur in many pieces, and are thus the property of the common-practice style as a whole. Such a pattern is frequently called a "schema." As the term is generally used in psychology, a schema is an entity defined as a cluster of features; not all features may be necessary to the schema (in some cases, *no* single feature is necessary), but a certain combination of features is generally typical. Perhaps the most important schema in common-practice music is the "perfect" cadence: a V-I progression, both chords in root position, with the tonic in the melody of the I chord. (A simple example is shown in figure 12.9a.) The perfect cadence serves an essential function in common-practice music; it is virtually required for a sense of closure at the end of a piece, and generally establishes sectional points of closure as well. The main point to be made about the perfect cadence is that it relies heavily on infrastructural levels for its definition, particularly harmony. Figure 12.9b shows the perfect cadences at the end of each movement of Bach's French Suite No. 6 in E major. As we would expect, each movement ends with a clear V-I progression in E major. However, each of these cadences is different, and none of them exactly matches the "textbook" cadence in figure 12.9a. The notes of the two harmonies are generally arpeggiated (presented successively rather than simultaneously); one or more notes of the chords may be omitted (for example, the third is omitted from both the B and E harmonies in the Bourée); and non-chord tones may be added. Identifying cadences by searching for an exact configuration of pitches, then, simply would not work. What all the cadences have in common with the stereotyped cadence (and with each other) is their harmonic structure: each of them consists of a B harmony followed by a E harmony. Of course, identifying these progressions as perfect cadences also requires knowledge that they represent V followed by I in the current key, E major, which in turn requires knowledge of key structure as well. Notice also that, in each of the Bach movements, the V is metrically weaker (in terms of its beginning point) than the I; this is a typical feature of the perfect cadence, though not a necessary one.

Infrastructural features are also necessary to the definition of other schemata. An example is the $\hat{1}$-$\hat{7}$-$\hat{4}$-$\hat{3}$ schema explored by Gjerdingen (1988): a common cliché of the Classical period. Gjerdingen defines the essential features of the schema as shown in figure 12.10. It features a I-V harmonic progression followed closely by a V-I; the first half must feature $\hat{1}$-$\hat{7}$ prominently in the melody (though other notes may also be used), and the second half $\hat{4}$-$\hat{3}$. Again, these features require knowledge

Figure 12.9
(A) A simple "perfect cadence." (B) Final cadences from the movements of Bach's French Suite No. 6 in E Major.

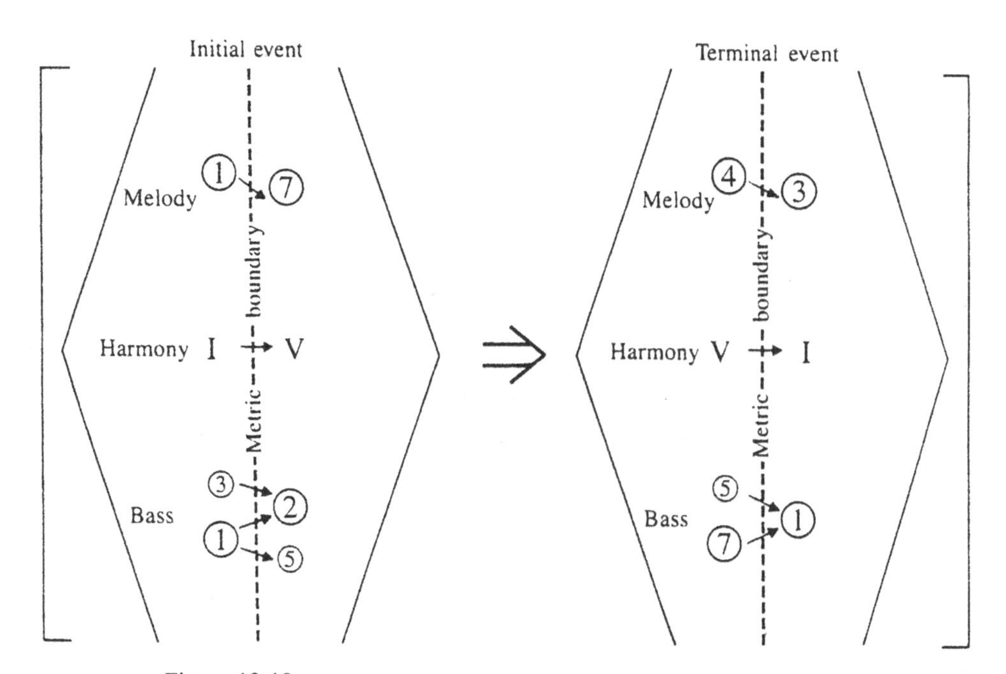

Figure 12.10
The $\hat{1}$-$\hat{7}$-$\hat{4}$-$\hat{3}$ schema. From Gjerdingen 1988. Reprinted by permission of the Society for Music Theory.

of both harmonic and key structure. The first and second chord of each half of the schema must be separated by a metric boundary. (In our terms, this appears to mean that a relatively strong beat must either separate the two chords or must coincide with the second chord—usually the latter.) Gjerdingen actually neglects some other features which it seems might also be important for the schema (perhaps because he considered them obvious). Surely the first and second chord of each half must be part of the same phrase; an instance of the schema with a clear phrase boundary after the first I chord would hardly seem characteristic. In the vast majority of cases Gjerdingen considers, the $\hat{1}$-$\hat{7}$ and $\hat{4}$-$\hat{3}$ gestures appear to be metrically parallel, although perhaps this is not absolutely necessary. (Also, the $\hat{1}$-$\hat{7}$-$\hat{4}$-$\hat{3}$ scale degrees must presumably belong to the same contrapuntal line, though that is implicit in the fact that they are in the melody.) In all these ways, then, infrastructural representations are necessary for perception of the $\hat{1}$-$\hat{7}$-$\hat{4}$-$\hat{3}$ schema.

Other examples could be cited of schemata which depend on infrastructural levels—for example, the "topics" of Classical-period music

(Ratner 1980; Agawu 1991). Topics are idiomatic gestures or clichés, which carry extramusical or expressive connotations; typically a topic will be presented in a single phrase or section of a piece, though in some cases it might pervade an entire movement. These include the "hunt" style, the "singing" style, the "march" style, and dance-related topics such as the minuet and polonaise. As Ratner (1980, 9–29) makes clear, topics frequently have essential metrical and harmonic characteristics.[8] The point to be emphasized, then, is that all of these schemata rely on infrastructural levels. Incidentally, such schemata also provide another important kind of *evidence* for infrastructural representations. For example, if we assume that listeners can recognize perfect cadences—and there is good evidence that they can (Rosner & Narmour 1992)—then it is reasonable to assume that they must be performing harmonic analysis as well.

12.4 Tension and Energy

Another higher-level function of the infrastructure has already been described, and requires only passing mention here. This is its role in musical tension. I argued in section 11.5 that a piece can convey varying levels of tension, depending on the degree to which its most preferred analysis satisfies the preference rules. A piece for which a high-scoring analysis (at some level of structure) can be found will seem calm and stable; a piece for which the best analysis found is fairly low-scoring will seem tense and daring. The clearest example is in harmony and key. The preference rule perspective accounts for the tension associated with a number of harmonic and tonal phenomena, including chromaticism, rapid or syncopated harmonic rhythm, long or accented ornamental dissonances, and sudden shifts in key or pitch collection. Low-scoring phenomena in meter, phrase structure, and contrapuntal structure can also create tension.

A further kind of meaning in music is perhaps obvious, but deserves mention. It is well-known that music has the power to convey varying levels of energy and arousal. Loud music conveys more energy than quiet music. Register is important too: a melody can convey energy by rising, and a large-scale melodic peak often conveys a climax of intensity and excitement. (I am not aware of experimental evidence for this, but it is surely true; consider the ascent to the high C in figure 2.7, or the rise to the high D in m. 47 of figure 3.10.)[9] These aspects are not strictly speaking infrastructural; the first relates to dynamics, which is not covered by the current model, and the second relates simply to the pitch-time representation (though it might be argued that what conveys energy is a rising melodic line, and this relies partly on contrapuntal structure). One aspect of musical energy which *is* captured by the infrastructure is what we might call musical speed. Faster music conveys more energy; but what exactly makes a piece seem fast? Clearly, the essential variable here is not simply notes per second. A piece with many notes can seem slow; think of a florid Chopin Nocturne or Beethoven slow movement, such as that shown in figure 5.14. Rather, the primary indicator of speed in music, I would argue, is *tempo*—that is, the speed of the tactus, the most salient level of meter. For this aspect of musical energy, then, metrical analysis is crucial.

12.5 The Functions of Harmony and Tonality

We have said that harmony and tonality are essential to the communication of conventional progressions such as cadences and topics; and they play an important role in musical tension. However, this hardly does justice to the expressive power of harmony and tonality in common-practice music. What else can we say about the function and meaning of these representations?

In the first place, tonal pieces have the power to convey a sense of a complex, multi-leveled journey through space. A sequence of harmonies can stay more or less in one area, or it can move far afield, either gradually or suddenly; at a larger level, a progression of keys can be formed in an analogous way. This in itself can create a kind of drama or narrative; moving from one key to another—particularly to a remote key—conveys a sense of a conflict demanding resolution. In section 5.2 I discussed the variety of spaces that have been proposed to model the experience of harmonic and tonal motion. I will not revisit this issue here; but the idea that there is a spatial aspect to the experience of tonal music seems to be a matter of general agreement.

Another important aspect of tonality is the distinction between major and minor. The expressive associations of major with happiness and minor with sadness are well-known. It was noted in chapter 7 that children display knowledge of these associations from a very early age. It is important to observe that the distinction between major and minor is really a continuum. While many passages are clearly in major or minor, many others involve some kind of "mixture" of the two modes. Consider the first movement of Beethoven's Sonata Op. 2 No. 1, discussed several times in earlier chapters. While the second key of the exposition is clearly A♭ major, there are prominent elements of A♭ minor as well: the ♭$\hat{6}$'s (F♭'s) in mm. 20–30 (figure 2.13) and the ♭$\hat{3}$'s (C♭'s) in the closing theme, mm. 41–8 (figure 2.11). These elements add a tinge of sadness to the otherwise hopeful mood of the section. Schubert's *Moment Musical* No. 6 provides another example (figure 11.7). The piece begins in A♭ major, but moves to the parallel minor in m. 17; after a move to F♭ major, the main theme returns in A♭ major, but elements of A♭ minor gradually creep in and eventually take over completely, leading to a tragic minor conclusion. This continuum between major and minor allows for an expression of many shades of feeling between pure sadness and pure happiness, a possibility exploited by many common-practice composers and Schubert in particular.

The current approach suggests a general framework for capturing this gradual variation in expressive feeling between pure major and pure minor. Imagine a section of a piece that uses an evenly distributed C major scale (that is, with equal numbers of C's, D's, E's, and so on). Another passage uses an evenly distributed C harmonic minor scale. For each passage, we can find a "center of gravity" (COG), representing the mean line-of-fifths position of all events (figure 12.11). (Something very similar to this was proposed in chapter 5, as a means of deciding on the optimal spelling of each event; however, that COG was more affected by more recent events and was constantly being updated, whereas this one is calculated just once and treats all events equally.) It can be seen in figure 12.11 that the COG for the minor mode is somewhat to the right of that for the major mode. More important, it is further to the right with respect to the tonic, C, which of course is at the same location in both cases. It can be seen that a pitch collection which mixes major and minor, such as C major with a flattened sixth degree, will receive a COG in between those for pure major and minor (this is shown in figure 12.11 as well). Possibly our sense of where a passage falls in the continuum between major and minor—and its associated expressive message—arises from a calculation of this kind. By hypothesis, a passage in which the pitch COG is further to the right, relative to the tonic, will seem "sadder," while one in which the pitch COG is further to the left will seem "happier." (Notice that the use of the *line* of fifths, as opposed to the circle, is crucial here. It is not clear how the center of gravity of points can be calculated on a circle; and in any case, it makes no sense to say that one point on a circle is further to the left than another.)

This model also sheds light on styles where the opposition between major and minor is not found, notably rock music. As discussed in chapter 9, rock music is largely modal, in that it uses the diatonic scale but often assumes a different tonic from that of the conventional "major

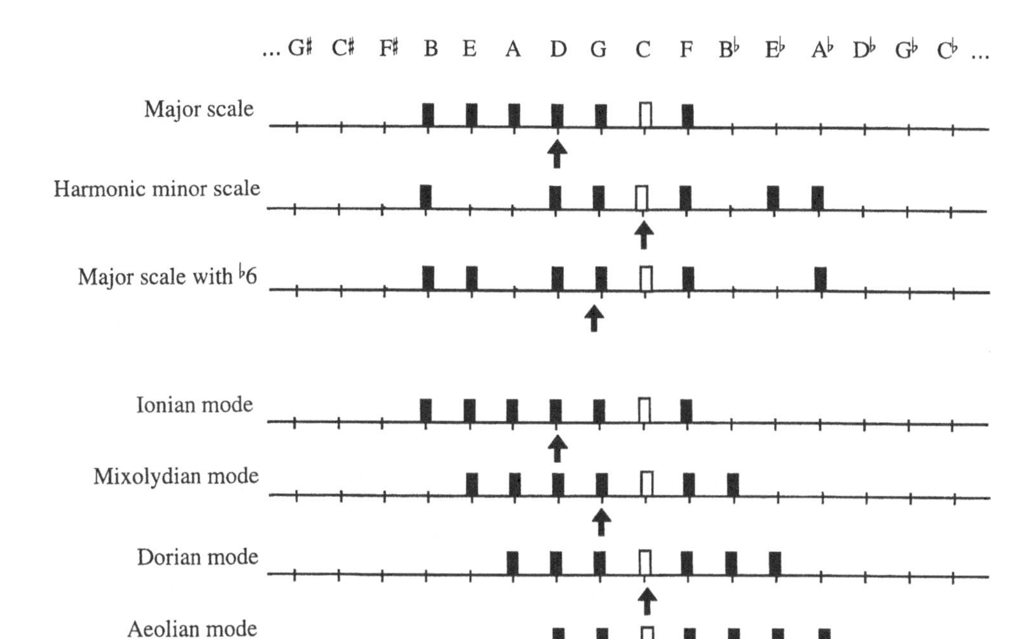

Figure 12.11
Common pitch collections represented on the line of fifths (assuming a tonal center of C). The arrows show where the "center of gravity" would be—the mean position of all pitch-events—in a piece using an even distribution of all pitch-classes in the collection.

scale." Figure 12.11 shows the four modes commonly used in rock—Ionian (the major scale), Mixolydian, Dorian, and Aeolian—along with their COGs (assuming, again, an even distribution of all the pitches of the mode). If we applied the rule just proposed for major and minor to the common modes of rock, considering the position of the COG relative to the tonic in each case, we would expect a continuum of expression, from happy to sad, of Ionian-Mixolydian-Dorian-Aeolian. (As with common-practice music, the term "continuum" is truly appropriate here, as many rock songs employ some kind of mixture of modes.) Though I will not attempt to demonstrate this statistically, I believe that this corresponds well with the emotional associations of modes in rock, as reflected in their lyrics. Particularly clear cases of this can be found in songs with modal shifts. When a song shifts from one mode to another, this is often accompanied by a change in the mood of the lyrics, and this

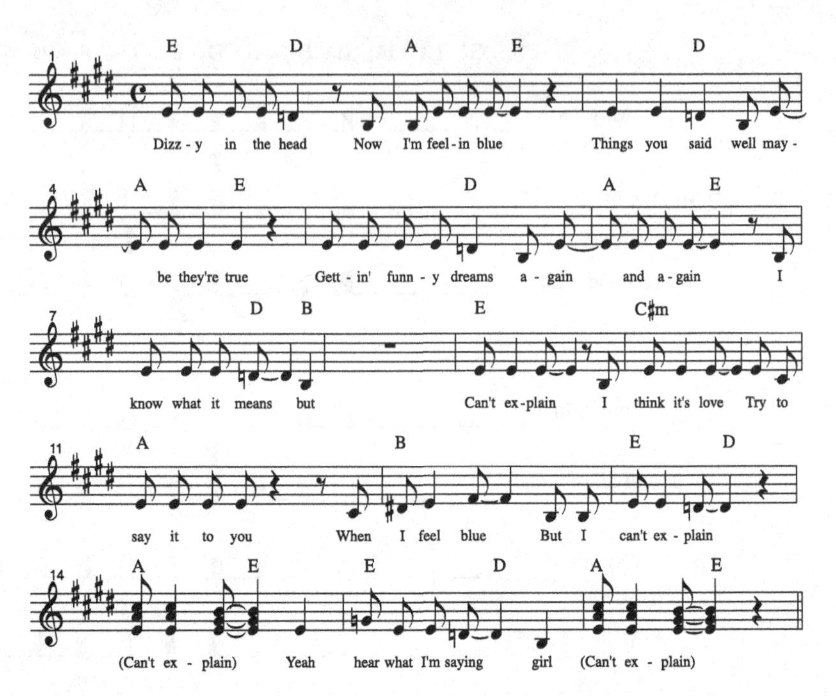

Figure 12.12
The Who, "I Can't Explain."

change often corresponds well with the "line-of-fifths" rule I have proposed. Consider the Who's "I Can't Explain" (figure 12.12). In the largely Mixolydian verse, the singer is confused and tormented. In the chorus, however (m. 9 in figure 12.12)—as the harmony shifts to Ionian—we find that the source of the singer's anguish is something fundamentally positive: he is in love. Similarly, in the Beatles' "A Hard Day's Night," the Dorian/Mixolydian mixture of the verse—describing the hard day—contrasts with the more sentimental Ionian harmony of the bridge, where the singer is home and "everything seems to be right." It is also interesting to observe that a number of bands of the 1980s and 1990s whose lyrics are consistently negative and depressing—such as the Police, U2, and Nirvana—display a strong fondness for the Aeolian mode, the most emotionally negative mode according to the current model. (See, for example, figure 9.18b.)

A final case should be discussed where the line of fifths appears to be used expressively in a rather different way: this is Monteverdi's opera *L'Orfeo*. Composed in 1607, the musical language of *L'Orfeo* is "proto-tonal": it employs a triadic harmonic language, and often implies clear tonal centers, but without adhering to the diatonic system of later common-practice music. Several authors, notably Dahlhaus (1990) and Chafe (1992), have argued that harmonies and tonal centers in *L'Orfeo* have strong emotional connotations. Perhaps the most convincing example is the connection between A minor (and to a lesser extent E major) and Orfeo's past or present sorrows, traced in detail by Chafe. The first act is generally joyful, as Orfeo and Euridice celebrate their love, the main keys being G major and D minor; but there are several moments in which Orfeo's earlier sorrows are recalled, most of which coincide with moves to A minor. These include the middle section of the shepherd's first speech, the shepherd's introduction of Orfeo ("Ma tu, gentil cantor"), and—most notably—the final strophic chorus in G major, in which the third line, modulating to A minor, almost always coincides with a reference to grief or suffering. This tonal contrast is made even more starkly in the second act, when the rejoicing of Orfeo and the shepherds is interrupted by the messenger who has come to inform Orfeo of Euridice's death; musically, the F major/C major of the shepherds is interrupted by the messenger's A minor. A rapid interchange occurs—the messenger singing in A minor or E major, the shepherds in F or C—as the messenger tries to break the awful news.

Chafe (1992) sees the significance of these tonal centers in *L'Orfeo* in terms of the hexachords they represent. In his view, *L'Orfeo* is built around contrasting hexachords of keys—in particular the two hexachords A, D, G, C, F, B♭ and E, A, D, G, C, F. However, this hexachordal explanation seems problematic, since it is often ambiguous which hexachord a particular chord or tonal center is implying. Rather, it seems more plausible to attach expressive meanings to tonal centers themselves; E and A are often sad, while D, G and C are generally either neutral or positive. F and B♭ are often joyful, particularly in Act III. (Orfeo's plea to Charon to allow him into hell, "Possente spirito," is mostly concerned with his suffering and travails, and is centered around G; but as he refers to Euridice—"O serene light of my eyes"—the music shifts to B♭ major.) This further suggests that the "happiness" of a harmony or tonal center is simply a function of its position on the line of fifths, with tonics in the flat direction being happy, those in the middle being neutral, and those in the sharp direction being sorrowful. (Again, this relies crucially on a linear space of tonics rather than a circular one; a given point on the circle—F♯/G♭, for example—might be regarded as a very "flat" or a very "sharp" key, depending on the direction from which it is approached.) The line of fifths thus emerges as a "happiness axis," but in a rather different way than in rock and common practice music; in *L'Orfeo*, it is the actual shifts in tonal center, rather than shifts in pitch collection relative to the tonal center, which govern the expressive tone.

Of course, shifts in tonal center are possible in rock and common-practice music as well, and often carry expressive meaning. Thus the tonal journey of a piece can be represented in terms of two cursors moving along the line of fifths, one representing the COG and the other representing the tonal center. My own feeling is that, in general, a shift in pitch collection relative to a tonal center conveys a change in perspective on an unchanging situation, whereas a shift in tonal center conveys an actual change in situation—or perhaps a shift in perspective from one character to another—with shifts in the flat direction generally being positive. In Schubert's *Moment Musical* No. 6 (figure 11.7), the major-minor shifts within the tonal center of A♭ suggest changing thoughts about the "status quo"; the shift to F♭ major in the middle section suggests a change or improvement in circumstances. In any case, the possibility afforded by the common-practice tonal system for creating complex trajectories of both tonal center and pitch collection is surely one of its great features.

12.6 Arbitrariness

At this point it is useful to introduce an idea from linguistics: the idea of arbitrariness (Lyons 1981, 19–21). Arbitrariness relates to the connection between the meaning of something and its internal structure. The relationship between meaning and form in words is generally held to be arbitrary. Every word is made up of a certain sequence of phonemes (or, in written language, letters): but you cannot predict the meaning of a word from the phonemes that constitute it. (Strictly speaking we should speak of morphemes here, not words, a morpheme being an indivisible unit of meaning. The meaning of "walking" could be figured out from knowledge of its component morphemes, "walk" and "ing"; but the meaning of these morphemes could not be figured out from any simpler principles. However, words often correspond to morphemes, so I will continue to speak of words in the interest of simplicity.) While the relationship between form and meaning in words is arbitrary, the relationship between form and meaning in sentences is not. The meaning of a sentence follows in a rule-governed way from the words that comprise it (the rules being both syntactic and semantic). It is for this reason that we can understand sentences we have not heard before; we cannot generally understand new words we have not heard before, at least not without benefit of context. This has important computational implications for language perception. The meaning of words cannot be figured out; they must simply be stored, in some kind of giant lexicon which matches the form of each word with its meaning. For sentences, however, no lexicon is necessary (or possible); the meaning of a sentence is figured out as it is heard.

It is instructive to compare this with the situation in music. Can we speak of arbitrariness at any level of musical structure? Does music comprise elements whose meaning is arbitrarily related to their form? (At this point it will be clear why I adopted a very broad definition of meaning, simply to mean the function of an element in the larger system.) We will examine several possibilities.

Hypothesis 1. The function of a chord is arbitrarily related to the pitches that comprise it.

It can be seen fairly easily that this hypothesis is false. Consider the functions of a chord. A chord implies a certain position in space (whatever kind of space is being assumed). As mentioned above, this allows tonal pieces to imply complex harmonic journeys, which in turn convey effects such as departure-return and conflict-resolution. Chords also have functions relative to keys, which in turn allow them to form schemata and other conventional progressions. A C major chord, by being I of C of major, has a unique capacity to establish closure in that key. Clearly, the relation between the functions of harmonies and the pitches they contain is not arbitrary, but highly rule-governed. The relation between pitches and roots is captured in the compatibility rule: the pitches C-E-G imply a root of C, because this is the preferred interpretation ($\hat{1}$-$\hat{3}$-$\hat{5}$) under the rule. (There are other factors in identifying chords—as discussed in chapter 6—but I have argued that these, too, can be captured in a few simple rules.) Once the root of a chord (and hence its line-of-fifths position) is known, its further properties—its role in harmonic journeys, its functions relative to keys and in harmonic schemata—can, likewise, be figured out from general principles. We could imagine a case where the relationship between the pitches and the functions of a chord was *not* rule-governed, so that it was necessary for us to store a large lexicon of harmonies, mapping their forms (i.e. their component pitches) on to arbitrary meanings. But this is not the case.

Figure 12.13

Hypothesis 2. The functions of a motive are arbitrarily related to the pitches that comprise it.

This brings us to the interesting question of what the function or meaning of a motive is. Consider the opening gesture of the *Pathetique* melody, shown in figure 12.13. This motive clearly implies a minor tonality, carrying a melancholy connotation; it opens with a rising staccato figure, which implies energy, followed by a legato descent back to the tonic, conveying tentative repose; the escape tones (the eighth notes in m. 1) provide a hint of tension and anxiety. Here again, it is quite clear that these functions of the motive are not arbitrarily related to the notes comprising it, but can be specified from a small number of simple rules. The proof of this, again, is that we are capable of understanding such meanings in motives that we have never heard before. (The rules governing the interpretation of motives may themselves be somewhat arbitrary—perhaps it is merely convention that rising contours imply rising energy, with no more general principle to support it; but there are general rules nonetheless.)

Hypothesis 3. The meaning of a musical schema is arbitrarily related to its form: the notes and infrastructural elements that comprise it.

I have defined the form of a schema in terms of both its notes and its infrastructural representations, since I argued above that infrastructural levels such as harmony are often important to a schema. This hypothesis is more problematic than the first two. Consider the fact that a V-I cadence implies closure in a key. Is this purely conventional—does its effect depend on our knowing (perhaps from hearing many pieces which end in cadences) that cadences imply closure? Or does it arise from more general principles about tension and stability? Under the current framework, V-I (or rather V7-I) is arguably the most stable possible chord progression within a key; it presents six of the seven diatonic pitches of a key, unambiguously establishing it and allowing it to receive a maximal score. I is more stable within the key than V is, given the higher values for members of the tonic triad in the key-profile, so it is natural that greater closure would be achieved by ending with I. Still, there may be something arbitrary in the convention that pieces have to end in exactly this way. What about other musical schemata? In the case of Classical topics, the relations between form and function are again clearly non-arbitrary at least in some cases: march music symbolizes the military because it is actually somewhat similar to military music. There may also be a degree of arbitrariness, however; to answer this question would require a greater historical knowledge than I possess. As for the $\hat{1}$-$\hat{7}$-$\hat{4}$-$\hat{3}$ schema discussed by Gjerdingen, the relation between form and function depends on knowing the functions of the schema—expressive, structural, and so on. Gjerdingen says little about this, and it may simply be a very difficult question to answer.[10]

Hypothesis 4. The association of major and minor with happy and sad is arbitrary.

One might argue that this association must be conventional, since some genres which employ something like major and minor modes (e.g., Indian classical music, which uses these scales among many others) do not assign them these expressive associations. One thing seems clear, however. If we accept that composers create intermediate expressive effects between major and minor in the manner described above, this is clearly not arbitrary. That is, it is not necessary for us to learn the connotations of many different "mixtures" of major and minor; there is a simple rule that predicts them. The same applies with the modes of rock music, and the many possible mixtures of these. The most that might be arbitrary—not explicable by any simpler principle—is the association of one direction on the line of fifths (in terms of the position of the pitch collection relative to the tonic) with happy and another with sad.

In short, there appears to be very little that is arbitrary about the relation between form and function in common-practice music. There is clearly no need for a lexicon of chords or motives, storing many arbitrary structural or expressive functions. In terms of schemata, the picture is less clear; in some cases, the relationship between the form of a schema (in terms of its configuration of pitches and infrastructural elements) and its function may be partly arbitrary and partly rule-governed. Even here, however, the difference between music and language in this respect is much more striking than the similarity. The fact that there is nothing in music comparable to the arbitrary relation found between form and meaning in words is an important difference between music and language which has not been fully appreciated.

Figure 12.14
Beethoven, Sonata Op. 10 No. 3, III, mm. 1–16.

12.7 Explaining Musical Details: An Experiment in Recomposition

It is something of a cliché to say that, with a piece of music (particularly with a great piece of music), every detail matters and to alter or remove a single note would be to significantly change (usually to harm) the overall effect. The preference rule approach allows us to bring some empirical rigor to this matter, by showing us the way that the details of a piece affect the infrastructural representations that are formed, and hence the higher-level representations as well. One way to examine this is through "recomposition"—selectively altering certain details of a piece, to see what kind of changes result in structure and effect. Figure 12.14 shows the first 16 measures of the minuet from Beethoven's Sonata Op. 10 No. 3; figure 12.15 shows a series of "recompositions" of part of the passage. In all cases, the aim was to make just a single change of one kind or another, so that the effect of this change could be examined.

In figure 12.15a, the third, fourth and fifth notes of the melody have been rearranged. The effect is subtle, but important: the dissonant D on the downbeat of m. 2 has now been replaced with a chord-tone C♯, robbing the passage of a small but important hint of harmonic tension. A slight effect on motivic structure occurs as well: the subtle connection between the first two notes of m. 2 and m. 5 is now broken. In the next two recompositions the consequences are mainly motivic. In figure 12.15b, the A in m. 4 has been shifted to the left by one beat, placing it on the downbeat of m. 4; now the first four measure-phrase is almost rhythmically identical to the second one, creating a strong link between

Figure 12.15
A series of partial recompositions of the Beethoven passage in figure 12.14. (A) mm. 1–3. (B) mm. 1–4. (C) mm. 1–4. (D) mm. 1–8. (E) mm. 1–4. (F) mm. 1–4. (G) mm. 9–13. (H) mm. 9–12. (I) mm. 1–4. (J) mm. 1–4.

Figure 12.15 (continued)

them. Since the third phrase (if altered in a similar way) and fourth feature the same rhythmic pattern, this would perhaps be a bit too much of one rhythmic idea. In figure 12.15c, this same A is shifted further to the left; now the strong-weak motive in m. 3 (D-A) is heard to relate to the similar descending strong-weak note pairs in mm. 2, 5 and 6.

In figure 12.15d, a single, radical change is introduced: the entire first eight measures are rewritten in 2/4. (Imagine that the tempo at the measure level is the same.) This brings up a general, important point: why does meter matter? Exactly what difference does it make (because we know that somehow it does make a difference) if a piece is in one meter or another? This example gives some possible answers. The dissonant D on the downbeat of m. 2 is now longer than before, as is the dissonant E-G dyad on the downbeat of m. 6, making them perhaps a little too destabilizing. The ii6 chord on the third beat of m. 6—already brief—is now briefer still, creating tension due to penalization by the strong beat rule (HPR 2). The grouping structure is also subtly affected; because the quarter-notes are longer than the corresponding notes in the original, there is more of a tendency to hear them as low-level group boundaries (PSPR 1), creating groups of eighth-eighth-quarter. This, in turn, tends to promote "eighth-eighth-quarter" as a rhythmic motive, one that occurs repeatedly in the passage.

Figure 12.15e has consequences mainly for contrapuntal structure. In the original version, the melody of the first eight measures has a slight tendency to break into two substreams, due to the shift in register from D to A in m. 4. (One can hear the upper stream continuing from the first phrase to the third; in the fourth phrase, the two sub-streams are perhaps brought together.) This is a case of the kind of ambiguity discussed in section 8.6; although the four phrases clearly constitute a single line, one is also aware of two separate contrapuntal strands. In figure 12.15e, the shift downwards on the downbeat of m. 3 tends to put this note in the lower stream rather than the upper one. The first phrase of the upper stream then ends on a metrically weak (and tonally inconclusive) note, creating some instability.

In figure 12.15f, the main consequence is for phrase structure. Since the long gap after the D of m. 3 has been filled in, there is no longer a strong tendency to hear a phrase boundary there (or anywhere else in the vicinity); thus the first eight measures tend to come across as a single phrase. However, the length of this phrase—20 notes—causes some tension due to violation of the phrase length rule (PSPR 2). This case brings to mind a point made earlier about the possible psychological origin of the phrase length rule. As discussed in section 3.4, a phrase seems to serve as a kind of perceptual present, a "chunk" of information that is available to awareness all at once; a long phrase such as the one in figure 12.15f (continuing with mm. 5–8 of the original) presents a lot of information in a single chunk, and is thus somewhat taxing for the listener.

Figure 12.15g, a recomposition of the third phrase (mm. 9–13), relates to harmonic and tonal structure. By replacing the A and C♯ in m. 12 with a B, we alter the A7 harmony to E minor; this is heard is as a continuation of the E minor harmony in the previous measure, making it quite emphatic. (The parallel between mm. 9–12 and mm. 1–4 is also increased.) In tonal terms, m. 12 is now slightly more compatible with E minor than in the original (since B is in the E minor tonic triad, while C♯ is not even in the E minor scale collection); the tonicization of E minor in mm. 9–11 is now felt to extend through m. 12 (at least until the last beat of m. 12). This—and the removal of the A7 chord in m. 12—slightly undermines the return to D major in the following measures, leaving the conflict with E minor not entirely resolved. Figure 12.15h shows another possible change to mm. 9–12, with the opposite effect: the D♯'s in m. 10 are changed to D's, and the B removed, creating a D6 chord instead of B4/3. This essentially neutralizes E as a tonal center (with the D♯'s gone, the entire passage remains comfortably within D major); it also significantly alters the harmonic journey of the passage, removing the B harmony which was the one move beyond the narrow line-of-fifths range of E to D.

Not all details are important. The effect of the change in figure 12.15i (the alteration of the tenor voice in m. 3) is negligible; the infrastructural representations (metrical, grouping, contrapuntal, TPC, harmonic and key) are almost completely unaffected, as is the motivic structure. Perhaps all it does is draw a slight degree of attention to the tenor voice, since there is now a slightly higher degree of activity in that voice. At the other extreme, many possible changes we might make—indeed, the vast majority of them—seem wholly unacceptable and well outside the bounds of the style. Here, too, the preference rule approach is of explanatory value (as discussed in section 11.2). The recomposition in figure 12.15j—that is to say, the wildly inappropriate high G♯ in m. 2—can be seen to be "low-scoring"—and hence "ungrammatical"—in several respects: in terms of harmony (because there is no high-scoring harmonic interpretation that accomodates the G♯ as well as the other nearby events), meter (because the G♯ is a long event on a weak beat), and counterpoint (because the G♯ does not fit well into the streams required by the surrounding events, and thus requires either a huge leap within an existing voice or the momentary addition of an extra voice). The note is also somewhat destabilizing in TPC and tonal terms, though not unusually so.

In this experiment, I have tried to show that the preference rule approach can be informative as to the consequences of microscopic compositional decisions and changes. It shows us how a piece of music is *fragile*: its effect can be profoundly changed by what seems superficially to be a small alteration.[11] Does the preference rule approach show us *why* Beethoven made the specific choices he did? Not really. It suggests that, perhaps, Beethoven chose figure 12.14 over figure 12.15a because he wanted the added degree of ornamental-dissonance tension that figure 12.14 provides, along with the subtle motivic connection to m. 5. However, it tells us nothing about why Beethoven's choice was better (as it surely was). In some cases, a melody can seem perfect with no ornamental-dissonance tension at all (see the minuet of Beethoven's Op. 7); and more motivic connections are not always better, as figure 12.15b suggests. In short, if our aim is to explain the quality of a piece—and I deliberately chose this piece, a particular favorite of mine, because I feel that there is a great deal of quality to be explained—a preference rule approach can only take us so far. It explains why the particular notes chosen lead to the infrastructural representations they do, but it says little about the much harder question of why a certain complex of infrastructural features is satisfying or enjoyable.

12.9 The Power of Common-Practice Music

Let us summarize the various functions of infrastructural levels that have been presented here. Through harmonic and tonal structure, the common-practice system has the ability to create complex, multi-leveled journeys. These journeys also carry complex emotional associations, related both to the tonal center and pitch collection used, which can in part be accounted for by the positions of these elements on the line of fifths. Motives—repeated patterns of rhythm (reinforced by meter) and pitch—are also an important aspect of common-practice music, and rely crucially on several aspects of the infrastructure. As other authors have noted, motives often serve as "agents," entities with feelings, desires, and the capacity for action. Combined with the expressive powers of the tonal system, this allows composers to create complex dramatic narratives. (On the other hand, other things besides motives can also serve as agents—chords, keys, instruments; as Maus [1988] has shown, the shifting sense of agency that one often finds in music is one of its unique and fascinating features.)

Metrical and harmonic analysis, along with motivic analysis, are also part of a complex process of searching for order and pattern in a piece—that is, of "making sense of it"—which has inherent appeal, analogous to many other intellectual activities. Music can convey fluctuating levels of energy, through the use of tempo, as well as through other means such as dynamics and register. Infrastructural representations—harmony and tonality in particular, but others as well—also have the power to convey varying levels of tension, by being more-or-less "high-scoring" by the preference rules used to generate them. Finally, configurations of infrastructural elements (as well as the input representation) can form higher-level patterns or "schemata," either those whose function is primarily structural (e.g., cadences), or those with extramusical or expressive associations (e.g., Classical "topics").

It is useful to consider the role of preference rule systems in these various aspects of music. In some cases, the kinds of higher-level entities and meanings I have described arise simply from the output of preference rule systems—that is, infrastructural representations—without directly involving preference rule systems themselves. For example, a Classical topic depends on a certain configuration of infrastructural representations, but how those infrastructural levels were in themselves derived is, perhaps, of little consequence for the perception of topics. In other respects, however, preference rule analysis is directly engaged with the higher-level effects of music. For example, the line of fifths—which I argued was important for the spatial and dramatic aspects of music—also plays a central role in the choice of harmonic and TPC labels (through the harmonic and pitch variance rules). Similarly, the kind of musical tension described earlier in this chapter results directly from the process of *forming* infrastructural representations, rather than from the output of that process; the idea of musical listening as a "search for order" is, likewise, heavily dependent on the analytical process itself.

None of these higher-level aspects of music are fully understood; to understand them better would require consideration of how they operate in many specific pieces. Moreover, this is only a partial survey of the functions and meanings arising from common-practice music. I have focused on meanings that in some way are connected with the infrastructural representations that have been our primary concern here. Nothing said here in any way denies that further meanings exist, and that they may be extremely important. Nevertheless, the functions discussed above begin to show us the tremendous range of resources available within the common-practice musical system; and in so doing, they begin to explain how music within this system is able to affect us and appeal to us as it does. Given this range of resources, I would argue, there is nothing fundamentally mysterious or inexplicable about the power of common-practice music.

Appendix: List of Rules

This appendix lists all the rules proposed for the six preference rule systems presented in Part I. Both well-formedness rules and preference rules are listed. The well-formedness rules are incorporated into a well-formedness definition (WFD), stating the constraints on what constitutes a well-formed structure.

I. Metrical Structure

WFD. A well-formed metrical structure consists of several levels of beats, such that

MWFR 1. Every beat at a given level must be a beat at all lower levels.

MWFR 2. Exactly one or two beats at a given level must elapse between each pair of beats at the next level up.

MPR 1 (Event Rule). Prefer a structure that aligns strong beats with event-onsets.

MPR 2 (Length Rule). Prefer a structure that aligns strong beats with onsets of longer events.

MPR 3 (Regularity Rule). Prefer beats at each level to be maximally evenly spaced.

MPR 4 (Grouping Rule). Prefer to locate strong beats near the beginning of groups.

MPR 5 (Duple Bias Rule). Prefer duple over triple relationships between levels.

The following rules are not included in the implemented model, but are considered important:

MPR 6 (Harmony Rule). Prefer to align strong beats with changes in harmony.

MPR 7 (Stress Rule). Prefer to align strong beats with onsets of louder events.

MPR 8 (Linguistic Stress Rule). Prefer to align strong beats with stressed syllables of text.

MPR 9 (Parallelism Rule). Prefer to assign parallel metrical structures to parallel segments. In cases where a pattern is immediately repeated, prefer to place the stronger beat on the first instance of the pattern rather than the second.

II. Phrase Structure

WFD. A well-formed phrase structure for a melody consists of a segmentation of the melody into non-overlapping phrases, such that every note is entirely contained in a phrase.

PSPR 1 (Gap Rule). Prefer to locate phrase boundaries at a) large inter-onset-intervals and b) large offset-to-onset intervals.

PSPR 2 (Phrase Length Rule). Prefer phrases to have roughly eight notes.

PSPR 3 (Metrical Parallelism Rule). Prefer to begin successive groups at parallel points in the metrical structure.

III. Contrapuntal Structure

WFD. A well-formed contrapuntal structure is a set of streams, subject to the following constraints:

CWFR 1. A stream must consist of a set of temporally contiguous squares on the plane.

CWFR 2. A stream may be only one square wide in the pitch dimension.

CWFR 3. Streams may not cross in pitch.

CWFR 4. Each note must be entirely included in a single stream.

CPR 1 (Pitch Proximity Rule). Prefer to avoid large leaps within streams.

CPR 2 (New Stream Rule). Prefer to minimize the number of streams.

CPR 3 (White Square Rule). Prefer to minimize the number of white squares in streams.

CPR 4 (Collision Rule). Prefer to avoid cases where a single square is included in more than one stream.

CPR 5 (Top Voice Rule). Prefer to maintain a single voice as the top voice: try to avoid cases where the top voice ends, or moves into an inner voice.

IV. Tonal-Pitch-Class Representation

WFD. A well-formed tonal-pitch-class representation is a labeling of each pitch-event with a single tonal-pitch-class label.

TPR 1 (Pitch Variance Rule). Prefer to label nearby events so that they are close together on the line of fifths.

TPR 2 (Voice-Leading Rule). Given two events that are adjacent in time and a half-step apart in pitch height: if the first event is remote from the current center of gravity, it should be spelled so that it is five steps away from the second on the line of fifths.

TPR 3 (Harmonic Feedback Rule). Prefer TPC representations which result in good harmonic representations.

V. Harmonic Structure

WFD. A well-formed harmonic structure is a complete segmentation of the piece into non-overlapping chord-spans.

HPR 1 (Compatibility Rule). In choosing roots for chord-spans, prefer certain TPC-root relationships over others, in the following order: $\hat{1}$, $\hat{5}$, $\hat{3}$, $\flat\hat{3}$, $\flat\hat{7}$, $\flat\hat{5}$, $\flat\hat{9}$, ornamental. (An ornamental relationship is any relationship besides those listed.)

HPR 2 (Strong Beat Rule). Prefer chord-spans that start on strong beats of the meter.

HPR 3 (Harmonic Variance Rule). Prefer roots that are close to the roots of nearby segments on the line of fifths.

HPR 4 (Ornamental Dissonance Rule). An event is an ornamental dissonance if it does not have a chord-tone relationship to the chosen root. Prefer ornamental dissonances that are a) closely followed by an event a step or half-step away in pitch height, and b) metrically weak.

VI. Key Structure

WFD. A well-formed key structure is a labeling of each segment of the piece with a key.

KPR 1 (Key-Profile Rule). For each segment, prefer a key which is compatible with the pitches in the segment, according to the (modified) key-profile formula.

KPR 2 (Modulation Rule). Prefer to minimize the number of key changes from one segment to the next.

Notes

Chapter 1

1. The term "tonal music" is motivated by the fact that this corpus of music shares a certain system of pitch organization. The term is problematic, since much other music—much popular music, for example—shares essentially the same system; moreover, there is another broader sense of "tonal," meaning music which has a focal pitch or tonal center. Nevertheless, "tonal music" is a useful and familiar term for the corpus I wish to study, and I will sometimes use it here.

2. While understanding human cognition is one goal of artificial intelligence, some work in the field has more practical applications. Given a process that humans perform—picking up an object, perceiving spoken language, or making medical diagnoses—one could study the process either with the aim of finding out how humans do it, or with the aim of devising a system to perform the process as well as possible in a practical situation. These are distinct goals, though they are certainly related. In the present case, computational analysis of music could be useful for a number of purposes. For example, in order to produce a sensible notation for something played on an electronic keyboard, one must determine the meter and barlines, the spelling of the notes, the correct division of the notes into different voices, the key signature, and so on. However, it is best to keep the psychological and the practical goals distinct; here my focus will be entirely on the former.

3. For reviews of this work, see Kauffman & Frisina 1992 and Marin & Perry 1999.

4. For discussions of this issue, see Chomsky 1980, 11–24, 189–97; Fodor & Pylyshyn 1988. Fodor and Pylyshyn observe that even in the debate between connectionist and symbolic approaches to cognition—a debate that is in some ways very fundamental—both sides agree on the necessity of mental representations. There have been, and continue to be, alternatives to the representational approach. One is behaviorism; another is the "direct perception" theory of J. J. Gibson (see Bruce and Green 1990, 381–9, for discussion).

5. For discussion of these points, see Putnam 1975; Fodor 1975, 27–53; Haugeland, 1981.

6. Not all the kinds of structure studied in this book are normally represented in music notation; for example, harmonic and key structure usually are not (except that the key signature shows the main key of a piece), and phrase structure often is not. And even when structures are shown in notation, they may not necessarily coincide with the structures that are heard; this sometimes happens in the case of contrapuntal structure, for example, as I will discuss later.

7. Lerdahl and Jackendoff's theory (discussed further below) has been criticized on exactly these grounds (Cohn 1985, 36–8). Perhaps the criticism has been due, in part, to the fact that the theory is largely concerned with pitch reduction—an area that seems particularly subjective and open to differences of opinion.

8. One could also mention articulation—that is, variations in the sounds of notes within an instrument, such as staccato versus legato and accent. However, these qualities are to some extent expressed in duration, and thus will be reflected in the input representation; for example, the fact that the first two chords in figure 1.1 are played staccato is clearly seen in the "piano roll." Other aspects of articulation are expressed in timbre and dynamics.

A further kind of information which might be considered is spatial location. We will discuss this further in chapter 4.

9. See Ashley 1989 for an insightful discussion of these issues. Ashley notes both the possibility of implementing preference rule systems and the importance of doing so.

10. All of the implementations discussed in this book—for the metrical, phrase structure, contrapuntal, harmonic-TPC, and key models—are publicly available at the website <www.link.cs.cmu.edu/music-analysis>. The programs are written in C and run on a UNIX platform. The website also contains information about the use of the programs, as well as input files for many of the musical excerpts discussed in this book.

Chapter 2

1. As I will explain below, it is often better to think of rhythms in terms of *inter-onset intervals*, rather than durations; but we will overlook this distinction for now.

2. I am not certain that exactly these results would be obtained; but it is only the general point that matters here, not the details.

3. We should be cautious in describing the perception of rhythms as "categorical." Categorical perception has a specific meaning in psychology; it refers to the fact that two stimuli in the same category are less easily discriminated than two stimuli, differing by an equal amount in physical terms, in different categories. The classic case is phonemes: two acoustically different "b's" are less easily distinguished than a "b" and "p" differing acoustically by the same amount (Garman 1990, 195–201). In the case of rhythms, however, discrimination is not the issue. It is certainly possible to discriminate two durational patterns that are understood as implying the same rhythm; for example, we can distinguish two performances

of the same passage played with different expressive timing. The point is that, even when we can discriminate them, we understand that their rhythms are in some sense the same.

4. The view I am embracing here is the prevailing one in music theory: most theorists hold that meter is primarily a local phenomenon, not usually extending more than two levels above the measure level. (See Lerdahl & Jackendoff 1983, 21–5; Benjamin 1984, 403–13; Schachter 1987, 16–17; and Lester 1986, 160–1. For an opposing view, see Kramer 1988, 98–102.)

5. Two partial exceptions are the systems of Longuet-Higgins and Steedman (1971) and Rosenthal (1992). Longuet-Higgins & Steedman tested their metrical algorithm on the 48 fugue subjects from Bach's *Well-Tempered Clavier*; Rosenthal tested his on a set of Mozart and Bach melodies.

6. In some cases, it is not obvious which notes *should* be considered extrametrical. Generally, extrametrical notes are notated in small noteheads (or with other special symbols); however, the triplet sixteenth notes in figure 2.7 might well be considered extrametrical. With some ornaments, such as some trills and mordents, the note falling on the beat is clearly part of the ornament; in such cases it could be argued that the note *following* the ornament is extrametrical.

7. Povel & Essens's (1985) model also incorporates this factor, as it prefers to locate beats on the first of a series of short notes.

8. There is a problem with the method just described for quantizing to pips: what if two notes are just 1 ms apart, but happen to be shifted over to different pips? To solve this, after quantization to pips, the program goes through all the pips, from left to right; each time it encounters two subsequent pips that both contain event-onsets, it shifts the onsets of the events in the second pip back to the first.

9. The note score for a pip is calculated as follows. Each note contributes a value which is either the maximum of its duration and its registral IOI, or 1.0, whichever is smaller (this value is known as the "effective length" of the note); this was to prevent notes of very long duration or registral IOI from adding too much weight to the score. The total note score for a pip is then the square root of its number of notes times their average effective length; this is similar to simply adding the effective lengths of the notes, but it prevents chords with many notes from having undue weight. Finally, a value of 0.2 is also added to the note score for each pip with any event-onsets at all, so that there is an extra bonus for putting beats on such pips.

10. As mentioned earlier, a bonus is added at level 4 for putting a beat at the first level 3 beat.

11. The workbook accompanies Kostka and Payne's textbook *Tonal Harmony* (1995a). The harmonic and key analyses are provided in an instructors' manual accompanying the workbook (Kostka 1995).

12. There are occasional cases where the metrical structure implied by notation is not totally clear. For example, in a passage featuring a triplet eighth notes in one hand and duple eighths in the other, it may be hard to decide which division is the primary one. Fortunately there were no such passages in the corpus. (In cases

where an occasional duple rhythm occurred against a steady triplet accompaniment, the triplet was treated as primary.) As for extrametrical notes (which only occurred in the unquantized input files), any analysis of such notes by the program was treated as correct. Even so, allowing extrametrical notes in the input still posed a challenge to the program, as I will discuss, since it sometimes caused errors in the analysis of metrical notes.

13. See Lester 1986 for discussion of the interaction between meter and harmony (pp. 66–7) and many interesting examples.

14. Looking at harmonic change, rather than bass notes, would give better results in the case of figure 2.11, since harmonic changes generally occur on strong beats. (At least, there are generally clear harmonic changes on the third quarter-note beat of each measure.)

15. Statistically, it has been found that performers (specifically pianists) tend to play metrically strong notes slightly louder than others; however, the difference is small. In a study by Drake and Palmer (1993), pianists were instructed to play simple rhythmic patterns; the notation of the patterns indicated the time signature and hence the metrical structure. In one typical pattern, the average intensity (loudness) of the metrically strongest events was about 58 units, while that of the weakest events was about 51 units (p. 356). While the units are arbitrary, the authors state that there is a typical performance range of 30 to 90 units; within this range, then, the variation due to metrical strength seems fairly small.

16. The importance of the "first-occurrence-strong" rule can be seen in cases where the second and third measures of piece present two occurrences of a pattern; this tends to make us hear the even-numbered measures as strong, contrary to the usual tendency. Instances include the following (in some of these cases the metrical unit in question is half a measure or two measures): Beethoven's String Quartet Op. 18 No. 3, first movement; Mozart, Serenade K. 388, second movement; Bach, *Well-Tempered Clavier* Book I, E Major Prelude; Haydn, String Quartet Op. 64 No. 2, first movement; Beethoven Symphony No. 7, III; Haydn Symphony No. 101, first movement (*presto*).

Chapter 3

1. This rule appears *not* to apply to auditory perception, as we will see in chapter 4.

2. See, for example, Koffka 1935, 431–48. See also the discussion of Gestalt theory in Lerdahl and Jackendoff 1983, 40–1.

3. Another quite different kind of musical grouping involves the grouping of events into a line, separate from other simultaneous lines in a polyphonic texture. This is the subject of chapter 4.

4. See Dowling & Harwood, 1986, 179–81. Evidence for this has also been found in studies of musical performance, as I will discuss in chapter 11.

5. These studies are somewhat problematic. The greater ease of learning Figure 3.2a may be due to the fact that the repeated pitch pattern in figure 3.2a (but not in figure 3.2b) is reinforced by a repeated rhythmic pattern, rather than to the effect of pauses as grouping cues. We return to this issue in section 12.2.

6. See Baker 1976 for an overview of Koch's theory.

7. It is worth noting that some theorists advocate an idea of the term "phrase" rather different from what I will assume here. In particular, Rothstein argues that a phrase should be defined primarily by tonal motion (1989, 5). This often leads to somewhat larger phrases than what is implied by the common use of the term; for example, in the "Blue Danube" Waltz, Rothstein's phrases are 16 measures in length. Rothstein acknowledges that his idea of phrase is somewhat prescriptive; he laments, for example, that four-measure segments in the "Blue Danube" are so often called phrases (1989, 5). As I will discuss later, though, it appears that tonal factors do play some role in phrase structure, even by the common understanding of the term.

8. The groups in hearing B are in fact an exact inversion of the opening motive, although I feel this is a weak factor perceptually.

9. See London (1997) for an insightful discussion of ambiguity in grouping. We will return to this issue in chapter 8.

10. These slurs are of course added by the transcriber of the song rather than the composer (in most cases the composer is unknown), but this is of little importance. Even structural information (phrasing slurs, bar lines, etc.) added by the composer is really only one "expert opinion," which one assumes generally corresponds to the way the piece is heard, but may not always do so.

11. As discussed earlier, Tenney and Polansky factor pitch and dynamic intervals into their scores as well as temporal intervals—and their temporal interval score includes only IOI, not OOI. However, it does not appear that Tenney and Polansky's complete model would produce good results here. Even if pitch were considered as well as IOI, for example, the fourth interval in figure 3.7b would still be a local maximum—since the third, fourth, and fifth intervals are all equal in pitch—and thus would still be (incorrectly) considered a phrase boundary.

12. The idea of two preference rule systems interacting in this way is not circular or incoherent, and in fact it is quite susceptible to computer implementation; for an example, see section 6.4. However, implementation will not be attempted in this case.

13. The stratified sample was prepared by Paul von Hippel.

14. One might question the decision to give the grouping program the "correct" metrical structures, as indicated in the transcriptions. It is possible that these metrical judgments were influenced by intuitions about grouping; in that case, there is a danger of circularity in giving the program this information. It seemed likely, however, that judgments of meter in the songs in this corpus were not very much affected by grouping, and depended mainly on other factors discussed in chapter 2—the event rule, the length rule, parallelism, and harmony. We should note, also, that the *phase* of the metrical structure (the exact placement of strong beats) does not influence the grouping program; only the period does. Still, this remains a possible problem with the test.

15. A clear break in the melody and a slight decrease in density in all voices—such as a shift from eighth notes to quarter notes—may have the same effect on the overall density of attack-points, but the former strongly suggests a phrase boundary while the latter does not.

Chapter 4

1. The renewed awareness of the contrapuntal basis of much Western music is partly due to the influence of the theorist Heinrich Schenker. I will return in section 8.6 to the connection between the current study and Schenkerian analysis.

2. Sequential integration is in fact somewhat broader than contrapuntal analysis; the former term includes sequential grouping of speech and other non-musical sounds. However, most of the research that Bregman reviews under the name of "sequential integration" uses stimuli which could well be considered musical: e.g., sequences of steady-state tones.

3. Van Noorden also found that, under some circumstances, listeners could choose to hear the two streams either segregating or fusing.

4. See Bregman 1990 for further discussion of the role of these factors and other experimental work on sequential integration.

5. We should note here that one could make a case for hierarchical grouping of streams. In the Mozart excerpt, for example (figure 4.1), one might suggest that the two left-hand streams from m. 12 onwards form a single larger stream. Such higher-level streams would, of course, contain multiple simultaneous notes. However, our only concern here is with the lowest level of streams.

6. The interaction of this rule with CWFR 4 also deserves comment: as long as a given note is entirely contained in a single stream, it is permissible for another stream to contain only part of the note.

7. It might occasionally be advantageous for a stream to move to a white square in order to avoid crossing or colliding with another stream. (I am indebted to Daniel Sleator for pointing this out.) In practice, however, this situation seems to arise very rarely.

Chapter 5

1. The line of fifths is not new. It is represented as one axis in several of the multi-dimensional spaces discussed earlier. It is also discussed more directly by Regener (1974) and Pople (1996); however, these authors make no argument for the line of fifths as a cognitive model for tonal music.

2. The problem is not choosing a location for a single element in isolation (presumably all A♭ points are equal, since the space is symmetrical), but rather choosing a location relative to the other elements already located in the space.

3. On instruments with variable tuning, such as stringed instruments, it is possible that different spellings of the same pitch are *acoustically* distinguished; and this in turn might affect our perceptual judgments. I am not aware of any acoustical or psychological evidence on this matter.

4. Longuet-Higgins and Steedman define "diatonic semitone" in terms of motion on their two-dimensional space. We could also define a "diatonic semitone" as an

interval of five steps on the line of fifths; this relates to TPR 2 in the model proposed below.

5. Krumhansl (1990, 67) provides data from various studies as to the frequency of scale degrees (i.e., pitches relative to the tonic) in tonal music. Considering only the diatonic scale degrees (which account for over 90% of the pitches), and assuming a tonic of C (2.0), the mean line-of-fifths position of all pitches in Krumhansl's corpus is 3.82: close to D.

6. This rule—that $\flat\hat{7}$ is generally preferred over $\sharp\hat{6}$, and $\sharp\hat{4}$ over $\flat\hat{5}$, regardless of voice-leading—is noted by Aldwell and Schachter (1989, 472–3).

The symbol "ˆ" above a number, which will be used frequently, indicates a scale or chord degree. For example, $\hat{3}$ represents either the third degree of a scale or the third of a chord; E is $\hat{3}$ of C. The major scale is assumed here; other scale steps can be indicated with ♯ or ♭ signs—for example, E♭ is $\flat\hat{3}$ of C, G♯ is $\sharp\hat{5}$ of C.

7. We should also consider how the current algorithm handles the spelling of events in minor passages. It is somewhat unclear what the conventional wisdom is on this matter. In major, the spelling of events within the scale of the current key is of course given by the usual spelling of that scale. In minor, however, there is not one scale collection, but three (the natural, harmonic, and ascending melodic). Aldwell and Schachter offer a rule for spelling chromatic scales in minor which seems to describe compositional practice fairly well (1989, 473). Any pitch that is in any of the three minor scales is spelled accordingly, regardless of voice-leading; that is, the spellings $\flat\hat{6}$-$\hat{6}$-$\flat\hat{7}$-$\hat{7}$ are always used. In addition, $\hat{3}$ is preferred over $\flat\hat{4}$, and $\sharp\hat{4}$ over $\flat\hat{5}$; thus only the spelling of $\flat\hat{2}/\sharp\hat{1}$ depends on voice-leading. The current model points to a possible, though imperfect, explanation for this. The scale degrees $\hat{1}$-$\hat{2}$-$\flat\hat{3}$-$\hat{3}$-$\hat{4}$-$\sharp\hat{4}$-$\hat{5}$-$\flat\hat{6}$-$\hat{6}$-$\flat\hat{7}$-$\hat{7}$ form a compact region of 11 steps on the line of fifths; as with major, it may be that the spelling of events within this range is enforced by the pitch variance rule, overruling voice-leading. I am unable to explain, however, why the range in minor should be slightly wider than in major (where the range of TPC's whose spelling is invariant is only 9 steps).

8. As with the contrapuntal program, this means that extrametrical notes cannot be handled.

9. At this point, we ensure that no change of spelling takes place within a single note. If two adjacent segments contain parts of the same note, the 12-step windows chosen for the two segments in an analysis must both imply the same spelling for that note.

10. When the program's output was adjusted to be completely correct (corresponding exactly with the Kostka-Payne corpus in relative terms, except in cases of enharmonic modulation), 84% of the TPC events were within the range of A♭ to C♯. However, this was not done in such a way as to maximize the number of TPC events within a certain range. By shifting the analyses of certain excerpts in their entirety by 12 steps, a completely correct (by the current criterion) TPC analysis of the corpus might be found which had an even higher proportion of events between the range of (say) A♭ to C♯.

11. In the context of B♭ major—the tonality of previous context—A♭ and G♯ are both chromatic also, strictly speaking. However, A♭ is a $\flat\hat{7}$ scale degree, and as such is not subject to the voice-leading rule.

While A♭ would not be considered "the" correct spelling for the note in question here, one might argue—as noted above—that it is the spelling that we first entertain upon hearing the note. The program's handling of such diachronic shifts in perception will be discussed further in chapter 8.

Chapter 6

1. We should note also that performing harmonic analysis in listening involves *hearing* the music. In a theory class, by contrast, students are often expected to do harmonic analysis from a *visual* representation of a piece—a score—which is quite different. One can learn to aurally imagine music from a score, of course, but this takes considerable practice.

2. In the discussion of grouping in chapter 3, I suggested that, in fact, harmony does not appear to be a central factor in melodic segmentation; and I will suggest in chapter 7 that harmonic information is generally not necessary in judgments of tonality either. Still, these experiments are valuable in providing further evidence for the psychological reality of harmony.

3. Two other less ambitious attempts to model tonal harmony using neural networks are Scarborough, Miller, & Jones 1991 and Laden & Keefe 1991.

4. The idea of virtual pitches was first formulated by Terhardt (1974).

5. The question of why these relationships are the most preferred ones (and why some are more preferred than others) is a complex one; psychoacoustic factors are undoubtedly relevant. According to Terhardt's "virtual pitch" theory (1974), the explanation has to do with the relative strength of partials of a complex tone. For example, the third partial of a sound (corresponding to $\hat{5}$ of a chord) tends to be stronger than the fifth partial (corresponding to $\hat{3}$); this is why a pitch is more readily heard as $\hat{5}$ than as $\hat{3}$. (This is the basis for Parncutt's theory of harmony [1989], discussed in section 6.2 above.)

6. Bharucha (1984, 494–5) actually expresses this in terms of scales: an anchored pitch is one that is followed by another pitch a step away in the current diatonic scale. The current formulation is somewhat different, but seems to work well in practice.

7. Recall that segments are determined by the lowest level of the metrical structure. In a quantized piece with perfectly regular meter, segments would all be the same length; but in a file generated from a live performance, or with irregular meter, they might vary in length.

8. The necessary metrical information was added to the input files. I simply added the metrical information that seemed correct; to a great extent, of course, the metrical structure is implied by the notation of the excerpts.

9. The German sixth chord in m. 36 is mishandled here; the A♯ is spelled as a B♭ (see section 5.6).

10. The harmonic and key programs can in fact be run in combination to produce a complete Roman numeral analysis, with mode, extension, and inversion information added; this is explained in section 7.6.

11. We should note that the harmonic variance rule—preferring successive harmonies to be located close together on the line of fifths—serves this function in a small way. It indirectly favors both circle-of-fifths progressions and cadences, since both involve movement to adjacent harmonies on the line of fifths. (See section 8.8 for further discussion.)

Chapter 7

1. Here, of course, I am referring to tonal music: music which conveys a sense of key or tonal center.

2. The original data was gathered for a variety of keys, but there was little variation between major keys (after adjusting for transposition), so the data were averaged over all major keys to produce a major key-profile which was then used for all major keys; the same was done for minor keys (Krumhansl 1990, 25, 27).

3. Normalizing the input vector values for their mean and variance has no effect on the algorithm's judgments, since the input vector is the same for all 24 keys. Normalizing the key-profile vectors does have a slight effect, however. Since the mean value for the minor key-profiles is a bit higher than for major key-profiles, removing this normalization tends to bias the algorithm towards minor keys, relative to the original K-S algorithm. However, this effect could be counteracted simply by adjusting the key-profile values. As we will see, the key-profile values seem to require significant adjustment in any case.

4. Actual statistical analyses of entire pieces show mixed results. Krumhansl (1990, 66–71) reports that some note counts in pieces show a high key-profile correlation with the appropriate keys, although she does not show that these note counts correlate more highly with the "correct" key than with any other. In other studies the results have been unpromising. Krumhansl (1990, 71–3) found the key-profile model predicted G major as the key of Schubert's *Moment Musical* No. 1, whereas the correct key is C major; Butler (1989, 226–8) found that the pitch-class distribution of Schubert's *Moment Musical* No. 2 in A♭ major yielded a poor match to the key-profile for that key. Of course, one might attribute these failures to the details of the algorithm, such as the fact that the key-profile values are not ideal. As I have said, however, I believe that using the key-profile model to predict global key is misguided.

5. The possible choice of keys was limited to a range of 28 steps (C♭♭ to Fx) on the line of fifths.

6. A similar set of tests was reported in Temperley 1999b. There, however, an NPC version of the model was used rather than a TPC version. Thus the results of the tests are slightly different. The results of the TPC and NPC versions will be compared below.

7. In 45 of the 48 *Well-Tempered-Clavier* cases, and 43 of the 46 Kostka-Payne cases, the procedure described here resulted in a segment level between 1.0 and 2.0 seconds; in the remaining cases, the segment length was slightly more than 2.0 seconds (these were slow triple meter pieces, where there was no level between 1.0 and 2.0 seconds). In cases where the excerpt began with an incomplete segment, this portion was treated as a separate segment if it was at least half the length of a regular segment; otherwise it was absorbed into the following seg-

ment. Partial segments at the end of the excerpt were always treated as complete segments.

8. In a few cases, an entire excerpt was spelled in a different cycle of the line of fifths than in the notation. This occurred because the spellings of the excerpts were generated by running them through the pitch spelling program and correcting them as necessary, and in some cases the pitch spelling program notated something in the wrong cycle of the line of fifths. In such cases, the "correct" key analysis was adjusted accordingly. Since both pitch spellings and key analyses are essentially relative rather than absolute, this did not appear to be problematic (see section 5.4 for discussion).

9. Vos & Van Geenen's assumption that only six of the fugue subjects modulate is not necessarily correct. Keller only mentions modulations in six cases, but it is not clear that he would have mentioned all cases. It appears that several others may modulate: Book II, No. 10, for example.

10. Krumhansl's tests of the K-S algorithm have already been described, as well as tests by various authors on the *Well-Tempered-Clavier* fugue subjects. Vos and Van Geenen also tested their model on fugue subjects by other composers. Holtzmann tested his model on a corpus of 22 other melodies, but it is not clear how these melodies were chosen. Longuet-Higgins and Steedman's, Holtzmann's, and Vos and Van Geenen's tests were all limited to monophonic excerpts. Winograd presents tests of three short pieces; Maxwell gives results for two pieces; Leman gives results for two pieces.

11. If the key-profile values were to be based on pitch-class distribution in pieces, this information would have to be gathered relative to the *local* key rather than the main key—unlike the pitch-class tallies by Krumhansl and Butler mentioned earlier (see note 4).

12. As a further comparison, the Kostka-Payne test was also run using the key-profile values from the original Krumhansl-Schmuckler algorithm. Various values of the change penalty were used; the best performance occurred with a change penalty of 12 (as in the tests of my own version). With the original key-profile values, 622 of 896 segments were judged correctly—a rate of 69.4%, compared to 83.8% in the test just described. (Since the original K-S profiles use NPCs, not TPCs, the appropriate comparison is with the NPC version of my algorithm.)

13. The time-point of each event was determined by its onset; duration was not considered.

14. The model was also tested using weighted, rather than flat, input vectors (so that the value for a pitch-class in a segment's local input vector was given by the total duration of events of that pitch-class in the segment). The level of performance was almost identical to the flat-input version; with a half-life of 4.0 seconds, it yielded a score of 69.9% correct.

Chapter 8

1. This is in keeping with the spirit of *GTTM*, which addressed itself to "the final state of [the listener's] understanding" of a piece, rather than the intermediate stages and provisional analyses leading to that understanding (Lerdahl & Jackendoff 1983, 3–4).

2. A widely discussed example of hypermetrical revision is the opening of Mozart's Symphony No. 40; see Lerdahl & Jackendoff 1983, 22–5, and Kramer 1988, 114–7.

3. A more extended investigation along these lines is offered by David Lewin (1986). Building on the ideas of phenomenology, Lewin explores the ways that the interpretation of an event can shift depending on the contextual perspective.

4. See, for example, Aldwell and Schachter's discussion of enharmonic modulation (1989, 560–4); as the authors note, this often involves the "reinterpretation" of the tonal implications of an event. (I discuss enharmonic modulation later in this section.) For a fascinating early discussion of tonal reinterpretation, see Weber 1832/1994.

5. This excerpt was analyzed by the program as part of the Kostka-Payne test reported in chapter 7. As noted there, all extrametrical notes were excluded from the Kostka-Payne excerpts. In cases such as the Chopin, this is unfortunate. One might argue that the grace notes in the excerpt have important tonal implications; in particular, the F's in m. 1 and elsewhere exert significant pressure towards B♭ major.

6. A subtle reinforcement of the 3/2 meter continues even past this point, but to my mind it is no longer strong enough to constitute a real ambiguity.

7. The idea of a hierarchical representation of tonal structure is fundamental to the work of Schenker (1935/1979) and others in the Schenkerian tradition. It is also reflected in Lerdahl and Jackendoff's Prolongational Reduction and Time-Span Reduction in *GTTM* (1983). However, neither Schenker's nor *GTTM*'s reductional structures are really hierarchical representations of harmony and key; they are something rather different.

8. The phenomena of diachronic and synchronic ambiguity might of course arise in combination. For example, assume that both hearings A and B of a passage were simultaneously present. It might still be that hearing A was slightly favored from one contextual perspective (say, when the passage was first heard), while hearing B was favored from another (in light of the following context). Describing such situations with any precision—let alone quantifying them—would be extremely complex, and I will not attempt it here.

9. Five of the seven roots can move by fifths to two other roots; two (VII and IV) can move to only one other root. This yields twelve possible "fifth" motions in all.

10. Of course, a certain NPC triad could be located at different points on the line of fifths, but we can assume—again due to the harmonic variance rule—that a triad will be located at the closest possible point to the previous triad. So if the first triad is C major, the second will preferably be interpreted as F major rather than E♯ major.

11. There is experimental evidence, as well, that musical expectations may not be much affected by actual knowledge of what is coming next. Schmuckler (1989, 144) found that subjects' judgments of the expectedness of different continuations were not greatly influenced by prior familiarity with the piece.

Chapter 9

1. The idea of regarding musical surfaces as transformations of underlying structures brings to mind Schenkerian approaches to rhythmic structure, such as those of Komar (1971) and Schachter (1980). These approaches involve adding rhythmic values to higher-level events in Schenkerian reductions, and sometimes entail the rhythmic alteration of events; for example, syncopated events may be shifted to metrically stronger positions, irregular phrases may be made regular (for example, a 5-measure phrase might be "normalized" to 4 measures), and so on. Some authors have argued that the idea of positing transformations as a general aspect of rhythmic structure is problematic (Lester 1986, 187–9; Lerdahl & Jackendoff 1983, 288). However, the kind of normalization that I propose here is extremely local and quite constrained in the way it may be applied.

2. For further discussion of some of the issues explored in this section, see Temperley 1999a.

3. For example, the melody of the Beatles' "Drive My Car" emphasizes the fourth degree of the tonic harmony; the opening melodic phrase of Pink Floyd's "Welcome to the Machine" ends on a striking $\hat{2}$ of the tonic chord.

4. Moore (1995, 189) has suggested that much rock reflects "the 'divorce' between melodic and surface harmonic schemes," in that a single melodic phrase can be repeated over different harmonies, with apparent disregard for the compatibility between the two. I find this claim somewhat overstated; a large majority of songs reflect a fairly close coordination between melody and harmony. However, one does sometimes find cases, such as those in figure 9.18, where melody and harmony appear to be truly independent.

5. Some might argue that a piece in A minor would use the C diatonic scale with A as the tonic; however, as suggested in section 7.4, the primary pitch collection of most minor common-practice music is actually the harmonic minor pitch collection. $\hat{6}$ and $\flat\hat{7}$ are sometimes used, but are much less common than $\flat\hat{6}$ and $\hat{7}$.

6. One does sometimes find $\sharp\hat{4}/\flat\hat{5}$ used as a melodic inflection in rock, moving either upwards to $\hat{5}$ or downwards to $\hat{4}$; examples of this usage are seen in figures 9.13b and 9.24. I would suggest that these are best regarded as chromatic passing tones; such tones are often found in common-practice music as well. I will give evidence below that $\sharp\hat{4}/\flat\hat{5}$ is generally not understood as a "legal" scale-degree.

7. While it is not uncommon for rock songs to modulate from one tonal center to another, we will mainly be concerned with songs that maintain a single tonal center throughout. However, modulation creates no fundamental problem for the current model. When a piece shifts tonal center, I would argue, the pitch collection for each section of the song is defined by the "supermode" for the tonal center of that section.

8. This point has been made by Moore (1992, 77). Moore suggests that tonicization in rock is often achieved through emphasis of a particular triad: through repetition, accentuation, or use at structurally important points. This relates quite closely to what I propose below.

9. For the moment, I am assuming a "neutral" model of pitch-class. If tonal-pitch-class distinctions are made, the situation becomes somewhat more compli-

cated. For example, A♭ is compatible with a tonic of C (as part of the Aeolian mode), but G♯ is not. The question of whether TPC distinctions are important in rock is a difficult one which I will not pursue here.

10. B is featured in some occurrences of the verse melody (such as the one in Figure 9.26, from the second verse), but not in others (e.g., the first verse).

Chapter 10

1. A distinction should be made between music in "strict rhythm" and music in "free rhythm"; this distinction is discussed by Nketia (1974, 168) and Agawu (1995, 73–4). It appears that, roughly speaking, "strict rhythm" refers to music with meter; "free rhythm," to music without meter. Much African music is in free rhythm; however, the work discussed here is almost entirely concerned with strict rhythm music.

2. At one point, Pantaleoni (1972a, 56–8) seems to question even the eighth-note pulse, arguing that Ewe dance music is understood directly in terms of the bell pattern, without involving any regular pulse at all. But he seems ambivalent about this (see Pantaleoni 1972b, 8), and clearly assumes an eighth-note pulse in his transcriptions.

3. Lerdahl and Jackendoff (1983, 70–4) propose that the tactus is invariably between 40 and 160 beats per minute, and often close to 70. Parncutt (1994) found experimentally that the preferred range for the tactus was around 80–90 beats per minute.

4. See Jones 1959 I, 41, 54, 84, 86, 124; Koetting 1970, 130–1. Other ethnic groups that use the standard pattern assume different beginning points, as some have pointed out (Pressing 1983a; Rahn 1996); but apparently the pattern always has a definite "beginning" point. (While there seems to be no doubt that the X position is metrically strong in Ewe music, there is some question as to whether it is truly the *beginning* of the pattern; see note 18.)

5. Two dissenting views should be mentioned. Arom (1991, 206–7, 229) accepts the tactus level in African music, but argues strongly against any higher levels. Kauffman (1980, 407–12) proposes that common rhythmic patterns in African music, such as 3-3-2 and the standard bell pattern, may actually be metrical patterns. The idea of irregular metrical structures—irregular (but regularly repeating) patterns of accents which are inferred by the listener and then imposed on subsequent events—is not out of the question; but since Kauffman is the only one to suggest it, and gives little evidence for it, I will not pursue it further here.

6. Jones's comments about the conflicting barlines are of little help. He emphasizes that rhythms which conflict with the underlying beat are not "syncopated" —rather, he claims that they are completely independent of this beat; it seems that the independent barlines are designed to convey this (Jones 1959 I, 20–1, 23, 32). Generally, syncopation simply refers to musical events which conflict (at least on the surface) with the prevailing beat; is there anything to be gained by saying that a line goes against the main beat, but is *not* syncopated? Chernoff (1979, 45) expresses a similar view to Jones, asserting that different lines of a drum ensemble piece must be given different meters. Agawu (1995, 71, 187–8, 200) is also somewhat inconsistent on the issue of meter. Despite his frequent

discussions of conflicts between accent and meter, he denies that the time signatures and barlines of his transcriptions represent an "accentual hierarchy"; rather they simply represent "grouping." By "accentual hierarchy," I assume he means a metrical framework with beats of varying strength. Surely Agawu is not claiming that such a framework is absent in African music; if so, what does he mean by "conflicts" between meter and surface events? (Agawu's claim that his barlines represent grouping will be discussed below.)

7. We should be cautious in how we interpret these authors' transcriptions. However, as we have seen, there is plenty of support in their writings—Jones included—for the existence of meter in African music. In general, then, it seems fair to take time signatures and barlines as indicators of the meter that is perceived.

8. It might be argued that the instruments here are heard, to some extent, as forming "resultant" lines rather than individual ones (Kubik 1962; Nketia 1974, 133–8); in this case, since a given resultant line may have notes on all or nearly all of the eighth-note beats, it may be loudness—resulting from several instruments playing together—more than the event rule that causes certain beats to be heard as strong.

9. Locke (1982) draws a useful distinction between "offbeat rhythms," which imply a metrical structure of the same period as the underlying meter but different in phase, and "cross rhythms," which imply a metrical structure of a different period.

10. Here I am echoing Agawu (1995, 4–5, 187–90), who argues eloquently against exaggerating the differences between Western and African music.

11. Some students of African music seem to fall victim to the common misconception that, in Western music, metrically accented notes are always (or at least normally) indicated as such by being played louder. In Western music, Chernoff remarks (1979, 42), "the rhythm is counted evenly and stressed on the main beat." Arom (1991, 202–4), similarly, seems to assume that meter requires dynamic accents. This misconception may account for the misgivings of some authors, notably Jones, about using barlines and time signatures in the traditional way. In an early article, Jones (1954, 27) explains his reluctance to use a 2/4 time signature in a piece; this would "imply the presence of alternate strong and weak accents. But in our example, and indeed in all cases where clapping is used, the claps are all of equal intensity" (see also Jones 1959 I, 17, 23). (This fear of implying dynamic accents might also explain Agawu's comments, noted earlier, that his barlines do not represent an "accentual hierarchy.") But Western metrical notation does not imply that notes on strong beats are dynamically accented. As noted in chapter 2, loudness appears to be a fairly minor factor in Western meter, and it is certainly not crucial; meter can readily be inferred in music for instruments permitting no dynamic variation.

12. Jones (1959 I, 19) discusses this as well, although his "Western" hearing is rather different.

13. Here again, I am speaking of "strict rhythm" rather than "free rhythm" music (see note 1).

14. Patterns such as figure 10.6 raise the issue of metrical ambiguity. We are assuming that an African listener infers a particular metrical structure for a piece. There may be cases of conflict, for example, between the underlying meter and the cross-rhythms of the master drummer; but either these cross-rhythms override the underlying meter, or they do not. But is it not possible that the listener maintains two, or even several, metrical interpretations at once? As noted in chapter 8, the potential for synchronic ambiguity in Western meter appears to be related to the degree of difference between the two structures in question. In a case of hemiola, all the levels are the same except for one intermediate level; thus entertaining two structures really means, simply, entertaining two candidates for this one level. In such cases metrical ambiguity does seem possible for Western listeners, and we might expect that the same would hold true of African listeners. Indeed, the cases where metrical ambiguity has been claimed in African rhythm are precisely cases of hemiola. Jones (1959 I, 102) suggests that, in pieces with pervasive hemiolas, African listeners are able to perceive both duple and triple meters simultaneously—for example, hearing the rhythm of figure 10.6 as being in 3/4 and 6/8 at once. See also Locke 1982, 223. Agawu (1995, 189–91, 193) takes issue with Jones, arguing that the 6/8 meter is primary in such cases.

15. "Position-finding" has not received much discussion with regard to rhythm. It is briefly mentioned by Rahn (1996, 79), who uses the term "individuation."

16. We should note that Blacking's claims apply only to the Venda; in fact he mentions that some other Africans (specifically Zulus) agreed with his own interpretations rather than the Venda ones (p. 158).

17. A similar case is seen in the five-eighth-note pattern in the master drum in figure 10.4. This is a more problematic case, since "quintuple" metrical structures are not accommodated by the current theory. We should note also that the metrical structures at issue in these examples are not—according to the consensus view—the primary metrical structures present, though they might be secondary structures heard momentarily, adding elements of tension or conflict with the primary meter.

18. Jones (1959 I, 54) asserts that, in some contexts, even the standard bell pattern itself is regarded as end-accented, that is, ending on the X position shown in figure 10.2. Similarly, Locke (1982, 224–5) notes that, in performance, the standard pattern is most commonly begun on the position just after the X position (see also Locke's transcription on p. 220); this, again, suggests an end-accented grouping. In most cases, however—as noted earlier—the X position is described as the beginning of the pattern.

19. An interesting case here is Pressing's study (1983b) of the Kidi patterns in Agbadza, an Ewe drumming piece. To his credit, Pressing distinguishes clearly between "phrasing" (grouping) and "polyrhythmic sampling" (metrical implications) (pp. 9–11). He shows the grouping structures for a number of Kidi patterns; he indicates the periods (or "cycles") of the meters implied by those patterns, but unfortunately not the placing of the strong beats (p. 6). Regarding grouping, Pressing notes that parallelism is an important factor (p. 9); he also notes that groups always begin with a bounce beat rather than a mute beat (the

bounce beats being much more prominent) (p. 9). Bounce beats are presumably strong phenomenal accents and would tend to be metrically strong (this is the main factor in the metrical implications of patterns, as Pressing implies [pp. 10–11]); it is interesting, then, that they tend to mark the *beginnings* of phrases, rather than the ends, as other authors argue is normative for African music. In cases of several bounce strokes in quick succession, however, the phrase always begins on the first of the series, rather than on the last (p. 6). This is predicted by *GTTM*'s similarity rule for grouping—a grouping boundary occurs at a change from one stroke type to another, grouping the similar ones together. (In metrical terms, we would probably expect the *last* bounce stroke of a series to be metrically strongest, since it is in a way the "longest.")

20. Several other concepts used in discussions of African rhythm should be briefly mentioned. Nketia and others have suggested that some rhythmic patterns are divisive, formed by divisions of a unit, whereas others are additive, formed by combining units together (Nketia 1974, 128–31; Pressing 1983b, 10–11). (See also Jones 1959 I, 20–1, on the additive nature of African rhythm.) It is unclear what these terms imply from a cognitive viewpoint. One could describe certain metrical structures—and rhythmic patterns that imply them—as additive or divisive: given a regular tactus level, an irregular higher level might be seen as forming an additive structure, while an irregular lower level (with successive tactus beats being divided differently) would be divisive. While there are some patterns in African music that imply such irregular structures, most African rhythmic patterns and polyphonic textures imply regular structures of several levels; in this case it is unclear why they should be regarded divisively or additively (for example, consider the "divisive" patterns shown in Nketia 1974, 130). Rhythmic patterns might be considered additive in other ways. Nketia points to the Ewe bell pattern as additive, arguing that it is based on a 5 + 7 structure (1974, 129–31). This pattern might be regarded as an additive "5 + 7" pattern in terms of its grouping structure, or in terms of its motivic structure (in that it involves a five-beat pattern followed by a seven-beat variant). However, its metrical structure is clearly not additive (at least in the contexts where it is normally used); strong beats do not occur at intervals of 5 and 7 beats, but rather, occur regularly every third beat.

The terms "polyrhythm" and "polymeter" are also sometimes used. The meaning of these terms is, again, unclear. If polyrhythm simply means the use of multiple rhythmic patterns simultaneously in a piece, then of course it characterizes much Western music as well as African. Arom (1991, 216, 272) uses it to mean the combination of rhythms "so as to create an interwoven effect." Most often, I think, it means using patterns which imply different meters taken individually (see Nketia 1974, 135–8; Pressing 1983b, 10–11). This is indeed an important feature of much African music, and is surely much less common in Western music, although not nonexistent. The same concept is also sometimes denoted by "polymeter." Waterman (1952, 212) defines polymeter as "the interplay of two or more metrical frameworks"; Chernoff (1979, 45) defines it as "the simultaneous use of different meters." What exactly is meant by this? Perhaps these authors are simply referring to simultaneous lines which (taken individually) imply different meters. Or are they implying that, when such lines are combined, several different meters are perceived simultaneously? This cannot be ruled out, although I have argued that this rarely occurs in Western music.

Chapter 11

1. For a useful review of recent work in the area of music performance, see Palmer 1997.

2. The same reasoning applies to the observation that the normative alignment of grouping and meter differs between African and Western music (see section 10.7). The fact that African perception prefers strong beats at the end of groups is reflected, presumably, in the fact that most African songs align metrically strong beats with group endings rather than group beginnings; this means that the metrical grouping rule can be satisfied along with the other rules. Western music, by contrast, is generally constructed so that strong beats naturally fall near the beginning of groups, allowing an analysis which yields "beginning-accented" groups while also satisfying other grouping and metrical rules.

3. Here again, I am speaking of African music in "strict rhythm" (see chapter 10, note 1). That some African music is in "free rhythm" is rather problematic for the current argument; we might have to argue that such music was perceived under a different rule system.

4. In the harmonic notations of jazz pieces—seen in "lead sheets" and also in the parts for accompanying instruments (piano, bass, guitar) in jazz ensembles—one finds a overwhelming majority of root position chords. (Inversions can be represented in jazz notation, using a chord symbol over a bass note, such as "C7/E," but this is relatively rare.) A "walking bass line" may use more than one note under a chord, but the root of the chord is usually emphasized.

5. An interesting analogy can be drawn here, as noted by Swain (1997, 141–67), with a phenomenon in language: what is called a "trading relationship." When change occurs in some aspect of a language, this may result in a loss of information which must be counteracted by change in some other aspect. For example, case information in English (for example, whether a noun is subject or object) used to be communicated through inflectional endings. As the inflections began to drop out, case information had to be conveyed in some other way; this resulted in the development of fixed rules of word order (subject-verb-object), which previously had varied rather freely. Swain finds interesting parallels to this phenomenon in music, although his examples relate more to what I would call schemata (see section 12.3) rather than infrastructural levels. For example, Swain notes that the stylized cadences of Renaissance music required both voice-leading and rhythmic features. With the development of tonal harmony, the harmonic identification of a cadence was clear enough that the strict rhythmic conventions of the Renaissance cadence were no longer necessary.

6. For an interesting quantitative approach to this question, see the discussion of Krumhansl's work in section 11.4 below.

7. Other practical problems arise with this scheme as well. In the way the scores are currently calculated, the score for a piece may be affected by things such as the number of notes in the piece: for example, under the metrical system, a piece with more notes will have a higher score. Scores for pieces would have to be

normalized for this in some way. With regard to the key model, the score for a segment is determined by the sum of the key-profile values for all the pitch-classes present in the segment; thus adding more pitch-classes to a segment will always increase the score. This seems counterintuitive; rather, for present purposes, it seems that the score should be given by the *mean* key-profile value for all the pitch-classes in a segment. In this way, a segment in which all the pitches present are diatonic (under some analysis) will score higher than one in which no purely diatonic analysis can be found. (See Temperley 1999b for development of this idea.)

8. In particular, one runs the risk of committing the "intentional fallacy"—attributing intentions to composers where one is not warranted to do so. For an interesting discussion of the intentional fallacy in musical discourse, see Haimo 1996.

9. See Benjamin 1984, 390–403. Lerdahl and Jackendoff's Metrical Well-Formedness Rules 3 and 4 (1983) stipulate that meter must be regular; however, the authors suggest that metrical structures may exist in some idioms which are not bound by these rules (pp. 96–9). Kramer (1988, 102) argues that metrical structures must reflect some degree of regularity, but also points out that irregularity at one level may not preclude regularity (in terms of relationships between levels) at both higher and lower levels.

10. This technique is sometimes called "cellular" organization. For discussion, see Kramer 1988, 221–85.

11. Von Hippel shows that, when the notes of a melody are randomly scrambled, the intervals between adjacent notes tend to be larger than in the original version, proving that there is a preference for small intervals. However, range constraints are also important: the probability of a certain interval following a note depends on where the note is in the range of the melody. Von Hippel proposes a model based on regression to the mean which accounts for both range constraints and the preference for small intervals. It is possible that taking range constraints into account would improve the performance of the model proposed in chapter 4.

12. In fact, Huron's statistical analyses of Bach's polyphonic works show that even *single* perfect intervals—fifths and octaves—are avoided to a surprising degree; this too, Huron argues, may arise from a desire to avoid tonal fusion.

13. On the other hand, parallel perfect intervals might *indirectly* hinder stream segregation, since the fusion of two voices into one would create discontinuities in the "piano-roll" representation. We should remember that the strict separation between spectral integration and sequential integration assumed here is certainly an oversimplification; in reality the two processes interact in complex ways.

14. Here the original Krumhansl-Schmuckler algorithm has an advantage over my modified version, in that it normalizes the scores for the size of the input vector values, thus solving the problem that passages with more notes receive higher scores (see note 7). On the other hand, it also normalizes for the *variance* in the input vector values. This is not desirable for current purposes; we want an input vector with low variance—i.e., with all pitch-classes used about equally often—to be judged as less tonal than one with high variance.

15. Also of interest is a study by Kessler, Hansen, and Shepard (1984), using Balinese subjects and materials.

16. Table 11.1 also shows the total scores for each segment in the right-hand column, though I am not sure how revealing these are. Adding together the tension of harmonic variance with the tension of ornamental dissonances is, perhaps, a case of "apples and oranges"—experientially, the two are very different (though I have suggested that considering total rule scores can be useful in other ways).

Table 11.1 omits the scores for one rule: the voice-leading rule (TPR 2). The preferred analysis incurred only one penalty from this rule: A penalty of 3.0 for the first E in the melody m. 12, which resolves incorrectly to the E♭ of the next measure.

17. One could also hear the A and C in the upper voices, along with the C-E-G in the cello, as part of an A minor seventh, but this would be penalized by the harmonic variance rule, as A is not close to the harmonies of nearby measures.

18. I have argued elsewhere that the sense of disorientation conveyed by much highly chromatic music may be due, in part, to the ambiguous and low-scoring TPC representations that are formed (Temperley 2000). For example, with a whole-tone or octatonic scale, several spelling interpretations are about equally favored; and all are low-scoring in TPC terms (relative to the diatonic scale, for example), since they all involve a wide dispersion of events on the line of fifths. Indeed, the degree of tension in such music might be considered to exceed the bounds of "grammaticality" within the common-practice idiom.

19. Measures 130–1 illustrate a phenomenon that has not yet been discussed: metrical parallelism between levels. The descending diminished-seventh motive has previously been presented as "strong-weak" at the half-note level; in light of this, when the motive occurs at the whole-note level, there is a tendency to hear it as "strong-weak" at that level as well, thus favoring m. 130 as strong.

20. Here again, getting these results would require some kind of normalization of the key scores, to adjust for the number of pitch-classes in each segment (see note 7 above).

21. More precisely, the hierarchical tension of an event is given by tracing its connections through superordinate events until the most superordinate event is reached, and adding all the resulting pitch-space distances.

Chapter 12

1. See Budd 1985 and Swain 1997 for useful surveys of this subject.

2. For a brief review of work on motivic structure, and numerous citations, see Temperley 1995, 143.

3. Povel and Essens (1985) devised a metric of the rhythmic "complexity" of a pattern, which proved a good predictor of subjects' ability to reproduce rhythmic patterns. However, this metric did not incorporate the factor of repetition (although Povel and Essens do discuss this factor).

4. Longuet-Higgins and Lee (1982) also propose viewing metrical structure in terms of a tree.

5. Again, this model pertains to motivic relations that are recognized in a direct, automatic way. Admittedly, many other relationships can be perceived with effort and attention. As I discuss elsewhere (Temperley 1995, 159–60), there are also other kinds of relationships that seem to be perceived fairly directly—such as similarities of contour—which are not covered by this model.

6. The idea of musical affect arising from denial of pattern completion—or completion of a pattern in an unexpected way—is central to the writings of Leonard Meyer (1956, 1973).

7. We should recall, also, that parallelism may affect grouping structure. Under the current framework, this occurs indirectly due to the influence of parallelism on meter, which then influences grouping (see section 3.4); the result is that grouping structure tends to be parallel with metrical structure, and hence with motivic patterns as well. As noted earlier, there is evidence that perceived motivic patterns must be supported by the grouping structure (so that motives correspond with groups, or at least are not interrupted by grouping boundaries). If so, the influence of parallelism on grouping can be seen as favoring a grouping structure which is likely to yield motivic connections.

8. It is important to consider which listeners are being described here. Ratner is concerned, presumably, with describing the perception of listeners in the Classical period. No doubt many of the topics he discusses are unlikely to be perceived by listeners today.

9. For a review of research on emotion and energy in music, see Dowling & Harwood 1986, 207–13. These authors cite evidence that loud music conveys more energy (210). Interestingly, the studies they review did not find that the ascending or descending contour of a melody greatly affected the expressive content. One might argue that the expressive impact of register relates to the relative effect of sections *within* a melody: the higher portion of a melody conveys more energy than the lower portion. On the other hand, it surely true that an ascending melody, taken as a whole, is different in expressive effect from a descending one; this deserves further study.

10. One might argue that positing schemata *implies* some kind of arbitrary or conventional meaning. If the function of the $\hat{1}$-$\hat{7}$-$\hat{4}$-$\hat{3}$ schema followed in a rule-governed way from its musical structure, why would we need to posit a schema? On the other hand, it might be argued that a schema is simply a common configuration of infrastructural elements which deserves recognition, even if it possesses no meaning beyond what is implied by its structure.

11. Not all music is fragile; nor is fragility a necessary or sufficient condition for good music (by my standards, and, I believe, most other people's). Consider Boulez's *Structures 1a*, a small but representative fragment of which is shown in figure 11.6. This is, to my mind, a highly effective and memorable piece. Yet I do not believe its effect, or the effect of a passage within it, would greatly be altered by altering a few notes. On the other hand, much mediocre common-practice music is as fragile as the Beethoven minuet, yet mediocre nonetheless.

References

Agawu, V. K. (1986). "Gi Dunu", "Nyekpadudo", and the study of West African rhythm. *Ethnomusicology*, 30, 64–83.

Agawu, V. K. (1991). *Playing with Signs: A Semiotic Interpretation of Classic Music*. Princeton, NJ: Princeton University Press.

Agawu, V. K. (1995). *African Rhythm*. Cambridge: Cambridge University Press.

Aldwell, E., & Schachter, C. (1989). *Harmony and Voice Leading*. Fort Worth, TX: Harcourt Brace Jovanovich.

Allen, P., & Dannenberg, R. (1990). Tracking musical beats in time. In *Proceedings of the 1990 International Computer Music Conference*, 140–3. San Francisco: International Computer Music Association.

Arom, S. (1991). *African Polyphony and Polyrhythm*. Translated by M. Thom, B. Tuckett, and R. Boyd. Cambridge: Cambridge University Press.

Ashley, R. (1989). Modeling musical listening: General considerations. *Contemporary Music Review*, 4, 295–310.

Baker, M. (1989a). An artificial intelligence approach to musical grouping analysis. *Contemporary Music Review*, 3, 43–68.

Baker, M. (1989b). A computational approach to modeling musical grouping structure. *Contemporary Music Review*, 4, 311–25.

Baker, N. (1976). Heinrich Koch and the theory of melody. *Journal of Music Theory*, 20, 1–48.

Benjamin, W. (1984). A theory of musical meter. *Music Perception*, 1, 355–413.

Berry, W. (1987). *Structural Functions in Music*. New York: Dover.

Bharucha, J. J. (1984). Anchoring effects in music: The resolution of dissonance. *Cognitive Psychology*, 16, 485–518.

Bharucha, J. J. (1987). Music cognition and perceptual facilitation: A connectionist framework. *Music Perception*, 5, 1–30.

Bharucha, J. J. (1991). Pitch, harmony and neural nets: A psychological perspective. In P. M. Todd & D. G. Loy (Eds.), *Music and Connectionism*, 84–99. Cambridge, MA: MIT Press.

Bharucha, J. J., & Krumhansl, C. (1983). The representation of harmonic structure in music. *Cognition*, 13, 63–102.

Bharucha, J. J., & Stoeckig, K. (1987). Priming of chords: Spreading activation or overlapping frequency spectra? *Perception & Psychophysics*, 41, 519–24.

Blacking, J. (1967). *Venda Children's Songs*. Johannesburg: Witwatersrand University Press.

Boltz, M., & Jones, M. R. (1986). Does rule recursion make melodies easier to reproduce? If not, what does? *Cognitive Psychology*, 18, 389–431.

Bregman, A. S. (1978). Auditory streaming is cumulative. *Journal of Experimental Psychology: Human Perception and Peformance*, 4, 380–7.

Bregman, A. S. (1990). *Auditory Scene Analysis*. Cambridge, MA: MIT Press.

Brown, H. (1988). The interplay of set content and temporal context in a functional theory of tonality perception. *Music Perception*, 5, 219–50.

Brown, H., Butler, D., & Jones, M. R. (1994). Musical and temporal influences on key discovery. *Music Perception*, 11, 371–407.

Browne, R. (1981). Tonal implications of the diatonic set. *In Theory Only*, 5(6, 7), 3–21.

Bruce, V., & Green, P. R. (1990). *Visual Perception: Physiology, Psychology and Ecology*. Hove, U.K.: Erlbaum.

Budd, M. (1985). *Music and the Emotions*. London: Routledge & Kegan Paul.

Butler, D. (1979). A further study of melodic channeling. *Perception & Psychophysics*, 25, 264–68.

Butler, D. (1989). Describing the perception of tonality in music: A critique of the tonal hierarchy theory and a proposal for a theory of intervallic rivalry. *Music Perception*, 6, 219–42.

Castellano, M. A., Bharucha, J. J., & Krumhansl, C. L. (1984). Tonal hierarchies in the music of North India. *Journal of Experimental Psychology: General*, 113, 394–412.

Chafe, C., Mont-Reynaud, B., & Rush, L. (1982). Toward an intelligent editor of digital audio: Recognition of musical constructs. *Computer Music Journal*, 6(1), 30–41.

Chafe, E. T. (1992). *Monteverdi's Tonal Language*. New York: Schirmer.

Chernoff, J. M. (1979). *African Rhythm and African Sensibility*. Chicago: Chicago University Press.

Chomsky, N. (1980). *Rules and Representations*. New York: Columbia University Press.

Clarke, E. (1989). Mind the gap: Formal structures and psychological processes in music. *Contemporary Music Review*, 3, 1–13.

Clough, J., & Douthett, J. (1991). Maximally even sets. *Journal of Music Theory*, 35, 93–173.

Cohn, R. (1985). Review of *A Generative Theory of Tonal Music*. *In Theory Only*, 8(6), 27–52.

Cohn, R. (1996). Maximally smooth cycles, hexatonic systems, and the analysis of late-romantic triadic progressions. *Music Analysis*, 15, 9–40.

Cook, N. (1987). The perception of large-scale tonal closure. *Music Perception*, 5, 197–206.

Cooper, G. B., & Meyer, L. B. (1960). *The Rhythmic Structure of Music*. Chicago: University of Chicago Press.

Cuddy, L. L., Cohen, A. J., & Mewhort, D. J. K. (1981). Perception of structure in short melodic sequences. *Journal of Experimental Psychology: Human Perception & Performance*, 7, 869–83.

Cuddy, L. L., Cohen, A. J., & Miller, J. (1979). Melody recognition: the experimental application of rules. *Canadian Journal of Psychology*, 33, 148–57.

Dahlhaus, C. (1990). *Studies on the Origin of Harmonic Tonality*. Translated by R. O. Gjerdingen. Princeton, NJ: Princeton University Press.

Deliege, I. (1987). Grouping conditions in listening to music: An approach to Lerdahl & Jackendoff's grouping preference rules. *Music Perception*, 4, 325–60.

Desain, P., & Honing, H. (1992). *Music, Mind & Machine*. Amsterdam: Thesis Publications.

Desain, P., Honing, H., vanThienen, H., & Windsor, L. (1998). Computational modeling of music cognition: Problem or solution? *Music Perception*, 16, 151–66.

Deutsch, D. (1975). Two-channel listening to musical scales. *Journal of the Acoustical Society of America*, 57, 1156–60.

Deutsch, D. (1980). The processing of structured and unstructured tonal sequences. *Perception & Psychophysics*, 28, 381–9.

Deutsch, D., & Feroe, J. (1981). The internal representation of pitch sequences in tonal music. *Psychological Review*, 88, 503–22.

Dowling, W. J. (1973a). The perception of interleaved melodies. *Cognitive Psychology*, 5, 322–37.

Dowling, W. J. (1973b). Rhythmic groups and subjective chunks in memory for melodies. *Perception & Psychophysics*, 14, 37–40.

Dowling, W. J. (1978). Scale and contour: Two components of a theory of memory for melodies. *Psychological Review*, 85, 341–54.

Dowling, W. J., & Harwood, D. L. (1986). *Music Cognition*. Orlando, FL: Academic Press.

Drake, C., & Palmer, C. (1993). Accent structures in music performance. *Music Perception*, 10, 343–78.

Fitzgibbons, P. J., Pollatsek, A., & Thomas, I. B. (1974). Detection of temporal gaps within and between perceptual tonal groups. *Perception & Psychophysics*, 16, 522–28.

Fodor, J. A. (1975). *The Language of Thought*. New York: Crowell.

Fodor, J. A. (1983). *Modularity of Mind*. Cambridge, MA: MIT Press.

Fodor, J. A., & Pylyshyn, Z. W. (1988). Connectionism and cognitive architectecture: A critical analysis. *Cognition*, 20, 3–71.

Foster, S., Schloss, W. A., & Rockmore, A. J. (1982). Toward an intelligent editor of digital audio: signal processing methods. *Computer Music Journal*, 6(1), 42–51.

Gabrielsson, A. (1973). Studies in rhythm. *Acta Universitatis Upsaliensis*, 7, 3–19.

Garman, M. (1990). *Psycholinguistics*. Cambridge: Cambridge University Press.

Gjerdingen, R. O. (1988). *A Classic Turn of Phrase: Music and the Psychology of Convention*. Philadelphia: University of Pennsylvania Press.

Gjerdingen, R. O. (1994). Apparent motion in music? *Music Perception*, 11, 335–70.

Haimo, E. (1996). Atonality, analysis, and the intentional fallacy. *Music Theory Spectrum*, 18, 167–99.

Halle, J., & Lerdahl, F. (1993). A generative textsetting model. *Current Musicology*, 55, 3–23.

Handel, S. (1973). Temporal segmentation of repeating auditory patterns. *Journal of Experimental Psychology*, 101, 46–54.

Handel, S., & Lawson, G. R. (1983). The contextual nature of rhythmic interpretation. *Perception & Psychophysics*, 34, 103–20.

Haugeland, J. (1981). Semantic engines: An introduction to mind design. In J. Haugeland (Ed.), *Mind Design*, 1–34. Cambridge, MA: MIT Press.

Holtzmann, S. R. (1977). A program for key determination. *Interface*, 6, 29–56.

Howell, D. (1997). *Statistical Methods for Psychology*. Belmont, CA: Wadsworth.

Huron, D. (1989). Voice segregation in selected polyphonic keyboard works by Johann Sebastian Bach. Ph.D diss., Nottingham University.

Huron, D. (under review). A derivation of the rules of voice-leading from perceptual principles. Submitted for publication.

Huron, D., & Parncutt, R. (1993). An improved model of tonality perception incorporating pitch salience and echoic memory. *Psychomusicology*, 12, 154–171.

Huron, D., & Royal, M. (1996). What is melodic accent? Converging evidence from musical practice. *Music Perception*, 13, 489–516.

Jackendoff, R. (1991). Musical parsing and musical affect. *Music Perception*, 9, 199–230.

Jones, A. M. (1954). African rhythm. *Africa*, 24, 26–47.

Jones, A. M. (1959). *Studies in African Music*. London: Oxford University Press.

Kamien, R. (1993). Conflicting metrical patterns in accompaniment and melody in works by Mozart and Beethoven: A preliminary study. *Journal of Music Theory*, 37, 311–350.

Kastner, M. P., & Crowder, R. G. (1990). Perception of the major/minor distinction: IV. Emotional connotations in young children. *Music Perception*, 8, 189–202.

Kauffman, R. (1980). African rhythm: A reassessment. *Ethnomusicology*, 24, 393–415.

Kauffman, W., & Frisina, R. (1992). The fusion of neuroscience and music. *Psychomusicology*, 11, 75–8.

Keller, H. (1976). *The Well-Tempered Clavier by Johann Sebastian Bach*. Translated by L. Gerdine. London: Norton.

Kessler, E. J., Hansen, C., & Shepard, R. N. (1984). Tonal schemata in the perception of music in Bali and the West. *Music Perception*, 2, 131–65.

Koetting, J. (1970). Analysis and notation of West African drum ensemble music. *Selected Reports in Ethnomusicology*, 1, 116–46.

Koffka, K. (1935). *Principles of Gestalt Psychology*. New York: Harcourt Brace.

Komar, A. J. (1971).*Theory of Suspensions*. Princeton, NJ: Princeton University Press.

Kostka, S. (1995). *Instructor's Manual to Accompany* Tonal Harmony. New York: McGraw-Hill.

Kostka, S., & Payne, D. (1995a). *Tonal Harmony*. New York: McGraw-Hill.

Kostka, S., & Payne, D. (1995b). *Workbook for* Tonal Harmony. New York: McGraw-Hill.

Kramer, J. (1988). *The Time of Music*. New York: Schirmer.

Krumhansl, C. L. (1990). *Cognitive Foundations of Musical Pitch*. New York: Oxford University Press.

Krumhansl, C., Bharucha, J. J., & Kessler, E. (1982). Perceived harmonic structure of chords in three related musical keys. *Journal of Experimental Psychology: Human Perception and Performance*, 8, 24–36.

Krumhansl, C. L., & Kessler, E. J. (1982). Tracing the dynamic changes in perceived tonal organization in a spatial representation of musical keys. *Psychological Review*, 89, 334–68.

Kubik, G. (1962). The phenomenon of inherent rhythms in East and Central African instrumental music. *African Music*, 3, 33–42.

Laden, B., & Keefe, D. H. (1991). The representation of pitch in a neural net model of chord classification. In P. M. Todd & D. G. Loy (Eds.), *Music and Connectionism*, 64–83. Cambridge, MA: MIT Press.

Large, E. W., & Kolen, J. F. (1994). Resonance and the perception of musical meter. *Connection Science*, 6, 177–208.

Lee, C. (1991). The perception of metrical structure: Experimental evidence and a model. In P. Howell, R. West, & I. Cross (Eds.), *Representing Musical Structure*, 59–127. London: Academic Press.

Leman, M. (1995). *Music and Schema Theory*. Berlin: Springer.

Lerdahl, F. (1988). Tonal pitch space. *Music Perception*, 5, 315–49.

Lerdahl, F. (1996). Calculating tonal tension. *Music Perception*, 13, 319–63.

Lerdahl, F., & Jackendoff, R. (1983). *A Generative Theory of Tonal Music*. Cambridge, MA: MIT Press.

Lester, J. (1986). *The Rhythms of Tonal Music*. Carbondale: Southern Illinois University Press.

Lewin, D. (1986). Music theory, phenomenology, and modes of perception. *Music Perception*, 3, 327–92.

Locke, D. (1982). Principles of offbeat timing and cross-rhythm in Southern Ewe dance drumming. *Ethnomusicology*, 26, 217–46.

London, J. (1997). Lerdahl and Jackendoff's strong reduction hypothesis and the limits of analytical description. *In Theory Only*, 13, 3–28.

Longuet-Higgins, H. C. (1962). Two letters to a musical friend. *Music Review*, 23, 244–8, 271–80.

Longuet-Higgins, H. C., & Lee, C. (1982). The perception of musical rhythms. *Perception*, 11, 115–28.

Longuet-Higgins, H. C., & Steedman, M. J. (1971). On interpreting Bach. *Machine Intelligence*, 6, 221–41.

Lyons, J. (1981). *Language and Linguistics*. Cambridge: Cambridge University Press.

Marin, O. S. M., & Perry, D. W. (1999). Neurological aspects of music perception and performance. In D. Deutsch (Ed.), *The Psychology of Music*, 653–724. San Diego, CA: Academic Press.

Marsden, A. (1992). Modelling the perception of musical voices: A case study in rule–based systems. In A. Marsden & A. Pople (Eds.), *Computer Representations and Models in Music*, 239–63. London: Academic Press.

Maus, F. (1988). Music as drama. *Music Theory Spectrum*, 10, 56–73.

Maxwell, H. J. (1992). An expert system for harmonic analysis of tonal music. In M. Balaban, K. Ebcioglu, & O. Laske (Eds.), *Understanding Music with AI*, 335–53. Cambridge, MA: MIT Press.

McCabe, S. L., & Denham, M. J. (1997). A model of auditory streaming. *Journal of the Acoustical Society of America*, 101, 1611–21.

Meyer, L. B. (1956). *Emotion and Meaning in Music.* Chicago: University of Chicago Press.

Meyer, L. B. (1973). *Explaining Music.* Berkeley: University of California Press.

Middleton, R. (1990). *Studying Popular Music.* Milton Keynes, U.K.: Open University Press.

Miller, G. A., & Heise, G. A. (1950). The trill threshold. *Journal of the Acoustical Society of America*, 22, 637–8.

Moore, A. (1992). Patterns of harmony. *Popular Music*, 11, 73–106.

Moore, A. (1993). *Rock: The Primary Text.* Buckingham, U.K.: Open University Press.

Moore, A. (1995). The so-called "flattened seventh" in rock. *Popular Music*, 14, 185–201.

Moorer, J. A. (1977). On the transcription of musical sound by computer. *Computer Music Journal*, 1(4), 32–8.

Narmour, E. (1990). *The Analysis and Cognition of Basic Melodic Structures: The Implication-Realization Model.* Chicago: University of Chicago Press.

Nketia, J. H. K. (1963). *African Music in Ghana.* Evanston, IL: Northwestern University Press.

Nketia, J. H. K. (1974). *The Music of Africa.* New York: Norton.

Ottman, R. (1986). *Music for Sight Singing.* Englewood Cliffs, NJ: Prentice-Hall.

Palmer, C. (1996a). Anatomy of a performance: Sources of musical expression. *Music Perception*, 13, 433–54.

Palmer, C. (1996b). On the assignment of structure in music performance. *Music Perception*, 14, 23–56.

Palmer, C. (1997). Music performance. *Annual Review of Psychology*, 48, 115–38.

Palmer, C., & Kelly, M. (1992). Linguistic prosody and musical meter in song. *Journal of Memory and Language*, 31, 525–42.

Palmer, C., & Krumhansl, C. (1987). Pitch and temporal contributions to musical phrase perception: Effects of harmony, performance timing, and familiarity. *Perception and Psychophysics*, 41, 505–18.

Palmer, C., & van de Sande, C. (1995). Range of planning in music performance. *Journal of Experimental Psychology: Human Perception and Performance*, 21, 947–62.

Pantaleoni, H. (1972a). Three principles of timing in Anlo dance drumming. *African Music*, 5, 50–63.

Pantaleoni, H. (1972b). Towards understanding the play of Sogo in Atsia. *Ethnomusicology*, 16, 1–37.

Parncutt, R. (1989). *Harmony: A Psychoacoustical Approach*. Berlin: Springer.

Parncutt, R. (1994). A perceptual model of pulse salience and metrical accent in musical rhythms. *Music Perception*, 11, 409–64.

Peel, J., & Slawson, W. (1984). Review of *A Generative Theory of Tonal Music*. *Journal of Music Theory*, 28, 271–94.

Piston, W. (1987). *Harmony*. New York: Norton.

Pople, A. (1996). Editorial: On coincidental collections. *Music Analysis*, 15, 1–7.

Povel, D.-J. (1981). Internal representation of simple temporal patterns. *Journal of Experimental Psychology: Human Perception and Performance*, 7, 3–18.

Povel, D.-J., & Essens, P. (1985). Perception of temporal patterns. *Music Perception*, 2, 411–40.

Pressing, J. (1983a). Cognitive isomorphisms between pitch and rhythm in world musics: West Africa, the Balkans and western tonality. *Studies in Music*, 17, 38–61.

Pressing, J. (1983b). Rhythmic design in the support drums of Agbadza. *African Music*, 6, 4–15.

Putnam, H. (1975). The nature of mental states. In *Mind, Language, and Reality: Philosophical Papers*, vol. 2, 429–40. London: Cambridge University Press.

Rahn, J. (1987). Asymmetrical ostinatos in sub–Saharan music: Time, pitch, and cycles reconsidered. *In Theory Only*, 9(7), 23–37.

Rahn, J. (1996). Turning the analysis around: Africa–derived rhythms and Europe–derived music theory. *Black Music Research Journal*, 16, 71–89.

Ratner, L. (1980). *Classic Music: Expression, Form, and Style*. New York: Schirmer.

Regener, E. (1974). On Allen Forte's theory of chords. *Perspectives of New Music*, 13, 191–212.

Riemann, H. ([1915]1992). Ideas for a study "on the imagination of tone." Translated by R. W. Wason and E. W. Marvin. *Journal of Music Theory*, 36, 81–117.

Rosen, C. (1971). *The Classical Style: Haydn, Mozart, Beethoven*. New York: Viking Press.

Rosenthal, D. (1992). Emulation of human rhythm perception. *Computer Music Journal*, 16(1), 64–76.

Rosner, B. S., & Narmour, E. (1992). Harmonic closure: Music theory and perception. *Music Perception*, 9, 383–412.

Rothstein, W. (1989). *Phrase Rhythm in Tonal Music.* New York: Schirmer.

Scarborough, D. L., Miller, B. O., & Jones, J. A. (1991). Connectionist models for tonal analysis. In P. M. Todd & D. G. Loy (Eds.), *Music and Connectionism.* 54–63. Cambridge, MA: MIT Press.

Schachter, C. E. (1976). Rhythm and linear analysis: A preliminary study. *Music Forum,* 4, 281–334.

Schachter, C. E. (1980). Rhythm and linear analysis: Durational reduction. *Music Forum,* 5, 197–232.

Schachter, C. E. (1987). Rhythm and linear analysis: Aspects of meter. *Music Forum,* 6, 1–59.

Schaffrath, H. (1995). *The Essen Folksong Collection.* Edited by D. Huron. Stanford, CA: Center for Computer–Assisted Research in the Humanities.

Schellenberg, E. G. (1997). Simplifying the implication-realization model of melodic expectancy. *Music Perception,* 14, 295–318.

Schenker, H. ([1935]1979). *Free Composition.* Translated and edited by E. Oster. New York: Longman.

Schenker, H. ([1926]1974). *Das Meisterwerk in der Musik,* vol. 2. Hildesheim: Georg Olms Verlag.

Schmuckler, M. (1989). Expectation in music: Investigation of melodic and harmonic processes. *Music Perception,* 7, 109–50.

Schoenberg, A. ([1954]1969). *Structural Functions of Harmony.* New York: Norton.

Shaffer, L. H., Clarke, E. F., & Todd, N. P. (1985). Metre and rhythm in piano playing. *Cognition,* 20, 61–77.

Shepard, R. N. (1982). Structural representations of musical pitch. In Diana Deutsch (Ed.), *The Psychology of Music,* 343–90. New York: Academic Press, 1982.

Sloboda, J. A. (1974). The eye-hand span: An approach to the study of sight–reading. *Psychology of Music,* 2, 4–10.

Sloboda, J. A. (1976). The effect of item position on the likelihood of identification by inference in prose reading and music reading. *Canadian Journal of Psychology,* 30, 228–36.

Sloboda, J. A. (1983). The communication of musical metre in piano performance. *Quarterly Journal of Experimental Psychology,* 35, 377–96.

Sloboda, J. A. (1985). *The Musical Mind.* Oxford: Clarendon Press.

Sloboda, J. A. (1988). *Generative Processes in Music.* Oxford: Clarendon Press.

Steedman, M. (1977). The perception of musical rhythm and meter. *Perception,* 6, 555–70.

Sundberg, J. (1988). Computer synthesis of music performance. In J. A. Sloboda (Ed.), *Generative Processes in Music,* 52–69. Oxford: Clarendon Press.

Swain, J. (1997). *Musical Languages*. New York: Norton.

Tagg, P. (1982). Analysing popular music: Theory, method, and practice. *Popular Music*, 2, 37–68.

Tan, N., Aiello, R., & Bever, T. G. (1981). Harmonic structure as a determinant of melodic organization. *Memory & Cognition*, 9, 533–9.

Tanguiane, A. S. (1993). *Artificial Perception and Music Recognition*. Berlin: Springer.

Tanguiane, A. S. (1994). A principle of correlativity of perception and its application to music recognition. *Music Perception*, 11, 465–502.

Temperley, D. (1995). Motivic perception and modularity. *Music Perception*, 13, 141–69.

Temperley, D. (1996). The perception of harmony and tonality: An algorithmic approach. Ph.D diss., Columbia University.

Temperley, D. (1999a). Syncopation in rock: A perceptual perspective. *Popular Music*, 18, 19–40.

Temperley, D. (1999b). What's key for key? The Krumhansl–Schmuckler key–finding algorithm reconsidered. *Music Perception*, 17, 65–100.

Temperley, D. (2000). The line of fifths. *Music Analysis*, 19, 289–319.

Temperley, D. (in-press-a). Hypermetrical ambiguity in sonata form closing themes. *In Theory Only*.

Temperley, D. (in-press-b). The question of purpose in music theory: Description, suggestion, and explanation. *Current Musicology*.

Temperley, D., & Sleator, D. (1999). Modeling meter and harmony: A preference-rule approach. *Computer Music Journal*, 23(1), 10–27.

Tenney, J., & Polanksy, L. (1980). Temporal Gestalt perception in music. *Journal of Music Theory*, 24, 205–41.

Terhardt, E. (1974). Pitch, consonance and harmony. *Journal of the Acoustical Society of America*, 55, 1061–9.

Thomassen, J. (1982). Melodic accent: Experiments and a tentative model. *Journal of the Acoustical Society of America*, 71, 1596–1605.

Thompson, W. F., & Cuddy, L. L. (1992). Perceived key movement in four-voice harmony and single voices. *Music Perception*, 9, 427–38.

Todd, N. P. M. (1985). A model of expressive timing in tonal music. *Music Perception*, 3, 33–58.

Todd, N. P. M. (1989). A computational model of rubato. *Contemporary Music Review*, 3, 69–88.

Todd, N. P. M. (1994). The auditory "primal sketch": A multiscale model of rhythmic grouping. *Journal of New Music Research*, 23, 25–70.

van Noorden, L. P. A. S. (1975). Temporal coherence in the perception of tone sequences. Ph.D diss., Eindhoven University of Technology.

von Hippel, P. (2000). Redefining pitch proximity: Tessitura and mobility as constraints on melodic intervals. *Music Perception*, 17, 315–27.

von Hippel, P., & Huron, D. (2000). Why do skips precede reversals? The effect of tessitura on melodic structure. *Music Perception*, 18, 59–85.

Vos, P. G., & Van Geenen, E. W. (1996). A parallel-processing key-finding model. *Music Perception*, 14, 185–224.

Waterman, R. A. (1948). "Hot" rhythm in Negro music. *Journal of the American Musicological Society*, 1, 24–37.

Waterman, R. A. (1952). African influence on the music of the Americas. In S. Tax (Ed.), *Acculturation in the Americas (Proceedings of the Nineteenth International Congress of Americanists)*, 207–218. Chicago: Chicago University Press.

Weber, G. ([1832]1994). A particularly remarkable passage in a string quartet in C by Mozart [K 465 ("Dissonance")]. Translated by I. Bent. In I. Bent (Ed.), *Music Analysis in the Nineteenth Century*, vol. 1, 157–83. Cambridge: Cambridge University Press.

Weber, G. (1851). *The Theory of Musical Composition.* Translated by James F. Warner. London: Cocks.

Wessel, D. L. (1979). Timbre space as a musical control structure. *Computer Music Journal*, 3(2), 45–52.

Winograd, T. (1968). Linguistics and the computer analysis of tonal harmony. *Journal of Music Theory*, 12, 2–49.

Author Index

Subject Index

Note: Page numbers in italics indicate main discussions.

Printed in the United States
By Bookmasters